ALSO BY BRIAN CHRISTIAN

*The Most Human Human: What Artificial Intelligence
Teaches Us About Being Alive*

Algorithms
to Live By

Algorithms to Live By

The Computer Science of Human Decisions

Brian Christian and
Tom Griffiths

ALLEN
LANE

ALLEN LANE
an imprint of Penguin Canada, a division of Penguin Random House Canada Limited

Penguin Canada, 320 Front Street West, Suite 1400, Toronto, Ontario M5V 3B6, Canada
Penguin Group (USA) LLC, 375 Hudson Street, New York, New York 10014, U.S.A.
Penguin Books Ltd, 80 Strand, London WC2R 0RL, England
Penguin Ireland, 25 St Stephen's Green, Dublin 2, Ireland (a division of Penguin Books Ltd)
Penguin Books Australia, 707 Collins Street, Melbourne, Victoria 3008, Australia
Penguin Books India, 11 Community Centre, Panchsheel Park, New Delhi – 110 017, India
Penguin Books New Zealand, 67 Apollo Drive, Rosedale, Auckland 0632, New Zealand
Penguin Books South Africa, 24 Sturdee Avenue, Rosebank,
Johannesburg 2196, South Africa
Penguin Books Ltd, Registered Offices: 80 Strand, London WC2R 0RL, England

Published in Allen Lane hardcover by Penguin Canada, 2016
Simultaneously published in the United States by Henry Holt and Company, LLC

1 2 3 4 5 6 7 8 9 10 (RRD)

Manufactured in the U.S.A.

Cover design: Lucy Kim
Cover photographs: Brain made of cables © Miguel Navarro/Getty Images;
USB type C connector © Chris Willson/Alamy Stock Photo
Interior design: Meryl Sussman Levavi

Library and Archives Canada Cataloguing in Publication

Christian, Brian, 1984-, author
Algorithms to live by : the computer science of human decisions
/ Brian Christian and Tom Griffiths.

Includes bibliographical references and index.
ISBN 978-0-670-06831-9 (bound)
ISBN 978-0-14-319647-1 (eBook)

1. Human behavior--Mathematical models. 2. Problem solving--
Mathematics. 3. Computer simulation. 4. Computer algorithms.
I. Griffiths, Tom, 1978-, author II. Title.

BF39.C47 2016   153.4'3   C2015-907884-9

www.penguinrandomhouse.ca

For our families

Contents

Algorithms
to Live By

Introduction

Algorithms to Live By

Imagine you're searching for an apartment in San Francisco—arguably the most harrowing American city in which to do so. The booming tech sector and tight zoning laws limiting new construction have conspired to make the city just as expensive as New York, and by many accounts more competitive. New listings go up and come down within minutes, open houses are mobbed, and often the keys end up in the hands of whoever can physically foist a deposit check on the landlord first.

Such a savage market leaves little room for the kind of fact-finding and deliberation that is theoretically supposed to characterize the doings of the rational consumer. Unlike, say, a mall patron or an online shopper, who can compare options before making a decision, the would-be San Franciscan has to decide instantly either way: you can take the apartment you are currently looking at, forsaking all others, or you can walk away, never to return.

Let's assume for a moment, for the sake of simplicity, that you care only about maximizing your chance of getting the very best apartment available. Your goal is reducing the twin, Scylla-and-Charybdis regrets of the "one that got away" and the "stone left unturned" to the absolute minimum. You run into a dilemma right off the bat: How are you to know that an apartment is indeed the best unless you have a baseline to judge it by? And how are you to establish that baseline unless you look at (and *lose*) a number of apartments? The more information you gather, the better you'll know the right opportunity when you see it—but the more likely you are to have already passed it by.

So what do you do? How do you make an informed decision when the very act of informing it jeopardizes the outcome? It's a cruel situation, bordering on paradox.

When presented with this kind of problem, most people will intuitively say something to the effect that it requires some sort of balance between looking and leaping—that you must look at enough apartments to establish a standard, then take whatever satisfies the standard you've established. This notion of balance is, in fact, precisely correct. What most people *don't* say with any certainty is what that balance is. Fortunately, there's an answer.

Thirty-seven percent.

If you want the best odds of getting the best apartment, spend 37% of your apartment hunt (eleven days, if you've given yourself a month for the search) noncommittally exploring options. Leave the checkbook at home; you're just calibrating. But after that point, be prepared to immediately commit—deposit and all—to the very first place you see that beats whatever you've already seen. This is not merely an intuitively satisfying compromise between looking and leaping. It is the *provably optimal* solution.

We know this because finding an apartment belongs to a class of mathematical problems known as "optimal stopping" problems. The 37% rule defines a simple series of steps—what computer scientists call an "algorithm"—for solving these problems. And as it turns out, apartment hunting is just one of the ways that optimal stopping rears its head in daily life. Committing to or forgoing a succession of options is a structure that appears in life again and again, in slightly different incarnations. How many times to circle the block before pulling into a parking space? How far to push your luck with a risky business venture before cashing out? How long to hold out for a better offer on that house or car?

The same challenge also appears in an even more fraught setting: dating. Optimal stopping is the science of serial monogamy.

Simple algorithms offer solutions not only to an apartment hunt but to all such situations in life where we confront the question of optimal stopping. People grapple with these issues every day—although surely poets have spilled more ink on the tribulations of courtship than of parking— and they do so with, in some cases, considerable anguish. But the anguish is unnecessary. Mathematically, at least, these are solved problems.

Every harried renter, driver, and suitor you see around you as you go through a typical week is essentially reinventing the wheel. They don't need a therapist; they need an algorithm. The therapist tells them to find the right, comfortable balance between impulsivity and overthinking.

The algorithm tells them the balance is thirty-seven percent.

◆

There is a particular set of problems that all people face, problems that are a direct result of the fact that our lives are carried out in finite space and time. What should we do, and leave undone, in a day or in a decade? What degree of mess should we embrace—and how much order is excessive? What balance between *new* experiences and *favored* ones makes for the most fulfilling life?

These might seem like problems unique to humans; they're not. For more than half a century, computer scientists have been grappling with, and in many cases solving, the equivalents of these everyday dilemmas. How should a processor allocate its "attention" to perform all that the user asks of it, with the minimum overhead and in the least amount of time? When should it switch between different tasks, and how many tasks should it take on in the first place? What is the best way for it to use its limited memory resources? Should it collect more data, or take an action based on the data it already has? Seizing the day might be a challenge for humans, but computers all around us are seizing milliseconds with ease. And there's much we can learn from how they do it.

Talking about algorithms for human lives might seem like an odd juxtaposition. For many people, the word "algorithm" evokes the arcane and inscrutable machinations of big data, big government, and big business: increasingly part of the infrastructure of the modern world, but hardly a source of practical wisdom or guidance for human affairs. But an algorithm is just a finite sequence of steps used to solve a problem, and algorithms are much broader—and older by far—than the computer. Long before algorithms were ever used by machines, they were used by people.

The word "algorithm" comes from the name of Persian mathematician al-Khwārizmī, author of a ninth-century book of techniques for doing mathematics by hand. (His book was called *al-Jabr wa'l-Muqābala*—and the "al-jabr" of the title in turn provides the source of our word "algebra.") The earliest known mathematical algorithms, however, predate even al-Khwārizmī's work: a four-thousand-year-old Sumerian clay tablet found near Baghdad describes a scheme for long division.

But algorithms are not confined to mathematics alone. When you cook bread from a recipe, you're following an algorithm. When you knit a sweater from a pattern, you're following an algorithm. When you put a sharp edge on a piece of flint by executing a precise sequence of strikes with the end of an antler—a key step in making fine stone tools—you're following an algorithm. Algorithms have been a part of human technology ever since the Stone Age.

◆

In this book, we explore the idea of *human algorithm design*—searching for better solutions to the challenges people encounter every day. Applying the lens of computer science to everyday life has consequences at many scales. Most immediately, it offers us practical, concrete suggestions for how to solve specific problems. Optimal stopping tells us when to look and when to leap. The explore/exploit tradeoff tells us how to find the balance between trying new things and enjoying our favorites. Sorting theory tells us how (and whether) to arrange our offices. Caching theory tells us how to fill our closets. Scheduling theory tells us how to fill our time.

At the next level, computer science gives us a vocabulary for understanding the deeper principles at play in each of these domains. As Carl Sagan put it, "Science is a way of thinking much more than it is a body of knowledge." Even in cases where life is too messy for us to expect a strict numerical analysis or a ready answer, using intuitions and concepts honed on the simpler forms of these problems offers us a way to understand the key issues and make progress.

Most broadly, looking through the lens of computer science can teach us about the nature of the human mind, the meaning of rationality, and the oldest question of all: how to live. Examining cognition as a means of solving the fundamentally computational problems posed by our environment can utterly change the way we think about human rationality.

The notion that studying the inner workings of computers might reveal how to think and decide, what to believe and how to behave, might strike many people as not only wildly reductive, but in fact misguided. Even if computer science did have things to say about how to think and how to act, would we want to listen? We look at the AIs and robots of science fiction, and it seems like theirs is not a life any of us would want to live.

In part, that's because when we think about computers, we think about coldly mechanical, deterministic systems: machines applying rigid deductive logic, making decisions by exhaustively enumerating the options, and grinding out the exact right answer no matter how long and hard they have to think. Indeed, the person who first imagined computers had something essentially like this in mind. Alan Turing defined the very notion of computation by an analogy to a human mathematician who carefully works through the steps of a lengthy calculation, yielding an unmistakably right answer.

So it might come as a surprise that this is not what modern computers are actually doing when they face a difficult problem. Straightforward arithmetic, of course, isn't particularly challenging for a modern computer. Rather, it's tasks like conversing with people, fixing a corrupted file, or winning a game of Go—problems where the rules aren't clear, some of the required information is missing, or finding exactly the right answer would require considering an astronomical number of possibilities—that now pose the biggest challenges in computer science. And the algorithms that researchers have developed to solve the hardest classes of problems have moved computers away from an extreme reliance on exhaustive calculation. Instead, tackling real-world tasks requires being comfortable with chance, trading off time with accuracy, and using approximations.

As computers become better tuned to real-world problems, they provide not only algorithms that people can borrow for their own lives, but a better standard against which to compare human cognition itself. Over the past decade or two, behavioral economics has told a very particular story about human beings: that we are irrational and error-prone, owing in large part to the buggy, idiosyncratic hardware of the brain. This self-deprecating story has become increasingly familiar, but certain questions remain vexing. Why are four-year-olds, for instance, still better than million-dollar supercomputers at a host of cognitive tasks, including vision, language, and causal reasoning?

The solutions to everyday problems that come from computer science tell a different story about the human mind. Life is full of problems that are, quite simply, *hard*. And the mistakes made by people often say more about the intrinsic difficulties of the problem than about the fallibility of human brains. Thinking algorithmically about the world, learning about the fundamental structures of the problems we face and about the properties of their solutions, can help us see how good we actually are, and better understand the errors that we make.

In fact, human beings turn out to consistently confront some of the hardest cases of the problems studied by computer scientists. Often, people need to make decisions while dealing with uncertainty, time constraints, partial information, and a rapidly changing world. In some of those cases, even cutting-edge computer science has not yet come up with efficient, always-right algorithms. For certain situations it appears that such algorithms might not exist at all.

Even where perfect algorithms haven't been found, however, the battle between generations of computer scientists and the most intractable real-world problems has yielded a series of insights. These hard-won precepts are at odds with our intuitions about rationality, and they don't sound anything like the narrow prescriptions of a mathematician trying to force the world into clean, formal lines. They say: Don't always consider all your options. Don't necessarily go for the outcome that seems best every time. Make a mess on occasion. Travel light. Let things wait. Trust your instincts and don't think too long. Relax. Toss a coin. Forgive, but don't forget. To thine own self be true.

Living by the wisdom of computer science doesn't sound so bad after all. And unlike most advice, it's backed up by proofs.

◆

Just as designing algorithms for computers was originally a subject that fell into the cracks between disciplines—an odd hybrid of mathematics and engineering—so, too, designing algorithms for humans is a topic that doesn't have a natural disciplinary home. Today, algorithm design draws not only on computer science, math, and engineering but on kindred fields like statistics and operations research. And as we consider how algorithms designed for machines might relate to human minds, we also need to look to cognitive science, psychology, economics, and beyond.

We, your authors, are familiar with this interdisciplinary territory. Brian studied computer science and philosophy before going on to graduate work in English and a career at the intersection of the three. Tom studied psychology and statistics before becoming a professor at UC Berkeley, where he spends most of his time thinking about the relationship between human cognition and computation. But nobody can be an expert in all of the fields that are relevant to designing better algorithms for humans. So as part of our quest for algorithms to live by, we talked to the people who came up with some of the most famous algorithms of the last fifty years. And we asked them, some of the smartest people in the world, how their research influenced the way they approached their own lives—from finding their spouses to sorting their socks.

The next pages begin our journey through some of the biggest challenges faced by computers and human minds alike: how to manage finite space, finite time, limited attention, unknown unknowns, incomplete information, and an unforeseeable future; how to do so with grace and confidence; and how to do so in a community with others who are all simultaneously trying to do the same. We will learn about the fundamental mathematical structure of these challenges and about how computers are engineered—sometimes counter to what we imagine—to make the most of them. And we will learn about how the mind works, about its distinct but deeply related ways of tackling the same set of issues and coping with the same constraints. Ultimately, what we can gain is not only a set of concrete takeaways for the problems around us, not only a new way to see the elegant structures behind even the hairiest human dilemmas, not only a recognition of the travails of humans and computers as deeply conjoined, but something even more profound: a new vocabulary for the world around us, and a chance to learn something truly new about ourselves.

1 | Optimal Stopping

When to Stop Looking

Though all Christians start a wedding invitation by solemnly declaring their marriage is due to special Divine arrangement, I, as a philosopher, would like to talk in greater detail about this . . .

—JOHANNES KEPLER

If you prefer Mr. Martin to every other person; if you think him the most agreeable man you have ever been in company with, why should you hesitate?

—JANE AUSTEN, *EMMA*

It's such a common phenomenon that college guidance counselors even have a slang term for it: the "turkey drop." High-school sweethearts come home for Thanksgiving of their freshman year of college and, four days later, return to campus single.

An angst-ridden Brian went to his own college guidance counselor his freshman year. His high-school girlfriend had gone to a different college several states away, and they struggled with the distance. They also struggled with a stranger and more philosophical question: how good a relationship did they have? They had no real benchmark of other relationships by which to judge it. Brian's counselor recognized theirs as a classic freshman-year dilemma, and was surprisingly nonchalant in her advice: "Gather data."

The nature of serial monogamy, writ large, is that its practitioners are confronted with a fundamental, unavoidable problem. When have you met enough people to know who your best match is? And what if acquiring the data costs you that very match? It seems the ultimate Catch-22 of the heart.

As we have seen, this Catch-22, this angsty freshman cri de coeur, is what mathematicians call an "optimal stopping" problem, and it may actually have an answer: 37%.

Of course, it all depends on the assumptions you're willing to make about love.

The Secretary Problem

In any optimal stopping problem, the crucial dilemma is not which option to *pick*, but how many options to even *consider*. These problems turn out to have implications not only for lovers and renters, but also for drivers, homeowners, burglars, and beyond.

The **37% Rule*** derives from optimal stopping's most famous puzzle, which has come to be known as the "secretary problem." Its setup is much like the apartment hunter's dilemma that we considered earlier. Imagine you're interviewing a set of applicants for a position as a secretary, and your goal is to maximize the chance of hiring the single best applicant in the pool. While you have no idea how to assign scores to individual applicants, you can easily judge which one you prefer. (A mathematician might say you have access only to the *ordinal* numbers—the relative ranks of the applicants compared to each other—but not to the *cardinal* numbers, their ratings on some kind of general scale.) You interview the applicants in random order, one at a time. You can decide to offer the job to an applicant at any point and they are guaranteed to accept, terminating the search. But if you pass over an applicant, deciding not to hire them, they are gone forever.

The secretary problem is widely considered to have made its first appearance in print—sans explicit mention of secretaries—in the February 1960 issue of *Scientific American*, as one of several puzzles posed in Martin Gardner's beloved column on recreational mathematics. But the origins of the problem are surprisingly mysterious. Our own initial search yielded little but speculation, before turning into unexpectedly physical detective work: a road trip down to the archive of Gardner's papers at Stanford, to haul out boxes of his midcentury correspondence. Reading paper

---

*We use boldface to indicate the algorithms that appear throughout the book.

correspondence is a bit like eavesdropping on someone who's on the phone: you're only hearing one side of the exchange, and must infer the other. In our case, we only had the replies to what was apparently Gardner's own search for the problem's origins fiftysome years ago. The more we read, the more tangled and unclear the story became.

Harvard mathematician Frederick Mosteller recalled hearing about the problem in 1955 from his colleague Andrew Gleason, who had heard about it from somebody else. Leo Moser wrote from the University of Alberta to say that he read about the problem in "some notes" by R. E. Gaskell of Boeing, who himself credited a colleague. Roger Pinkham of Rutgers wrote that he first heard of the problem in 1955 from Duke University mathematician J. Shoenfield, "and I believe he said that he had heard the problem from someone at Michigan."

"Someone at Michigan" was almost certainly someone named Merrill Flood. Though he is largely unheard of outside mathematics, Flood's influence on computer science is almost impossible to avoid. He's credited with popularizing the traveling salesman problem (which we discuss in more detail in chapter 8), devising the prisoner's dilemma (which we discuss in chapter 11), and even with possibly coining the term "software." It's Flood who made the first known discovery of the 37% Rule, in 1958, and he claims to have been considering the problem since 1949—but he himself points back to several other mathematicians.

Suffice it to say that wherever it came from, the secretary problem proved to be a near-perfect mathematical puzzle: simple to explain, devilish to solve, succinct in its answer, and intriguing in its implications. As a result, it moved like wildfire through the mathematical circles of the 1950s, spreading by word of mouth, and thanks to Gardner's column in 1960 came to grip the imagination of the public at large. By the 1980s the problem and its variations had produced so much analysis that it had come to be discussed in papers as a subfield unto itself.

As for secretaries—it's charming to watch each culture put its own anthropological spin on formal systems. We think of chess, for instance, as medieval European in its imagery, but in fact its origins are in eighth-century India; it was heavy-handedly "Europeanized" in the fifteenth century, as its shahs became kings, its viziers turned to queens, and its elephants became bishops. Likewise, optimal stopping problems have had a number of incarnations, each reflecting the predominating concerns of its time. In the nineteenth century such problems were typified by baroque lotteries and by women choosing male suitors; in the early twentieth century by holidaying motorists searching for hotels and by male suitors choosing women; and in the paper-pushing, male-dominated mid-twentieth century, by male bosses choosing female assistants. The first explicit mention of it by name as the "secretary problem" appears to be in a 1964 paper, and somewhere along the way the name stuck.

Whence 37%?

In your search for a secretary, there are two ways you can fail: stopping early and stopping late. When you stop too early, you leave the best applicant undiscovered. When you stop too late, you hold out for a better applicant who doesn't exist. The optimal strategy will clearly require finding the right balance between the two, walking the tightrope between looking too much and not enough.

If your aim is finding the very best applicant, settling for nothing less, it's clear that as you go through the interview process you shouldn't even consider hiring somebody who isn't the best you've seen so far. However, simply being the best yet isn't enough for an offer; the very first applicant, for example, will of course be the best yet by definition. More generally, it stands to reason that the rate at which we encounter "best yet" applicants will go down as we proceed in our interviews. For instance, the second applicant has a 50/50 chance of being the best we've yet seen, but the fifth applicant only has a 1-in-5 chance of being the best so far, the sixth has a 1-in-6 chance, and so on. As a result, best-yet applicants will become steadily more impressive as the search continues (by definition, again, they're better than all those who came before)—but they will also become more and more infrequent.

Okay, so we know that taking the *first* best-yet applicant we encounter (a.k.a. the first applicant, period) is rash. If there are a hundred applicants, it also seems hasty to make an offer to the *next* one who's best-yet, just because she was better than the first. So how do we proceed?

Intuitively, there are a few potential strategies. For instance, making an offer the third time an applicant trumps everyone seen so far—or maybe the fourth time. Or perhaps taking the next best-yet applicant to come along after a long "drought"—a long streak of poor ones.

But as it happens, neither of these relatively sensible strategies comes out on top. Instead, the optimal solution takes the form of what we'll call the **Look-Then-Leap Rule**: You set a predetermined amount of time for "looking"—that is, exploring your options, gathering data—in which you categorically don't choose anyone, no matter how impressive. After that point, you enter the "leap" phase, prepared to instantly commit to anyone who outshines the best applicant you saw in the look phase.

We can see how the Look-Then-Leap Rule emerges by considering how the secretary problem plays out in the smallest applicant pools. With just one applicant the problem is easy to solve—hire her! With two applicants, you have a 50/50 chance of success no matter what you do. You can hire the first applicant (who'll turn out to be the best half the time), or dismiss the first and by default hire the second (who is also best half the time).

Add a third applicant, and all of a sudden things get interesting. The odds if we hire at random are one-third, or 33%. With two applicants we could do no better than chance; with three, can we? It turns out we can, and it all comes down to what we do with the second interviewee. When we see the first applicant, we have no *information*—she'll always appear to be the best yet. When we see the third applicant, we have no *agency*—we have to make an offer to the final applicant, since we've dismissed the others. But when we see the second applicant, we have a little bit of both: we know whether she's better or worse than the first, and we have the freedom to either hire or dismiss her. What happens when we just hire her if she's better than the first applicant, and dismiss her if she's not? This turns out to be the best possible strategy when facing three applicants; using this approach it's possible, surprisingly, to do just as well in the three-applicant problem as with two, choosing the best applicant exactly half the time.*

Enumerating these scenarios for four applicants tells us that we should still begin to leap as soon as the second applicant; with five applicants in the pool, we shouldn't leap before the third.

As the applicant pool grows, the exact place to draw the line between looking and leaping settles to 37% of the pool, yielding the 37% Rule: look

---

*With this strategy we have a 33% risk of dismissing the best applicant and a 16% risk of never meeting her. To elaborate, there are exactly six possible orderings of the three applicants: 1-2-3, 1-3-2, 2-1-3, 2-3-1, 3-1-2, and 3-2-1. The strategy of looking at the first applicant and then leaping for whoever surpasses her will succeed in three of the six cases (2-1-3, 2-3-1, 3-1-2) and will fail in the other three—twice by being overly choosy (1-2-3, 1-3-2) and once by not being choosy enough (3-2-1).

Number of Applicants	Take the Best Applicant After	Chance of Getting the Best
3	1 (33.33%)	50%
4	1 (25%)	45.83%
5	2 (40%)	43.33%
6	2 (33.33%)	42.78%
7	2 (28.57%)	41.43%
8	3 (37.5%)	40.98%
9	3 (33.33%)	40.59%
10	3 (30%)	39.87%
20	7 (35%)	38.42%
30	11 (36.67%)	37.86%
40	15 (37.5%)	37.57%
50	18 (36%)	37.43%
100	37 (37%)	37.10%
1000	369 (36.9%)	36.81%

How to optimally choose a secretary.

at the first 37% of the applicants,* choosing none, then be ready to leap for anyone better than all those you've seen so far.

As it turns out, following this optimal strategy ultimately gives us a 37% chance of hiring the best applicant; it's one of the problem's curious mathematical symmetries that the strategy itself and its chance of success work out to the very same number. The table above shows the optimal strategy for the secretary problem with different numbers of applicants, demonstrating how the chance of success—like the point to switch from looking to leaping—converges on 37% as the number of applicants increases.

A 63% failure rate, when following the *best possible* strategy, is a sobering fact. Even when we act optimally in the secretary problem, we will still fail most of the time—that is, we won't end up with the single best applicant in the pool. This is bad news for those of us who would frame romance as a search for "the one." But here's the silver lining. Intuition would suggest that our chances of picking the single best applicant should steadily

---

*Just a hair under 37%, actually. To be precise, the mathematically optimal proportion of applicants to look at is $1/e$—the same mathematical constant e, equivalent to $2.71828\ldots$, that shows up in calculations of compound interest. But you don't need to worry about knowing e to twelve decimal places: anything between 35% and 40% provides a success rate extremely close to the maximum. For more of the mathematical details, see the notes at the end of the book.

decrease as the applicant pool grows. If we were hiring at random, for instance, then in a pool of a hundred applicants we'd have a 1% chance of success, and in a pool of a million applicants we'd have a 0.0001% chance. Yet remarkably, the math of the secretary problem doesn't change. If you're stopping optimally, your chance of finding the single best applicant in a pool of a hundred is 37%. And in a pool of a million, believe it or not, your chance is still 37%. Thus the bigger the applicant pool gets, the more valuable knowing the optimal algorithm becomes. It's true that you're unlikely to find the needle the majority of the time, but optimal stopping is your best defense against the haystack, no matter how large.

Lover's Leap

> *The passion between the sexes has appeared in every age to be so nearly the same that it may always be considered, in algebraic language, as a given quantity.*
>
> —THOMAS MALTHUS

> *I married the first man I ever kissed. When I tell this to my children they just about throw up.*
>
> —BARBARA BUSH

Before he became a professor of operations research at Carnegie Mellon, Michael Trick was a graduate student, looking for love. "It hit me that the problem has been studied: it is the Secretary Problem! I had a position to fill [and] a series of applicants, and my goal was to pick the best applicant for the position." So he ran the numbers. He didn't know how many women he could expect to meet in his lifetime, but there's a certain flexibility in the 37% Rule: it can be applied to either the number of applicants or the *time* over which one is searching. Assuming that his search would run from ages eighteen to forty, the 37% Rule gave age 26.1 years as the point at which to switch from looking to leaping. A number that, as it happened, was exactly Trick's age at the time. So when he found a woman who was a better match than all those he had dated so far, he knew exactly what to do. He leapt. "I didn't know if she was Perfect (the assumptions of the model don't allow me to determine that), but there was no doubt that she met the qualifications for this step of the algorithm. So I proposed," he writes.

"And she turned me down."

Mathematicians have been having trouble with love since at least the seventeenth century. The legendary astronomer Johannes Kepler is today perhaps best remembered for discovering that planetary orbits are elliptical and for being a crucial part of the "Copernican Revolution" that included Galileo and Newton and upended humanity's sense of its place in the heavens. But Kepler had terrestrial concerns, too. After the death of his first wife in 1611, Kepler embarked on a long and arduous quest to remarry, ultimately courting a total of eleven women. Of the first four, Kepler liked the fourth the best ("because of her tall build and athletic body") but did not cease his search. "It would have been settled," Kepler wrote, "had not both love and reason forced a fifth woman on me. This one won me over with love, humble loyalty, economy of household, diligence, and the love she gave the stepchildren."

"However," he wrote, "I continued."

Kepler's friends and relations went on making introductions for him, and he kept on looking, but halfheartedly. His thoughts remained with number five. After eleven courtships in total, he decided he would search no further. "While preparing to travel to Regensburg, I returned to the fifth woman, declared myself, and was accepted." Kepler and Susanna Reuttinger were wed and had six children together, along with the children from Kepler's first marriage. Biographies describe the rest of Kepler's domestic life as a particularly peaceful and joyous time.

Both Kepler and Trick—in opposite ways—experienced firsthand some of the ways that the secretary problem oversimplifies the search for love. In the classical secretary problem, applicants always accept the position, preventing the rejection experienced by Trick. And they cannot be "recalled" once passed over, contrary to the strategy followed by Kepler.

In the decades since the secretary problem was first introduced, a wide range of variants on the scenario have been studied, with strategies for optimal stopping worked out under a number of different conditions. The possibility of rejection, for instance, has a straightforward mathematical solution: propose early and often. If you have, say, a 50/50 chance of being rejected, then the same kind of mathematical analysis that yielded the 37% Rule says you should start making offers after just a *quarter* of your search. If turned down, keep making offers to every best-yet person you see until somebody accepts. With such a strategy, your chance of overall success—that is, proposing and being accepted by the best applicant in the pool—will also be 25%. Not such terrible odds, perhaps, for a scenario that combines the obstacle of rejection with the general difficulty of establishing one's standards in the first place.

Kepler, for his part, decried the "restlessness and doubtfulness" that pushed him to keep on searching. "Was there no other way for my uneasy heart to be content with its fate," he bemoaned in a letter to a confidante, "than by realizing the impossibility of the fulfillment of so many other desires?" Here, again, optimal stopping theory provides some measure of consolation. Rather than being signs of moral or psychological degeneracy, restlessness and doubtfulness actually turn out to be part of the best strategy for scenarios where second chances are possible. If you can recall previous applicants, the optimal algorithm puts a twist on the familiar Look-Then-Leap Rule: a longer noncommittal period, and a fallback plan.

For example, assume an immediate proposal is a sure thing but belated proposals are rejected half the time. Then the math says you should keep looking noncommittally until you've seen 61% of applicants, and then only leap if someone in the remaining 39% of the pool proves to be the best yet. If you're still single after considering all the possibilities—as Kepler was—then go back to the best one that got away. The symmetry between strategy and outcome holds in this case once again, with your chances of ending up with the best applicant under this second-chances-allowed scenario also being 61%.

For Kepler, the difference between reality and the classical secretary problem brought with it a happy ending. In fact, the twist on the classical problem worked out well for Trick, too. After the rejection, he completed his degree and took a job in Germany. There, he "walked into a bar, fell in love with a beautiful woman, moved in together three weeks later, [and] invited her to live in the United States 'for a while.'" She agreed—and six years later, they were wed.

Knowing a Good Thing When You See It: Full Information

The first set of variants we considered—rejection and recall—altered the classical secretary problem's assumptions that timely proposals are always accepted, and tardy proposals, never. For these variants, the best approach remained the same as in the original: look noncommittally for a time, then be ready to leap.

But there's an even more fundamental assumption of the secretary problem that we might call into question. Namely, in the secretary problem we know *nothing* about the applicants other than how they compare to one another. We don't have an objective or preexisting sense of what makes for a good or a bad applicant; moreover, when we compare two of them, we know which of the two is better, but not by how much. It's this fact that gives rise to the unavoidable "look" phase, in which we risk passing up a superb early applicant while we calibrate our expectations and standards. Mathematicians refer to this genre of optimal stopping problems as "no-information games."

This setup is arguably a far cry from most searches for an apartment, a partner, or even a secretary. Imagine instead that we had some kind of objective criterion—if every secretary, for instance, had taken a typing exam scored by percentile, in the fashion of the SAT or GRE or LSAT. That is, every applicant's score will tell us where they fall among all the typists who took the test: a 51st-percentile typist is just above average, a 75th-percentile typist is better than three test takers out of four, and so on.

Suppose that our applicant pool is representative of the population at large and isn't skewed or self-selected in any way. Furthermore, suppose we decide that typing speed is the only thing that matters about our applicants. Then we have what mathematicians call "full information," and everything changes. "No buildup of experience is needed to set a standard," as the seminal 1966 paper on the problem put it, "and a profitable choice can sometimes be made immediately." In other words, if a 95th-percentile applicant happens to be the first one we evaluate, we know it instantly and can confidently hire her on the spot—that is, of course, assuming we don't think there's a 96th-percentile applicant in the pool.

And there's the rub. If our goal is, again, to get the single best person for the job, we still need to weigh the likelihood that there's a stronger applicant out there. However, the fact that we have full information gives us everything we need to calculate those odds directly. The chance that our next applicant is in the 96th percentile or higher will always be 1 in 20, for instance. Thus the decision of whether to stop comes down entirely to how many applicants we have left to see. Full information means that we don't need to look before we leap. We can instead use the **Threshold Rule**, where we immediately accept an applicant if she is above a certain percentile. We don't need to look at an initial group of candidates to set this threshold—but we do, however, need to be keenly aware of how much looking remains available.

The math shows that when there are a lot of applicants left in the pool, you should pass up even a very good applicant in the hopes of finding someone still better than that—but as your options dwindle, you should be prepared to hire anyone who's simply better than average. It's a familiar, if not exactly inspiring, message: in the face of slim pickings, lower your standards. It also makes clear the converse: with more fish in the sea, raise them. In both cases, crucially, the math tells you exactly by how much.

The easiest way to understand the numbers for this scenario is to start at the end and think backward. If you're down to the last applicant, of course, you are necessarily forced to choose her. But when looking at the next-to-last applicant, the question becomes: is she above the 50th percentile? If yes, then hire her; if not, it's worth rolling the dice on the last applicant instead, since *her* odds of being above the 50th percentile are 50/50 by definition. Likewise, you should choose the third-to-last applicant if she's above the 69th percentile, the fourth-to-last applicant if she's above the 78th, and so on, being more choosy the more applicants are left. No matter what, never hire someone who's below average unless you're totally out of options. (And since you're still interested only in finding the very best person in the applicant pool, never hire someone who isn't the best you've seen so far.)

The chance of ending up with the single best applicant in this full-information version of the secretary problem comes to 58%—still far from a guarantee, but considerably better than the 37% success rate offered by the

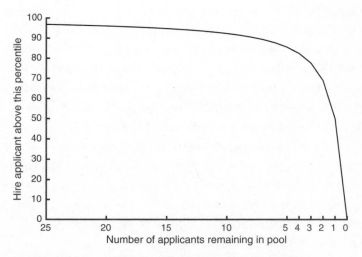

Optimal stopping thresholds in the full-information secretary problem.

37% Rule in the no-information game. If you have all the facts, you can succeed more often than not, even as the applicant pool grows arbitrarily large.

The full-information game thus offers an unexpected and somewhat bizarre takeaway. *Gold digging is more likely to succeed than a quest for love.* If you're evaluating your partners based on any kind of objective criterion—say, their income percentile—then you've got a lot more information at your disposal than if you're after a nebulous emotional response ("love") that might require both experience and comparison to calibrate.

Of course, there's no reason that net worth—or, for that matter, typing speed—needs to be the thing that you're measuring. Any yardstick that provides full information on where an applicant stands relative to the population at large will change the solution from the Look-Then-Leap Rule to the Threshold Rule and will dramatically boost your chances of finding the single best applicant in the group.

There are many more variants of the secretary problem that modify its other assumptions, perhaps bringing it more in line with the real-world challenges of finding love (or a secretary). But the lessons to be learned from optimal stopping aren't limited to dating or hiring. In fact, trying to make the best choice when options only present themselves one by one is also the basic structure of selling a house, parking a car, and quitting when you're ahead. And they're all, to some degree or other, solved problems.

When to Sell

If we alter two more aspects of the classical secretary problem, we find ourselves catapulted from the realm of dating to the realm of real estate. Earlier we talked about the process of renting an apartment as an optimal stopping problem, but *owning* a home has no shortage of optimal stopping either.

Imagine selling a house, for instance. After consulting with several real estate agents, you put your place on the market; a new coat of paint, some landscaping, and then it's just a matter of waiting for the offers to come in. As each offer arrives, you typically have to decide whether to accept it or turn it down. But turning down an offer comes at a cost—another week (or month) of mortgage payments while you wait for the next offer, which isn't guaranteed to be any better.

Selling a house is similar to the full-information game. We know the objective dollar value of the offers, telling us not only which ones are better than which, but also by how much. What's more, we have information about the broader state of the market, which enables us to at least roughly predict the range of offers to expect. (This gives us the same "percentile" information about each offer that we had with the typing exam above.) The difference here, however, is that our goal isn't actually to secure the single best offer—it's to make the most money through the process overall. Given that waiting has a cost measured in dollars, a good offer today beats a slightly better one several months from now.

Having this information, we don't need to look noncommittally to set a threshold. Instead, we can set one going in, ignore everything below it, and take the first option to exceed it. Granted, if we have a limited amount of savings that will run out if we don't sell by a certain time, or if we expect to get only a limited number of offers and no more interest thereafter, then we should lower our standards as such limits approach. (There's a reason why home buyers look for "motivated" sellers.) But if neither concern leads us to believe that our backs are against the wall, then we can simply focus on a cost-benefit analysis of the waiting game.

Here we'll analyze one of the simplest cases: where we know for certain the price range in which offers will come, and where all offers within that range are equally likely. If we don't have to worry about the offers (or our savings) running out, then we can think purely in terms of what we can expect to gain or lose by waiting for a better deal. If we decline the current offer, will the chance of a better one, multiplied by how *much* better we expect it to be, more than compensate for the cost of the wait? As it turns out, the math here is quite clean, giving us an explicit function for stopping price as a function of the cost of waiting for an offer.

This particular mathematical result doesn't care whether you're selling a mansion worth millions or a ramshackle shed. The only thing it cares about is the difference between the highest and lowest offers you're likely to receive. By plugging in some concrete figures, we can see how this algorithm offers us a considerable amount of explicit guidance. For instance, let's say the range of offers we're expecting runs from $400,000 to $500,000. First, if the cost of waiting is trivial, we're able to be almost infinitely choosy. If the cost of getting another offer is only a dollar, we'll maximize our earnings by waiting for someone willing to offer us $499,552.79 and not a dime less. If waiting costs $2,000 an offer, we should hold out for an even $480,000. In a slow market where waiting costs $10,000 an offer, we should take anything over $455,279. Finally, if

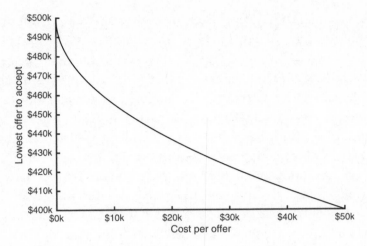

Optimal stopping thresholds in the house-selling problem.

waiting costs half or more of our expected range of offers—in this case, $50,000—then there's no advantage whatsoever to holding out; we'll do best by taking the very first offer that comes along and calling it done. Beggars can't be choosers.

The critical thing to note in this problem is that our threshold depends *only* on the cost of search. Since the chances of the next offer being a good one—and the cost of finding out—never change, our stopping price has no reason to ever get lower as the search goes on, regardless of our luck. We set it once, before we even begin, and then we quite simply hold fast.

The University of Wisconsin–Madison's Laura Albert McLay, an optimization expert, recalls turning to her knowledge of optimal stopping problems when it came time to sell her own house. "The first offer we got was great," she explains, "but it had this huge cost because they wanted us to move out a month before we were ready. There was another competitive offer . . . [but] we just kind of held out until we got the right one." For many sellers, turning down a good offer or two can be a nerve-racking proposition, especially if the ones that immediately follow are no better. But McLay held her ground and stayed cool. "That would have been really, really hard," she admits, "if I didn't know the math was on my side."

This principle applies to any situation where you get a series of offers and pay a cost to seek or wait for the next. As a consequence, it's relevant to cases that go far beyond selling a house. For example, economists have used this algorithm to model how people look for jobs, where it handily explains the otherwise seemingly paradoxical fact of unemployed workers and unfilled vacancies existing at the same time.

In fact, these variations on the optimal stopping problem have another, even more surprising property. As we saw, the ability to "recall" a past opportunity was vital in Kepler's quest for love. But in house selling and job hunting, even if it's possible to reconsider an earlier offer, and even if that offer is guaranteed to still be on the table, you should nonetheless *never* do so. If it wasn't above your threshold then, it won't be above your threshold now. What you've paid to keep searching is a sunk cost. Don't compromise, don't second-guess. And don't look back.

When to Park

I find that the three major administrative problems on a campus are sex for the students, athletics for the alumni, and parking for the faculty.
—CLARK KERR, PRESIDENT OF UC BERKELEY, 1958–1967

Another domain where optimal stopping problems abound—and where looking back is also generally ill-advised—is the car. Motorists feature in some of the earliest literature on the secretary problem, and the framework of constant forward motion makes almost every car-trip decision into a stopping problem: the search for a restaurant; the search for a bathroom; and, most acutely for urban drivers, the search for a parking space. Who better to talk to about the ins and outs of parking than the man described by the *Los Angeles Times* as "the parking rock star," UCLA Distinguished Professor of Urban Planning Donald Shoup? We drove down from Northern California to visit him, reassuring Shoup that we'd be leaving plenty of time for unexpected traffic. "As for planning on 'unexpected traffic,' I think you should plan on expected traffic," he replied. Shoup is perhaps best known for his book *The High Cost of Free Parking*, and he has done much to advance the discussion and understanding of what really happens when someone drives to their destination.

We should pity the poor driver. The ideal parking space, as Shoup models it, is one that optimizes a precise balance between the "sticker price" of the space, the time and inconvenience of walking, the time taken seeking the space (which varies wildly with destination, time of day, etc.), and the gas burned in doing so. The equation changes with the number of passengers in the car, who can split the monetary cost of a space but not the search time or the walk. At the same time, the driver needs to consider that the area with the most parking supply may also be the area with the most demand; parking has a game-theoretic component, as you try to outsmart the other drivers on the road while they in turn are trying to outsmart you.* That said, many of the challenges of parking boil down to a single number: the occupancy rate. This is the proportion of all parking spots that are currently occupied. If the occupancy rate is low, it's easy to find a good parking spot. If it's high, finding anywhere at all to park is a challenge.

Shoup argues that many of the headaches of parking are consequences of cities adopting policies that result in extremely high occupancy rates. If the cost of parking in a particular location is too low (or—horrors!— nothing at all), then there is a high incentive to park there, rather than to park a little farther away and walk. So everybody tries to park there, but most of them find the spaces are already full, and people end up wasting time and burning fossil fuel as they cruise for a spot.

Shoup's solution involves installing digital parking meters that are capable of adaptive prices that rise with demand. (This has now been implemented in downtown San Francisco.) The prices are set with a target occupancy rate in mind, and Shoup argues that this rate should be somewhere around 85%—a radical drop from the nearly 100%-packed curbs of most major cities. As he notes, when occupancy goes from 90% to 95%, it accommodates only 5% more cars but *doubles* the length of everyone's search.

The key impact that occupancy rate has on parking strategy becomes clear once we recognize that parking is an optimal stopping problem. As you drive along the street, every time you see the occasional empty spot you have to make a decision: should you take this spot, or go a little closer to your destination and try your luck?

Assume you're on an infinitely long road, with parking spots evenly spaced, and your goal is to minimize the distance you end up walking to your destination. Then the solution is the Look-Then-Leap Rule. The optimally stopping driver should pass up all vacant spots occurring more than a certain distance from the destination and then take the first space that appears thereafter. And the distance at which to switch from looking to leaping depends on the proportion of spots that are likely to be filled—the occupancy rate. The table on the next page gives the distances for some representative proportions.

---

*More on the computational perils of game theory in chapter 11.

With this occupancy rate (%)	Wait until this many spaces away, then take the next free spot
0	0
50	1
75	3
80	4
85	5
90	7
95	14
96	17
97	23
98	35
99	69
99.9	693

How to optimally find parking.

If this infinite street has a big-city occupancy rate of 99%, with just 1% of spots vacant, then you should take the first spot you see starting at almost 70 spots—more than a quarter mile—from your destination. But if Shoup has his way and occupancy rates drop to just 85%, you don't need to start seriously looking until you're half a block away.

Most of us don't drive on perfectly straight, infinitely long roads. So as with other optimal stopping problems, researchers have considered a variety of tweaks to this basic scenario. For instance, they have studied the optimal parking strategy for cases where the driver can make U-turns, where fewer parking spaces are available the closer one gets to the destination, and where the driver is in competition against rival drivers also heading to the same destination. But whatever the exact parameters of the problem, more vacant spots are always going to make life easier. It's something of a policy reminder to municipal governments: parking is not as simple as having a resource (spots) and maximizing its utilization (occupancy). Parking is also a *process*—an optimal stopping problem—and it's one that consumes attention, time, and fuel, and generates both pollution and congestion. The right policy addresses the whole problem. And, counterintuitively, empty spots on highly desirable blocks can be the sign that things are working correctly.

We asked Shoup if his research allows him to optimize his own commute, through the Los Angeles traffic to his office at UCLA. Does arguably the world's top expert on parking have some kind of secret weapon?

He does: "I ride my bike."

When to Quit

In 1997, *Forbes* magazine identified Boris Berezovsky as the richest man in Russia, with a fortune of roughly $3 billion. Just ten years earlier he had been living on a mathematician's salary from the USSR Academy of Sciences. He made his billions by drawing on industrial relationships he'd formed through his research to found a company that facilitated interaction between foreign carmakers and the Soviet car manufacturer AvtoVAZ. Berezovky's company then became a large-scale dealer for the cars that AvtoVAZ produced, using a payment installment scheme to take advantage of hyperinflation in the ruble. Using the funds from this partnership he bought partial ownership of AvtoVAZ itself, then the ORT Television network, and finally the Sibneft oil company. Becoming one of a new class of oligarchs, he participated in politics, supporting Boris Yeltsin's reelection in 1996 and the choice of Vladimir Putin as his successor in 1999.

But that's when Berezovsky's luck turned. Shortly after Putin's election, Berezovsky publicly objected to proposed constitutional reforms that would expand the power of the president. His continued public criticism of Putin led to the deterioration of their relationship. In October 2000, when Putin was asked about Berezovsky's criticisms, he replied, "The state has a cudgel in its hands that you use to hit just once, but on the head. We haven't used this cudgel yet. . . . The day we get really angry, we won't hesitate." Berezovsky left Russia permanently the next month, taking up exile in England, where he continued to criticize Putin's regime.

How did Berezovsky decide it was time to leave Russia? Is there a way, perhaps, to think mathematically about the advice to "quit while you're ahead"? Berezovsky in particular might have considered this very question himself, since the topic he had worked on all those years ago as a mathematician was none other than optimal stopping; he authored the first (and, so far, the only) book entirely devoted to the secretary problem.

The problem of quitting while you're ahead has been analyzed under several different guises, but perhaps the most appropriate to Berezovsky's case—with apologies to Russian oligarchs—is known as the "burglar problem." In this problem, a burglar has the opportunity to carry out a sequence of robberies. Each robbery provides some reward, and there's a chance of getting away with it each time. But if the burglar is caught, he gets arrested and loses all his accumulated gains. What algorithm should he follow to maximize his expected take?

The fact that this problem has a solution is bad news for heist movie screenplays: when the team is trying to lure the old burglar out of retirement for one last job, the canny thief need only crunch the numbers. Moreover, the results are pretty intuitive: the number of robberies you should carry out is roughly equal to the chance you get away, divided by the chance you get caught. If you're a skilled burglar and have a 90% chance of pulling off each robbery (and a 10% chance of losing it all), then retire after 90/10 = 9 robberies. A ham-fisted amateur with a 50/50 chance of success? The first time you have nothing to lose, but don't push your luck more than once.

Despite his expertise in optimal stopping, Berezovsky's story ends sadly. He died in March 2013, found by a bodyguard in the locked bathroom of his house in Berkshire with a ligature around his neck. The official conclusion of a postmortem examination was that he had committed suicide, hanging himself after losing much of his wealth through a series of high-profile legal cases involving his enemies in Russia. Perhaps he should have stopped sooner—amassing just a few tens of millions of dollars, say, and not getting into politics. But, alas, that was not his style. One of his mathematician friends, Leonid Boguslavsky, told a story about Berezovsky from when they were both young researchers: on a water-skiing trip to a lake near Moscow, the boat they had planned to use broke down. Here's how David Hoffman tells it in his book *The Oligarchs*:

> While their friends went to the beach and lit a bonfire, Boguslavsky and Berezovsky headed to the dock to try to repair the motor. . . . Three hours later, they had taken apart and reassembled the motor. It was still dead. They had missed most of the party, yet Berezovsky insisted they *had* to keep trying. "We tried this and that," Boguslavsky recalled. Berezovsky would not give up.

Surprisingly, not giving up—ever—also makes an appearance in the optimal stopping literature. It might not seem like it from the wide range of problems we have discussed, but there are sequential decision-making problems for which there *is* no optimal stopping rule. A simple example is the game of "triple or nothing." Imagine you have $1.00, and can play the following game as many times as you want: bet all your money, and have a 50% chance of receiving triple the amount and a 50% chance of losing your entire stake. How many times should you play? Despite its simplicity, there is no optimal stopping rule for this problem, since each time you play, your average gains are a little higher. Starting with $1.00, you will get $3.00 half the time and $0.00 half the time, so on average you expect to end the first round with $1.50 in your pocket. Then, if you were lucky in the first round, the two possibilities from the $3.00 you've just won are $9.00 and $0.00—for an average return of $4.50 from the second bet. The math shows that you should *always* keep playing. But if you follow this strategy, you will eventually lose everything. Some problems are better avoided than solved.

Always Be Stopping

> *I expect to pass through this world but once. Any good therefore that I can do, or any kindness that I can show to any fellow creature, let me do it now. Let me not defer or neglect it, for I shall not pass this way again.*
>
> —STEPHEN GRELLET

> *Spend the afternoon. You can't take it with you.*
>
> —ANNIE DILLARD

We've looked at specific cases of people confronting stopping problems in their lives, and it's clear that most of us encounter these kinds of problems, in one form or another, daily. Whether it involves secretaries, fiancé(e)s, or apartments, life is full of optimal stopping. So the irresistible question is whether—by evolution or education or intuition—we actually do follow the best strategies.

At first glance, the answer is no. About a dozen studies have produced the same result: people tend to stop early, leaving better applicants unseen. To get a better sense for these findings, we talked to UC Riverside's Amnon Rapoport, who has been running optimal stopping experiments in the laboratory for more than forty years.

The study that most closely follows the classical secretary problem was run in the 1990s by Rapoport and his collaborator Darryl Seale. In this study people went through numerous repetitions of the secretary problem, with either 40 or 80 applicants each time. The overall rate at which people found the best possible applicant was pretty good: about 31%, not far from the optimal 37%. Most people acted in a way that was consistent with the Look-Then-Leap Rule, but they leapt sooner than they should have more than four-fifths of the time.

Rapoport told us that he keeps this in mind when solving optimal stopping problems in his own life. In searching for an apartment, for instance, he fights his own urge to commit quickly. "Despite the fact that by nature I am very impatient and I want to take the first apartment, I try to control myself!"

But that impatience suggests another consideration that isn't taken into account in the classical secretary problem: the role of time. After all, the whole time you're searching for a secretary, you don't have a secretary. What's more, you're spending the day conducting interviews instead of getting your own work done.

This type of cost offers a potential explanation for why people stop early when solving a secretary problem in the lab. Seale and Rapoport showed that if the cost of seeing each applicant is imagined to be, for instance, 1% of the value of finding the best secretary, then the optimal strategy would perfectly align with where people actually switched from looking to leaping in their experiment.

The mystery is that in Seale and Rapoport's study, there wasn't a cost for search. So why might people in the laboratory be acting like there was one?

Because for people there's *always* a time cost. It doesn't come from the design of the experiment. It comes from people's lives.

The "endogenous" time costs of searching, which aren't usually captured by optimal stopping models, might thus provide an explanation for why human decision-making routinely diverges from the prescriptions of those models. As optimal stopping researcher Neil Bearden puts it, "After searching for a while, we humans just tend to get bored. It's not irrational to get bored, but it's hard to model that rigorously."

But this doesn't make optimal stopping problems less important; it actually makes them more important, because the flow of time turns *all* decision-making into optimal stopping.

"The theory of optimal stopping is concerned with the problem of choosing a time to take a given action," opens the definitive textbook on optimal stopping, and it's hard to think of a more concise description of the human condition. We decide the right time to buy stocks and the right time to sell them, sure; but also the right time to open the bottle of wine we've been keeping around for a special occasion, the right moment to interrupt someone, the right moment to kiss them.

Viewed this way, the secretary problem's most fundamental yet most unbelievable assumption—its strict seriality, its inexorable one-way march—is revealed to be the nature of time itself. As such, the explicit premise of the optimal stopping problem is the implicit premise of what it is to be alive. It's this that forces us to decide based on possibilities we've not yet seen, this that forces us to embrace high rates of failure even when acting optimally. No choice recurs. We may get *similar* choices again, but never that exact one. Hesitation—inaction—is just as irrevocable as action. What the motorist, locked on the one-way road, is to space, we are to the fourth dimension: we truly pass this way but once.

Intuitively, we think that rational decision-making means exhaustively enumerating our options, weighing each one carefully, and then selecting the best. But in practice, when the clock—or the ticker—is ticking, few aspects of decision-making (or of thinking more generally) are as important as this one: when to stop.

Explore/Exploit

The Latest vs. the Greatest

Your stomach rumbles. Do you go to the Italian restaurant that you know and love, or the new Thai place that just opened up? Do you take your best friend, or reach out to a new acquaintance you'd like to get to know better? This is too hard—maybe you'll just stay home. Do you cook a recipe that you know is going to work, or scour the Internet for new inspiration? Never mind, how about you just order a pizza? Do you get your "usual," or ask about the specials? You're already exhausted before you get to the first bite. And the thought of putting on a record, watching a movie, or reading a book—*which one?*—no longer seems quite so relaxing.

Every day we are constantly forced to make decisions between options that differ in a very specific dimension: do we try new things or stick with our favorite ones? We intuitively understand that life is a balance between novelty and tradition, between the latest and the greatest, between taking risks and savoring what we know and love. But just as with the look-or-leap dilemma of the apartment hunt, the unanswered question is: what balance?

In the 1974 classic *Zen and the Art of Motorcycle Maintenance*, Robert Pirsig decries the conversational opener "What's new?"—arguing that the question, "if pursued exclusively, results only in an endless parade of trivia and fashion, the silt of tomorrow." He endorses an alternative as vastly superior: "What's best?"

But the reality is not so simple. Remembering that every "best" song and restaurant among your favorites began humbly as something merely "new" to you is a reminder that there may be yet-unknown bests still out there—and thus that the new is indeed worthy of at least some of our attention.

Age-worn aphorisms acknowledge this tension but don't solve it. "Make new friends, but keep the old / Those are silver, these are gold," and "There is no life so rich and rare / But one more friend could enter there" are true enough; certainly their scansion is unimpeachable. But they fail to tell us anything useful about the *ratio* of, say, "silver" and "gold" that makes the best alloy of a life well lived.

Computer scientists have been working on finding this balance for more than fifty years. They even have a name for it: the explore/exploit tradeoff.

Explore/Exploit

In English, the words "explore" and "exploit" come loaded with completely opposite connotations. But to a computer scientist, these words have much more specific and neutral meanings. Simply put, exploration is *gathering* information, and exploitation is *using* the information you have to get a known good result.

It's fairly intuitive that never exploring is no way to live. But it's also worth mentioning that never exploiting can be every bit as bad. In the computer science definition, exploitation actually comes to characterize many of what we consider to be life's best moments. A family gathering together on the holidays is exploitation. So is a bookworm settling into a reading chair with a hot cup of coffee and a beloved favorite, or a band playing their greatest hits to a crowd of adoring fans, or a couple that has stood the test of time dancing to "their song."

What's more, exploration can be a curse.

Part of what's nice about music, for instance, is that there are constantly new things to listen to. Or, if you're a music journalist, part of what's terrible about music is that there are *constantly* new things to listen to. Being a music journalist means turning the exploration dial all the way to 11, where it's nothing but new things all the time. Music lovers might imagine working in music journalism to be paradise, but when you constantly have to explore the new you can never enjoy the fruits of your connoisseurship—a particular kind of hell. Few people know this experience as deeply as Scott Plagenhoef, the former editor in chief of *Pitchfork*. "You try to find spaces when you're working to listen to something that you just want to listen to," he says of a critic's life. His desperate urges to stop wading through unheard tunes of dubious quality and just listen to what he loved were so strong that Plagenhoef would put only new music on his iPod, to make himself physically incapable of abandoning his duties in those moments when he just really, really, really wanted to listen to the Smiths. Journalists are martyrs, exploring so that others may exploit.

In computer science, the tension between exploration and exploitation takes its most concrete form in a scenario called the "multi-armed bandit problem." The odd name comes from the colloquial term for a casino slot machine, the "one-armed bandit." Imagine walking into a casino full of different slot machines, each one with its own odds of a payoff. The rub, of course, is that you aren't told those odds in advance: until you start playing, you won't have any idea which machines are the most lucrative ("loose," as slot-machine aficionados call it) and which ones are just money sinks.

Naturally, you're interested in maximizing your total winnings. And it's clear that this is going to involve some combination of pulling the arms on different machines to test them out (exploring), and favoring the most promising machines you've found (exploiting).

To get a sense for the problem's subtleties, imagine being faced with only two machines. One you've played a total of 15 times; 9 times it paid out, and 6 times it didn't. The other you've played only twice, and it once paid out and once did not. Which is more promising?

Simply dividing the wins by the total number of pulls will give you the machine's "expected value," and by this method the first machine clearly comes out ahead. Its 9–6 record makes for an expected value of 60%, whereas the second machine's 1–1 record yields an expected value of only 50%. But there's more to it than that. After all, just two pulls aren't really very many. So there's a sense in which we just don't yet *know* how good the second machine might actually be.

Choosing a restaurant or an album is, in effect, a matter of deciding which arm to pull in life's casino. But understanding the explore/exploit tradeoff isn't just a way to improve decisions about where to eat or what to listen to. It also provides fundamental insights into how our goals should change as we age, and why the most rational course of action isn't always trying to choose the best. And it turns out to be at the heart of, among other things, web design and clinical trials—two topics that normally aren't mentioned in the same sentence.

People tend to treat decisions in isolation, to focus on finding each time the outcome with the highest expected value. But decisions are almost never isolated, and expected value isn't the end of the story. If you're thinking not just about the *next* decision, but about *all* the decisions you are going to make about the same options in the future, the explore/exploit tradeoff is crucial to the process. In this way, writes mathematician Peter Whittle, the bandit problem "embodies in essential form a conflict evident in all human action."

So which of those two arms should you pull? It's a trick question. It completely depends on something we haven't discussed yet: how long you plan to be in the casino.

Seize the Interval

"Carpe diem," urges Robin Williams in one of the most memorable scenes of the 1989 film *Dead Poets Society*. "Seize the day, boys. Make your lives extraordinary."

It's incredibly important advice. It's also somewhat self-contradictory. Seizing a day and seizing a lifetime are two entirely different endeavors. We have the expression "Eat, drink, and be merry, for tomorrow we die," but perhaps we should also have its inverse: "Start learning a new language or an instrument, and make small talk with a stranger, because life is long, and who knows what joy could blossom over many years' time." When balancing favorite experiences and new ones, nothing matters as much as the interval over which we plan to enjoy them.

"I'm more likely to try a new restaurant when I move to a city than when I'm leaving it," explains data scientist and blogger Chris Stucchio, a veteran of grappling with the explore/exploit tradeoff in both his work and his life. "I mostly go to restaurants I know and love now, because I know I'm going to be leaving New York fairly soon. Whereas a couple years ago I moved to Pune, India, and I just would eat friggin' everywhere that didn't look like it was gonna kill me. And as I was leaving the city I went back to all my old favorites, rather than trying out new stuff.... Even if I find a slightly better place, I'm only going to go there once or twice, so why take the risk?"

A sobering property of trying new things is that the value of exploration, of finding a new favorite, can only go down over time, as the remaining opportunities to savor it dwindle. Discovering an enchanting café on your last night in town doesn't give you the opportunity to return.

The flip side is that the value of exploitation can only go *up* over time. The loveliest café that you know about today is, by definition, at least as lovely as the loveliest café you knew about last month. (And if you've found another favorite since then, it might just be more so.) So explore when you will have time to use the resulting knowledge, exploit when you're ready to cash in. The interval makes the strategy.

Interestingly, since the interval makes the strategy, then by observing the strategy we can also infer the interval. Take Hollywood, for instance: Among the ten highest-grossing movies of 1981, only two were sequels. In 1991, it was three. In 2001, it was five. And in 2011, *eight* of the top ten highest-grossing films were sequels. In fact, 2011 set a record for the greatest percentage of sequels among major studio releases. Then 2012 immediately broke that record; the next year would break it again. In December 2012, journalist Nick Allen looked ahead with palpable fatigue to the year to come:

> Audiences will be given a sixth helping of X-Men plus *Fast and Furious 6*, *Die Hard 5*, *Scary Movie 5* and *Paranormal Activity 5*. There will also be *Iron Man 3*, *The Hangover 3*, and second outings for *The Muppets*, *The Smurfs*, *GI Joe* and *Bad Santa*.

From a studio's perspective, a sequel is a movie with a guaranteed fan base: a cash cow, a sure thing, an exploit. And an overload of sure things signals a short-termist approach, as with Stucchio on his way out of town. The sequels are more likely than brand-new movies to be hits this year, but where will the beloved franchises of the future come from? Such a sequel deluge is not only lamentable (certainly critics think so); it's also somewhat poignant. By entering an almost purely exploit-focused phase, the film industry seems to be signaling a belief that it is near the end of its interval.

A look into the economics of Hollywood confirms this hunch. Profits of the largest film studios declined by 40% between 2007 and 2011, and ticket sales have declined in seven of the past ten years. As the *Economist* puts it, "Squeezed between rising costs and falling revenues, the big studios have responded by trying to make more films they think will be hits: usually sequels, prequels, or anything featuring characters with name recognition." In other words, they're pulling the arms of the best machines they've got before the casino turns them out.

Win-Stay

Finding optimal algorithms that tell us exactly how to handle the multi-armed bandit problem has proven incredibly challenging. Indeed, as Peter Whittle recounts, during World War II efforts to solve the question "so sapped the energies and minds of Allied analysts . . . that the suggestion was made that the problem be dropped over Germany, as the ultimate instrument of intellectual sabotage."

The first steps toward a solution were taken in the years after the war, when Columbia mathematician Herbert Robbins showed that there's a simple strategy that, while not perfect, comes with some nice guarantees.

Robbins specifically considered the case where there are exactly two slot machines, and proposed a solution called the **Win-Stay, Lose-Shift** algorithm: choose an arm at random, and keep pulling it as long as it keeps paying off. If the arm doesn't pay off after a particular pull, then switch to the other one. Although this simple strategy is far from a complete solution, Robbins proved in 1952 that it performs reliably better than chance.

Following Robbins, a series of papers examined the "stay on a winner" principle further. Intuitively, if you were already willing to pull an arm, and it has just paid off, that should only increase your estimate of its value, and you should be only more willing to pull it again. And indeed, win-stay turns out to be an element of the optimal strategy for balancing exploration and exploitation under a wide range of conditions.

But lose-shift is another story. Changing arms each time one fails is a pretty rash move. Imagine going to a restaurant a hundred times, each time having a wonderful meal. Would one disappointment be enough to induce you to give up on it? Good options shouldn't be penalized too strongly for being imperfect.

More significantly, Win-Stay, Lose-Shift doesn't have any notion of the interval over which you are optimizing. If your favorite restaurant disappointed you the last time you ate there, that algorithm always says you should go to another place—even if it's your last night in town.

Still, Robbins's initial work on the multi-armed bandit problem kicked off a substantial literature, and researchers made significant progress over the next few years. Richard Bellman, a mathematician at the RAND

Corporation, found an exact solution to the problem for cases where we know in advance exactly how many options and opportunities we'll have in total. As with the full-information secretary problem, Bellman's trick was essentially to work backward, starting by imagining the final pull and considering which slot machine to choose given all the possible outcomes of the previous decisions. Having figured that out, you'd then turn to the second-to-last opportunity, then the previous one, and the one before that, all the way back to the start.

The answers that emerge from Bellman's method are ironclad, but with many options and a long casino visit it can require a dizzying—or impossible—amount of work. What's more, even if we are able to calculate all possible futures, we of course don't always know exactly how many opportunities (or even how many options) we'll have. For these reasons, the multi-armed bandit problem effectively stayed unsolved. In Whittle's words, "it quickly became a classic, and a byword for intransigence."

The Gittins Index

As so often happens in mathematics, though, the particular is the gateway to the universal. In the 1970s, the Unilever corporation asked a young mathematician named John Gittins to help them optimize some of their drug trials. Unexpectedly, what they got was the answer to a mathematical riddle that had gone unsolved for a generation.

Gittins, who is now a professor of statistics at Oxford, pondered the question posed by Unilever. Given several different chemical compounds, what is the quickest way to determine which compound is likely to be effective against a disease? Gittins tried to cast the problem in the most general form he could: multiple options to pursue, a different probability of reward for each option, and a certain amount of effort (or money, or time) to be allocated among them. It was, of course, another incarnation of the multi-armed bandit problem.

Both the for-profit drug companies and the medical profession they serve are constantly faced with the competing demands of the explore/exploit tradeoff. Companies want to invest R & D money into the discovery of new drugs, but also want to make sure their profitable current product lines are flourishing. Doctors want to prescribe the best existing treatments so that patients get the care they need, but also want to encourage experimental studies that may turn up even better ones.

In both cases, notably, it's not entirely clear what the relevant interval ought to be. In a sense, drug companies and doctors alike are interested in the *indefinite* future. Companies want to be around theoretically forever, and on the medical side a breakthrough could go on to help people who haven't even been born yet. Nonetheless, the present has a higher priority: a cured patient today is taken to be more valuable than one cured a week or a year from now, and certainly the same holds true of profits. Economists refer to this idea, of valuing the present more highly than the future, as "discounting."

Unlike previous researchers, Gittins approached the multi-armed bandit problem in those terms. He conceived the goal as maximizing payoffs not for a fixed interval of time, but for a future that is endless yet discounted.

Such discounting is not unfamiliar to us from our own lives. After all, if you visit a town for a ten-day vacation, then you should be making your restaurant decisions with a fixed interval in mind; but if you live in the town, this doesn't make as much sense. Instead, you might imagine the value of payoffs decreasing the further into the future they are: you care more about the meal you're going to eat tonight than the meal you're going to eat tomorrow, and more about tomorrow's meal than one a year from now, with the specifics of how much more depending on your particular "discount function." Gittins, for his part, made the assumption that the value assigned to payoffs decreases geometrically: that is, each restaurant visit you make is worth a constant fraction of the last one. If, let's say, you believe there is a 1% chance you'll get hit by a bus on any given day, then you should value tomorrow's dinner at 99% of the value of tonight's, if only because you might never get to eat it.

Working with this geometric-discounting assumption, Gittins investigated a strategy that he thought "at least would be a pretty good approximation": to think about each arm of the multi-armed bandit separately from the others, and try to work out the value of that arm on its own. He did this by imagining something rather ingenious: a bribe.

In the popular television game show *Deal or No Deal*, a contestant chooses one of twenty-six briefcases, which contain prizes ranging from a penny to a million dollars. As the game progresses, a mysterious character called the Banker will periodically call in and offer the contestant various sums of money to *not* open the chosen briefcase. It's up to the contestant to decide at what price they're willing to take a sure thing over the uncertainty of the briefcase prize.

Gittins (albeit many years before the first episode of *Deal or No Deal* aired) realized that the multi-armed bandit problem is no different. For every slot machine we know little or nothing about, there is some guaranteed payout rate which, if offered to us in lieu of that machine, will make us quite content never to pull its handle again. This number—which Gittins called the "dynamic allocation index," and which the world now knows as the **Gittins index**—suggests an obvious strategy on the casino floor: always play the arm with the highest index.*

In fact, the index strategy turned out to be more than a good approximation. It completely solves the multi-armed bandit with geometrically discounted payoffs. The tension between exploration and exploitation resolves into the simpler task of maximizing a single quantity that accounts for both. Gittins is modest about the achievement—"It's not quite Fermat's Last Theorem," he says with a chuckle—but it's a theorem that put to rest a significant set of questions about the explore/exploit dilemma.

Now, actually calculating the Gittins index for a specific machine, given its track record and our discounting rate, is still fairly involved. But once the Gittins index for a particular set of assumptions is known, it can be used for any problem of that form. Crucially, it doesn't even matter how many arms are involved, since the index for each arm is calculated separately.

In the table on the next page we provide the Gittins index values for up to nine successes and failures, assuming that a payoff on our next pull is worth 90% of a payoff now. These values can be used to resolve a variety of everyday multi-armed bandit problems. For example, under these assumptions you should, in fact, choose the slot machine that has a track record of 1–1 (and an expected value of 50%) over the one with a track record of 9–6 (and an expected value of 60%). Looking up the relevant coordinates in the table shows that the lesser-known machine has an index of 0.6346, while the more-played machine scores only a 0.6300. Problem solved: try your luck this time, and explore.

Looking at the Gittins index values in the table, there are a few other interesting observations. First, you can see the win-stay principle at work: as you go from left to right in any row, the index scores always increase. So

---

*The basic summary of this section: git while the Gittins's good.

Wins

	0	1	2	3	4	5	6	7	8	9
0	.7029	.8001	.8452	.8723	.8905	.9039	.9141	.9221	.9287	.9342
1	.5001	.6346	.7072	.7539	.7869	.8115	.8307	.8461	.8588	.8695
2	.3796	.5163	.6010	.6579	.6996	.7318	.7573	.7782	.7956	.8103
3	.3021	.4342	.5184	.5809	.6276	.6642	.6940	.7187	.7396	.7573
4	.2488	.3720	.4561	.5179	.5676	.6071	.6395	.6666	.6899	.7101
5	.2103	.3245	.4058	.4677	.5168	.5581	.5923	.6212	.6461	.6677
6	.1815	.2871	.3647	.4257	.4748	.5156	.5510	.5811	.6071	.6300
7	.1591	.2569	.3308	.3900	.4387	.4795	.5144	.5454	.5723	.5960
8	.1413	.2323	.3025	.3595	.4073	.4479	.4828	.5134	.5409	.5652
9	.1269	.2116	.2784	.3332	.3799	.4200	.4548	.4853	.5125	.5373

(Losses, labeled vertically on the left axis)

Gittins index values as a function of wins and losses, assuming that a payoff next time is worth 90% of a payoff now.

if an arm is ever the correct one to pull, and that pull is a winner, then (following the chart to the right) it can only make more sense to pull the same arm again. Second, you can see where lose-shift would get you into trouble. Having nine initial wins followed by a loss gets you an index of 0.8695, which is still higher than most of the other values in the table—so you should probably stay with that arm for at least another pull.

But perhaps the most interesting part of the table is the top-left entry. A record of 0–0—an arm that's a complete unknown—has an expected value of 0.5000 but a Gittins index of 0.7029. In other words, something you have no experience with whatsoever is more attractive than a machine that you know pays out seven times out of ten! As you go down the diagonal, notice that a record of 1–1 yields an index of 0.6346, a record of 2–2 yields 0.6010, and so on. If such 50%-successful performance persists, the index does ultimately converge on 0.5000, as experience confirms that the machine is indeed nothing special and takes away the "bonus" that spurs further exploration. But the convergence happens fairly slowly; the exploration bonus is a powerful force. Indeed, note that even a failure on the very first pull, producing a record of 0–1, makes for a Gittins index that's still above 50%.

We can also see how the explore/exploit tradeoff changes as we change the way we're discounting the future. The following table presents exactly the same information as the preceding one, but assumes that a payoff next

Wins

		0	1	2	3	4	5	6	7	8	9
	0	.8699	.9102	.9285	.9395	.9470	.9525	.9568	.9603	.9631	.9655
	1	.7005	.7844	.8268	.8533	.8719	.8857	.8964	.9051	.9122	.9183
	2	.5671	.6726	.7308	.7696	.7973	.8184	.8350	.8485	.8598	.8693
	3	.4701	.5806	.6490	.6952	.7295	.7561	.7773	.7949	.8097	.8222
Losses	4	.3969	.5093	.5798	.6311	.6697	.6998	.7249	.7456	.7631	.7781
	5	.3415	.4509	.5225	.5756	.6172	.6504	.6776	.7004	.7203	.7373
	6	.2979	.4029	.4747	.5277	.5710	.6061	.6352	.6599	.6811	.6997
	7	.2632	.3633	.4337	.4876	.5300	.5665	.5970	.6230	.6456	.6653
	8	.2350	.3303	.3986	.4520	.4952	.5308	.5625	.5895	.6130	.6337
	9	.2117	.3020	.3679	.4208	.4640	.5002	.5310	.5589	.5831	.6045

Gittins index values as a function of wins and losses, assuming that a payoff next time is worth 99% of a payoff now.

time is worth 99% of one now, rather than 90%. With the future weighted nearly as heavily as the present, the value of making a chance discovery, relative to taking a sure thing, goes up even more. Here, a totally untested machine with a 0–0 record is worth a guaranteed 86.99% chance of a payout!

The Gittins index, then, provides a formal, rigorous justification for preferring the unknown, provided we have some opportunity to exploit the results of what we learn from exploring. The old adage tells us that "the grass is always greener on the other side of the fence," but the math tells us why: the unknown has a chance of being better, even if we actually expect it to be no different, or if it's just as likely to be worse. The untested rookie is worth more (early in the season, anyway) than the veteran of seemingly equal ability, precisely because we know less about him. Exploration in itself has value, since trying new things increases our chances of finding the best. So taking the future into account, rather than focusing just on the present, drives us toward novelty.

The Gittins index thus provides an amazingly straightforward solution to the multi-armed bandit problem. But it doesn't necessarily close the book on the puzzle, or help us navigate *all* the explore/exploit tradeoffs of everyday life. For one, the Gittins index is optimal only under some strong assumptions. It's based on geometric discounting of future reward, valuing each pull at a constant fraction of the previous one, which is something that a variety of experiments in behavioral economics and psychology suggest people don't do. And if there's a cost to switching among options, the Gittins strategy is no longer optimal either. (The grass on the other side of the fence may look a bit greener, but that doesn't necessarily warrant climbing the fence—let alone taking out a second mortgage.) Perhaps even more importantly, it's hard to compute the Gittins index on the fly. If you carry around a table of index values you can optimize your dining choices, but the time and effort involved might not be worth it. ("Wait, I can resolve this argument. That restaurant was good 29 times out of 35, but this other one has been good 13 times out of 16, so the Gittins indices are . . . Hey, where did everybody go?")

In the time since the development of the Gittins index, such concerns have sent computer scientists and statisticians searching for simpler and more flexible strategies for dealing with multi-armed bandits. These strategies are easier for humans (and machines) to apply in a range of situations than crunching the optimal Gittins index, while still providing comparably good performance. They also engage with one of our biggest human fears regarding decisions about which chances to take.

Regret and Optimism

> Regrets, I've had a few. But then again, too few to mention.
>
> —FRANK SINATRA

> For myself I am an optimist. It does not seem to be much use being anything else.
>
> —WINSTON CHURCHILL

If the Gittins index is too complicated, or if you're not in a situation well characterized by geometric discounting, then you have another option: focus on *regret*. When we choose what to eat, who to spend time with, or what city to live in, regret looms large—presented with a set of good options, it is easy to torture ourselves with the consequences of making the wrong choice. These regrets are often about the things we failed to do, the options we never tried. In the memorable words of management theorist Chester Barnard, "To try and fail is at least to learn; to fail to try is to suffer the inestimable loss of what might have been."

Regret can also be highly motivating. Before he decided to start

Amazon.com, Jeff Bezos had a secure and well-paid position at the investment company D. E. Shaw & Co. in New York. Starting an online bookstore in Seattle was going to be a big leap—something that his boss (that's D. E. Shaw) advised him to think about carefully. Says Bezos:

> The framework I found, which made the decision incredibly easy, was what I called—which only a nerd would call—a "regret minimization framework." So I wanted to project myself forward to age 80 and say, "Okay, now I'm looking back on my life. I want to have minimized the number of regrets I have." I knew that when I was 80 I was not going to regret having tried this. I was not going to regret trying to participate in this thing called the Internet that I thought was going to be a really big deal. I knew that if I failed I wouldn't regret that, but I knew the one thing I might regret is not ever having tried. I knew that that would haunt me every day, and so, when I thought about it that way it was an incredibly easy decision.

Computer science can't offer you a life with no regret. But it can, potentially, offer you just what Bezos was looking for: a life with *minimal* regret.

Regret is the result of comparing what we actually did with what would have been best in hindsight. In a multi-armed bandit, Barnard's "inestimable loss" can in fact be measured precisely, and regret assigned a number: it's the difference between the total payoff obtained by following a particular strategy and the total payoff that theoretically could have been obtained by just pulling the best arm every single time (had we only known from the start which one it was). We can calculate this number for different strategies, and search for those that minimize it.

In 1985, Herbert Robbins took a second shot at the multi-armed bandit problem, some thirty years after his initial work on Win-Stay, Lose-Shift. He and fellow Columbia mathematician Tze Leung Lai were able to prove several key points about regret. First, assuming you're not omniscient, your total amount of regret will probably never stop increasing, even if you pick the best possible strategy—because even the best strategy isn't perfect every time. Second, regret will increase at a slower rate if you pick the best strategy than if you pick others; what's more, with a good strategy regret's rate of growth will go down over time, as you learn more about the problem and are able to make better choices. Third, and most specifically, the minimum possible regret—again assuming non-omniscience—is regret that increases at a *logarithmic* rate with every pull of the handle.

Logarithmically increasing regret means that we'll make as many mistakes in our first ten pulls as in the following ninety, and as many in our first year as in the rest of the decade combined. (The first decade's mistakes, in turn, are as many as we'll make for the rest of the century.) That's some measure of consolation. In general we can't realistically expect someday to never have any more regrets. But if we're following a regret-minimizing algorithm, every year we can expect to have fewer new regrets than we did the year before.

Starting with Lai and Robbins, researchers in recent decades have set about looking for algorithms that offer the guarantee of minimal regret. Of the ones they've discovered, the most popular are known as **Upper Confidence Bound** algorithms.

Visual displays of statistics often include so-called error bars that extend above and below any data point, indicating uncertainty in the measurement; the error bars show the range of plausible values that the quantity being measured could actually have. This range is known as the "confidence interval," and as we gain more data about something the confidence interval will shrink, reflecting an increasingly accurate assessment. (For instance, a slot machine that has paid out once out of two pulls will have a wider confidence interval, though the same expected value, as a machine that has paid out five times on ten pulls.) In a multi-armed bandit problem, an Upper Confidence Bound algorithm says, quite simply, to pick the option for which the top of the confidence interval is highest.

Like the Gittins index, therefore, Upper Confidence Bound algorithms assign a single number to each arm of the multi-armed bandit. And that number is set to the highest value that the arm could reasonably have, based on the information available so far. So an Upper Confidence Bound algorithm doesn't care which arm *has* performed best so far; instead, it chooses the arm that *could* reasonably perform best in the future. If you have never been to a restaurant before, for example, then for all you know it could be great. Even if you have gone there once or twice, and tried a couple of their dishes, you might not have enough information to rule out the possibility that it could yet prove better than your regular favorite. Like the Gittins index, the Upper Confidence Bound is always greater than the expected value, but by less and less as we gain more experience with a particular option. (A restaurant with a single mediocre review still retains a *potential* for greatness that's absent in a restaurant with hundreds of such reviews.) The recommendations given by Upper Confidence Bound algorithms will be similar to those provided by the Gittins index, but they are significantly easier to compute, and they don't require the assumption of geometric discounting.

Upper Confidence Bound algorithms implement a principle that has been dubbed "optimism in the face of uncertainty." Optimism, they show, can be perfectly rational. By focusing on the best that an option *could* be, given the evidence obtained so far, these algorithms give a boost to possibilities we know less about. As a consequence, they naturally inject a dose of exploration into the decision-making process, leaping at new options with enthusiasm because any one of them could be the next big thing. The same principle has been used, for instance, by MIT's Leslie Kaelbling, who builds "optimistic robots" that explore the space around them by boosting the value of uncharted terrain. And it clearly has implications for human lives as well.

The success of Upper Confidence Bound algorithms offers a formal justification for the benefit of the doubt. Following the advice of these algorithms, you should be excited to meet new people and try new things—to assume the best about them, in the absence of evidence to the contrary. In the long run, optimism is the best prevention for regret.

Bandits Online

In 2007, Google product manager Dan Siroker took a leave of absence to join the presidential campaign of then senator Barack Obama in Chicago. Heading the "New Media Analytics" team, Siroker brought one of Google's web practices to bear on the campaign's bright-red DONATE button. The result was nothing short of astonishing: $57 million of additional donations were raised as a direct result of his work.

What exactly did he do to that button?

He A/B tested it.

A/B testing works as follows: a company drafts several different versions of a particular webpage. Perhaps they try different colors or images, or different headlines for a news article, or different arrangements of items on the screen. Then they randomly assign incoming users to these various pages, usually in equal numbers. One user may see a red button, while another user may see a blue one; one may see DONATE and another may see CONTRIBUTE. The relevant metrics (e.g., click-through rate or average revenue per visitor) are then monitored. After a period of time, if statistically significant effects are observed, the "winning" version is typically locked into place—or becomes the control for another round of experiments.

In the case of Obama's donation page, Siroker's A/B tests were revealing. For first-time visitors to the campaign site, a DONATE AND GET A GIFT button turned out to be the best performer, even after the cost of sending the gifts was taken into account. For longtime newsletter subscribers who had never given money, PLEASE DONATE worked the best, perhaps appealing to their guilt. For visitors who had already donated in the past, CONTRIBUTE worked best at securing follow-up donations—the logic being perhaps that the person had already "donated" but could always "contribute" more. And in all cases, to the astonishment of the campaign team, a simple black-and-white photo of the Obama family outperformed any other photo or video the team could come up with. The net effect of all these independent optimizations was gigantic.

If you've used the Internet basically at all over the past decade, then you've been a part of someone else's explore/exploit problem. Companies want to discover the things that make them the most money while simultaneously making as much of it as they can—explore, exploit. Big tech firms such as Amazon and Google began carrying out live A/B tests on their users starting in about 2000, and over the following years the Internet has become the world's largest controlled experiment. What are these companies exploring and exploiting? In a word, you: whatever it is that makes you move your mouse and open your wallet.

Companies A/B test their site navigation, the subject lines and timing of their marketing emails, and sometimes even their actual features and pricing. Instead of "the" Google search algorithm and "the" Amazon checkout flow, there are now untold and unfathomably subtle permutations. (Google infamously tested forty-one shades of blue for one of its toolbars in 2009.) In some cases, it's unlikely that any pair of users will have the exact same experience.

Data scientist Jeff Hammerbacher, former manager of the Data group at Facebook, once told *Bloomberg Businessweek* that "the best minds of my generation are thinking about how to make people click ads." Consider it the millennials' *Howl*—what Allen Ginsberg's immortal "I saw the best minds of my generation destroyed by madness" was to the Beat Generation. Hammerbacher's take on the situation was that this state of affairs "sucks." But regardless of what one makes of it, the web is allowing for an experimental science of the click the likes of which had never even been dreamed of by marketers of the past.

We know what happened to Obama in the 2008 election, of course. But what happened to his director of analytics, Dan Siroker? After the inauguration, Siroker returned west, to California, and with fellow Googler Pete Koomen co-founded the website optimization firm Optimizely. By the 2012 presidential election cycle, their company counted among its clients both the Obama re-election campaign *and* the campaign of Republican challenger Mitt Romney.

Within a decade or so after its first tentative use, A/B testing was no longer a secret weapon. It has become such a deeply embedded part of how business and politics are conducted online as to be effectively taken for granted. The next time you open your browser, you can be sure that the colors, images, text, perhaps even the prices you see—and certainly the ads—have come from an explore/exploit algorithm, tuning itself to your clicks. In this particular multi-armed bandit problem, you're not the gambler; you're the jackpot.

The process of A/B testing itself has become increasingly refined over time. The most canonical A/B setup—splitting the traffic evenly between two options, running the test for a set period of time, and thereafter giving all the traffic to the winner—might not necessarily be the best algorithm for solving the problem, since it means half the users are stuck getting the inferior option as long as the test continues. And the rewards for finding a better approach are potentially very high. More than 90% of Google's approximately $50 billion in annual revenue currently comes from paid advertising, and online commerce comprises hundreds of billions of dollars a year. This means that explore/exploit algorithms effectively power, both economically and technologically, a significant fraction of the Internet itself. The best algorithms to use remain hotly contested, with rival statisticians, engineers, and bloggers endlessly sparring about the optimal way to balance exploration and exploitation in every possible business scenario.

Debating the precise distinctions among various takes on the explore/exploit problem may seem hopelessly arcane. In fact, these distinctions turn out to matter immensely—and it's not just presidential elections and the Internet economy that are at stake.

It's also human lives.

Clinical Trials on Trial

Between 1932 and 1972, several hundred African-American men with syphilis in Macon County, Alabama, went deliberately untreated by medical professionals, as part of a forty-year experiment by the US Public Health Service known as the Tuskegee Syphilis Study. In 1966, Public Health Service employee Peter Buxtun filed a protest. He filed a second protest in 1968. But it was not until he broke the story to the press—it appeared in the *Washington Star* on July 25, 1972, and was the front-page story in the *New York Times* the next day—that the US government finally halted the study.

What followed the public outcry, and the subsequent congressional hearing, was an initiative to formalize the principles and standards of medical ethics. A commission held at the pastoral Belmont Conference Center in Maryland resulted in a 1979 document known as the Belmont Report. The Belmont Report lays out a foundation for the ethical practice of medical experiments, so that the Tuskegee experiment—an egregious, unambiguously inappropriate breach of the health profession's duty to its patients—might never be repeated. But it also notes the difficulty, in many other cases, of determining exactly where the line should be drawn.

"The Hippocratic maxim 'do no harm' has long been a fundamental principle of medical ethics," the report points out. "[The physiologist] Claude Bernard extended it to the realm of research, saying that one should not injure one person regardless of the benefits that might come to others. However, even avoiding harm requires learning what is harmful; and, in the process of obtaining this information, persons may be exposed to risk of harm."

The Belmont Report thus acknowledges, but does not resolve, the tension that exists between acting on one's best knowledge and gathering more. It also makes it clear that gathering knowledge can be so valuable that some aspects of normal medical ethics can be suspended. Clinical testing of new drugs and treatments, the report notes, often requires risking harm to some patients, even if steps are taken to minimize that risk.

> The principle of beneficence is not always so unambiguous. A difficult ethical problem remains, for example, about research [on childhood diseases]

that presents more than minimal risk without immediate prospect of direct benefit to the children involved. Some have argued that such research is inadmissible, while others have pointed out that this limit would rule out much research promising great benefit to children in the future. Here again, as with all hard cases, the different claims covered by the principle of beneficence may come into conflict and force difficult choices.

One of the fundamental questions that has arisen in the decades since the Belmont Report is whether the standard approach to conducting clinical trials really does minimize risk to patients. In a conventional clinical trial, patients are split into groups, and each group is assigned to receive a different treatment for the duration of the study. (Only in exceptional cases does a trial get stopped early.) This procedure focuses on decisively resolving the question of which treatment is better, rather than on providing the best treatment to each patient in the trial itself. In this way it operates exactly like a website's A/B test, with a certain fraction of people receiving an experience during the experiment that will eventually be proven inferior. But doctors, like tech companies, are gaining some information about which option is better *while* the trial proceeds—information that could be used to improve outcomes not only for future patients beyond the trial, but for the patients currently in it.

Millions of dollars are at stake in experiments to find the optimal configuration of a website, but in clinical trials, experimenting to find optimal treatments has direct life-or-death consequences. And a growing community of doctors and statisticians think that we're doing it wrong: that we should be treating the selection of treatments as a multi-armed bandit problem, and trying to get the better treatments to people even while an experiment is in progress.

In 1969, Marvin Zelen, a biostatistician who is now at Harvard, proposed conducting "adaptive" trials. One of the ideas he suggested was a randomized "play the winner" algorithm—a version of Win-Stay, Lose-Shift, in which the chance of using a given treatment is increased by each win and decreased by each loss. In Zelen's procedure, you start with a hat that contains one ball for each of the two treatment options being studied. The treatment for the first patient is selected by drawing a ball at random from the hat (the ball is put back afterward). If the chosen treatment is a success, you put another ball for that treatment into the hat—now you have three balls, two of which are for the successful treatment. If it fails, then you put another ball for the *other* treatment into the hat, making it more likely you'll choose the alternative.

Zelen's algorithm was first used in a clinical trial sixteen years later, for a study of extracorporeal membrane oxygenation, or "ECMO"—an audacious approach to treating respiratory failure in infants. Developed in the 1970s by Robert Bartlett of the University of Michigan, ECMO takes blood that's heading for the lungs and routes it instead out of the body, where it is oxygenated by a machine and returned to the heart. It is a drastic measure, with risks of its own (including the possibility of embolism), but it offered a possible approach in situations where no other options remained. In 1975 ECMO saved the life of a newborn girl in Orange County, California, for whom even a ventilator was not providing enough oxygen. That girl has now celebrated her fortieth birthday and is married with children of her own. But in its early days the ECMO technology and procedure were considered highly experimental, and early studies in adults showed no benefit compared to conventional treatments.

From 1982 to 1984, Bartlett and his colleagues at the University of Michigan performed a study on newborns with respiratory failure. The team was clear that they wanted to address, as they put it, "the ethical issue of withholding an unproven but potentially lifesaving treatment," and were "reluctant to withhold a lifesaving treatment from alternate patients simply to meet conventional random assignment technique." Hence they turned to Zelen's algorithm. The strategy resulted in one infant being assigned the "conventional" treatment and dying, and eleven infants in a row being assigned the experimental ECMO treatment, all of them surviving. Between April and November of 1984, after the end of the official study, ten additional infants met the criteria for ECMO treatment. Eight were treated with ECMO, and all eight survived. Two were treated conventionally, and both died.

These are eye-catching numbers, yet shortly after the University of Michigan study on ECMO was completed it became mired in controversy. Having so few patients in a trial receive the conventional treatment deviated significantly from standard methodology, and the procedure itself was highly invasive and potentially risky. After the publication of the paper, Jim Ware, professor of biostatistics at the Harvard School of Public Health, and his medical colleagues examined the data carefully and concluded that they "did not justify routine use of ECMO without further study." So Ware and his colleagues designed a second clinical trial, still trying to balance the acquisition of knowledge with the effective treatment of patients but using a less radical design. They would randomly assign patients to either ECMO or the conventional treatment until a prespecified number of deaths was observed in one of the groups. Then they would switch all the patients in the study to the more effective treatment of the two.

In the first phase of Ware's study, four of ten infants receiving conventional treatment died, and all nine of nine infants receiving ECMO survived. The four deaths were enough to trigger a transition to the second phase, where all twenty patients were treated with ECMO and nineteen survived. Ware and colleagues were convinced, concluding that "it is difficult to defend further randomization ethically."

But some had already concluded this *before* the Ware study, and were vocal about it. The critics included Don Berry, one of the world's leading experts on multi-armed bandits. In a comment that was published alongside the Ware study in *Statistical Science*, Berry wrote that "randomizing patients to non-ECMO therapy as in the Ware study was unethical. . . . In my view, the Ware study should not have been conducted."

And yet even the Ware study was not conclusive for all in the medical community. In the 1990s yet another study on ECMO was conducted, enrolling nearly two hundred infants in the United Kingdom. Instead of using adaptive algorithms, this study followed the traditional methods, splitting the infants randomly into two equal groups. The researchers justified the experiment by saying that ECMO's usefulness "is controversial because of varying interpretation of the available evidence." As it turned out, the difference between the treatments wasn't as pronounced in the United Kingdom as it had been in the two American studies, but the results were nonetheless declared "in accord with the earlier preliminary findings that a policy of ECMO support reduces the risk of death." The cost of that knowledge? Twenty-four more infants died in the "conventional" group than in the group receiving ECMO treatment.

The widespread difficulty with accepting results from adaptive clinical trials might seem incomprehensible. But consider that part of what the advent of statistics did for medicine, at the start of the twentieth century, was to transform it from a field in which doctors had to persuade each other in ad hoc ways about every new treatment into one where they had clear guidelines about what sorts of evidence were and were not persuasive. Changes to accepted standard statistical practice have the potential to upset this balance, at least temporarily.

After the controversy over ECMO, Don Berry moved from the statistics department at the University of Minnesota to the MD Anderson Cancer Center in Houston, where he has used methods developed by studying multi-armed bandits to design clinical trials for a variety of cancer treatments. While he remains one of the more vocal critics of randomized clinical trials, he is by no means the only one. In recent years, the ideas he's been fighting for are finally beginning to come into the mainstream. In February 2010, the FDA released a "guidance" document, "Adaptive Design Clinical Trials for Drugs and Biologics," which suggests—despite a long history of sticking to an option they trust—that they might at last be willing to explore alternatives.

The Restless World

Once you become familiar with them, it's easy to see multi-armed bandits just about everywhere we turn. It's rare that we make an isolated decision, where the outcome doesn't provide us with any information that we'll use to make other decisions in the future. So it's natural to ask, as we did with optimal stopping, how well people generally tend to solve these problems—a question that has been extensively explored in the laboratory by psychologists and behavioral economists.

In general, it seems that people tend to over-explore—to favor the new disproportionately over the best. In a simple demonstration of this phenomenon, published in 1966, Amos Tversky and Ward Edwards conducted experiments where people were shown a box with two lights on it and told that each light would turn on a fixed (but unknown) percentage of the time. They were then given 1,000 opportunities either to observe which light came on, or to place a bet on the outcome without getting to observe it. (Unlike a more traditional bandit problem setup, here one could not make a "pull" that was both wager and observation at once; participants would not learn whether their bets had paid off until the end.) This is pure exploration vs. exploitation, pitting the gaining of information squarely against the use of it. For the most part, people adopted a sensible strategy of observing for a while, then placing bets on what seemed like the best outcome—but they consistently spent a lot more time observing than they should have. How much more time? In one experiment, one light came on 60% of the time and the other 40% of the time, a difference neither particularly blatant nor particularly subtle. In that case, people chose to observe 505 times, on average, placing bets the other 495 times. But the math says they should have started to bet after just 38 observations—leaving 962 chances to cash in.

Other studies have produced similar conclusions. In the 1990s, Robert Meyer and Yong Shi, researchers at Wharton, ran a study where people were given a choice between two options, one with a known payoff chance and one unknown—specifically two airlines, an established carrier with a known on-time rate and a new company without a track record yet. Given the goal of maximizing the number of on-time arrivals over some period of time, the mathematically optimal strategy is to initially only fly the new airline, as long as the established one isn't clearly better. If at any point it's apparent that the well-known carrier is better—that is, if the Gittins index of the new option falls below the on-time rate of the familiar carrier—then you should switch hard to the familiar one and never look back. (Since in this setup you can't get any more information about the new company once you stop flying it, there is no opportunity for it to redeem itself.) But in the experiment, people tended to use the untried airline too little when it was good and too much when it was bad. They also didn't make clean breaks away from it, often continuing to alternate, particularly when neither airline was departing on time. All of this is consistent with tending to over-explore.

Finally, psychologists Mark Steyvers, Michael Lee, and E.-J. Wagenmakers have run an experiment with a four-armed bandit, asking a group of people to choose which arm to play over a sequence of fifteen opportunities. They then classified the strategies that participants seemed to use. The results suggested that 30% were closest to the optimal strategy, 47% most resembled Win-Stay, Lose-Shift, and 22% seemed to move at random between selecting a new arm and playing the best arm found so far. Again, this is consistent with over-exploring, as Win-Stay, Lose-Shift and occasionally trying an arm at random are both going to lead people to try things other than the best option late in the game, when they should be purely exploiting.

So, while we tend to commit to a new secretary too soon, it seems like we tend to stop trying new airlines too late. But just as there's a cost to not having a secretary, there's a cost to committing too soon to a particular airline: the world might change.

The standard multi-armed bandit problem assumes that the probabilities with which the arms pay off are fixed over time. But that's not necessarily true of airlines, restaurants, or other contexts in which people have to make repeated choices. If the probabilities of a payoff on the different arms change over time—what has been termed a "restless bandit"—the problem becomes much harder. (So much harder, in fact, that there's no tractable algorithm for completely solving it, and it's believed there never will be.) Part of this difficulty is that it is no longer simply a matter of exploring for a while and then exploiting: when the world can change, continuing to explore can be the right choice. It might be worth going back to that disappointing restaurant you haven't visited for a few years, just in case it's under new management.

In his celebrated essay "Walking," Henry David Thoreau reflected on how he preferred to do his traveling close to home, how he never tired of his surroundings and always found something new or surprising in the Massachusetts landscape. "There is in fact a sort of harmony discoverable between the capabilities of the landscape within a circle of ten miles' radius, or the limits of an afternoon walk, and the threescore years and ten of human life," he wrote. "It will never become quite familiar to you."

To live in a restless world requires a certain restlessness in oneself. So long as things continue to change, you must never fully cease exploring.

Still, the algorithmic techniques honed for the standard version of the multi-armed bandit problem are useful even in a restless world. Strategies like the Gittins index and Upper Confidence Bound provide reasonably good approximate solutions and rules of thumb, particularly if payoffs don't change very much over time. And many of the world's payoffs are arguably more static today than they've ever been. A berry patch might be ripe one week and rotten the next, but as Andy Warhol put it, "A Coke is a Coke." Having instincts tuned by evolution for a world in constant flux isn't necessarily helpful in an era of industrial standardization.

Perhaps most importantly, thinking about versions of the multi-armed bandit problem that do have optimal solutions doesn't just offer algorithms, it also offers insights. The conceptual vocabulary derived from the classical form of the problem—the tension of explore/exploit, the importance of the interval, the high value of the 0–0 option, the minimization of regret—gives us a new way of making sense not only of specific problems that come before us, but of the entire arc of human life.

Explore . . .

While laboratory studies can be illuminating, the interval of many of the most important problems people face is far too long to be studied in the lab. Learning the structure of the world around us and forming lasting social relationships are both lifelong tasks. So it's instructive to see how the general pattern of early exploration and late exploitation appears over the course of a lifetime.

One of the curious things about human beings, which any developmental psychologist aspires to understand and explain, is that we take years to become competent and autonomous. Caribou and gazelles must be prepared to run from predators the day they're born, but humans take more than a year to make their first steps. Alison Gopnik, professor of developmental psychology at UC Berkeley and author of *The Scientist in the Crib*, has an explanation for why human beings have such an extended period of dependence: "it gives you a developmental way of solving the exploration/exploitation tradeoff." As we have seen, good algorithms for playing multi-armed bandits tend to explore more early on, exploiting the resulting knowledge later. But as Gopnik points out, "the disadvantage of that is that you don't get good payoffs when you are in the exploration stage." Hence childhood: "Childhood gives you a period in which you can just explore possibilities, and you don't have to worry about payoffs because payoffs are being taken care of by the mamas and the papas and the grandmas and the babysitters."

Thinking about children as simply being at the transitory exploration stage of a lifelong algorithm might provide some solace for parents of preschoolers. (Tom has two highly exploratory preschool-age daughters, and hopes they are following an algorithm that has minimal regret.) But it also provides new insights about the rationality of children. Gopnik points out that "if you look at the history of the way that people have thought about children, they have typically argued that children are cognitively deficient in various ways—because if you look at their exploit capacities, they look terrible. They can't tie their shoes, they're not good at long-term planning, they're not good at focused attention. Those are all things that kids are really awful at." But pressing buttons at random, being very interested in new toys, and jumping quickly from one thing to another are all things that kids are really great at. And those are exactly what they should be doing if their goal is exploration. If you're a baby, putting every object in the house into your mouth is like studiously pulling all the handles at the casino.

More generally, our intuitions about rationality are too often informed by exploitation rather than exploration. When we talk about decision-making, we usually focus just on the immediate payoff of a single decision—and if you treat every decision as if it were your last, then indeed only exploitation makes sense. But over a lifetime, you're going to make a lot of decisions. And it's actually rational to emphasize exploration—the new rather than the best, the exciting rather than the safe, the random rather than the considered—for many of those choices, particularly earlier in life.

What we take to be the caprice of children may be wiser than we know.

. . . And Exploit

> I had reached a juncture in my reading life that is familiar to those who have been there: in the allotted time left to me on earth, should I read more and more new books, or should I cease with that vain consumption—vain because it is endless—and begin to reread those books that had given me the intensest pleasure in my past.
>
> —LYDIA DAVIS

At the other extreme from toddlers we have the elderly. And thinking about aging from the perspective of the explore/exploit dilemma also provides some surprising insights into how we should expect our lives to change as time goes on.

Laura Carstensen, a professor of psychology at Stanford, has spent her career challenging our preconceptions about getting older. Particularly, she has investigated exactly how, and why, people's social relationships change as they age. The basic pattern is clear: the size of people's social networks (that is, the number of social relationships they engage in) almost invariably decreases over time. But Carstensen's research has transformed how we should think about this phenomenon.

The traditional explanation for the elderly having smaller social networks is that it's just one example of the decrease in quality of life that comes with aging—the result of diminished ability to contribute to social relationships, greater fragility, and general disengagement from society. But Carstensen has argued that, in fact, the elderly have fewer social relationships by choice. As she puts it, these decreases are "the result of life-long selection processes by which people strategically and adaptively cultivate their social networks to maximize social and emotional gains and minimize social and emotional risks."

What Carstensen and her colleagues found is that the shrinking of social networks with aging is due primarily to "pruning" peripheral relationships and focusing attention instead on a core of close friends and family members. This process seems to be a deliberate choice: as people approach the end of their lives, they want to focus more on the connections that are the most meaningful.

In an experiment testing this hypothesis, Carstensen and her collaborator Barbara Fredrickson asked people to choose who they'd rather spend thirty minutes with: an immediate family member, the author of a book they'd recently read, or somebody they had met recently who seemed to share their interests. Older people preferred the family member; young people were just as excited to meet the author or make a new friend. But in a critical twist, if the young people were asked to imagine that they were about to move across the country, they preferred the family member too. In another study, Carstensen and her colleagues found the same result in the other direction as well: if older people were asked to imagine that a medical breakthrough would allow them to live twenty years longer, their preferences became indistinguishable from those of young people. The point is that these differences in social preference are not about age as such—they're about where people perceive themselves to be on the *interval* relevant to their decision.

Being sensitive to how much time you have left is exactly what the computer science of the explore/exploit dilemma suggests. We think of the young as stereotypically fickle; the old, stereotypically set in their ways. In fact, both are behaving completely appropriately with respect to their intervals. The deliberate honing of a social network down to the most meaningful relationships is the rational response to having less time to enjoy them.

Recognizing that old age is a time of exploitation helps provide new perspectives on some of the classic phenomena of aging. For example, while going to college—a new social environment filled with people you haven't met—is typically a positive, exciting time, going to a retirement home—a new social environment filled with people you haven't met—can be painful. And that difference is partly the result of where we are on the explore/exploit continuum at those stages of our lives.

The explore/exploit tradeoff also tells us how to think about advice from our elders. When your grandfather tells you which restaurants are good, you should listen—these are pearls gleaned from decades of searching. But when he only goes to the same restaurant at 5:00 p.m. every day, you should feel free to explore other options, even though they'll likely be worse.

Perhaps the deepest insight that comes from thinking about later life as a chance to exploit knowledge acquired over decades is this: life should get better over time. What an explorer trades off for knowledge is pleasure. The Gittins index and the Upper Confidence Bound, as we've seen, inflate the appeal of lesser-known options beyond what we actually expect, since pleasant surprises can pay off many times over. But at the same time, this means that exploration *necessarily* leads to being let down on most occasions. Shifting the bulk of one's attention to one's favorite things should increase quality of life. And it seems like it does: Carstensen has found that older people are generally more satisfied with their social networks, and often report levels of emotional well-being that are higher than those of younger adults.

So there's a lot to look forward to in being that late-afternoon restaurant regular, savoring the fruits of a life's explorations.

3 | Sorting

Making Order

Nowe if the word, which thou art desirous to finde, begin with (a) then looke in the beginning of this Table, but if with (v) looke towards the end. Againe, if thy word beginne with (ca) looke in the beginning of the letter (c) but if with (cu) then looke toward the end of that letter. And so of all the rest. &c.

—ROBERT CAWDREY, *A TABLE ALPHABETICALL* (1604)

Before Danny Hillis founded the Thinking Machines corporation, before he invented the famous Connection Machine parallel supercomputer, he was an MIT undergraduate, living in the student dormitory, and horrified by his roommate's socks.

What horrified Hillis, unlike many a college undergraduate, wasn't his roommate's hygiene. It wasn't that the roommate didn't *wash* the socks; he did. The problem was what came next.

The roommate pulled a sock out of the clean laundry hamper. Next he pulled another sock out at random. If it didn't match the first one, he tossed it back in. Then he continued this process, pulling out socks one by one and tossing them back until he found a match for the first.

With just 10 different pairs of socks, following this method will take on average *19* pulls merely to complete the first pair, and 17 more pulls to complete the second. In total, the roommate can expect to go fishing in the hamper 110 times just to pair 20 socks.

It was enough to make any budding computer scientist request a room transfer.

Now, just how socks *should* be sorted is a good way get computer scientists talking at surprising length. A question about socks posted to the programming website Stack Overflow in 2013 prompted some twelve thousand words of debate.

"Socks confound me!" confessed legendary cryptographer and Turing Award–winning computer scientist Ron Rivest to the two of us when we brought up the topic.

He was wearing sandals at the time.

The Ecstasy of Sorting

Sorting is at the very heart of what computers do. In fact, in many ways it was sorting that brought the computer into being.

In the late nineteenth century, the American population was growing by 30% every decade, and the number of "subjects of inquiry" in the US Census had gone from just five in 1870 to more than two hundred in 1880. The tabulation of the 1880 census took eight years—just barely finishing by the time the 1890 census began. As a writer at the time put it, it was a wonder "the clerks who toiled at the irritating slips of tally paper . . . did not go blind and crazy." The whole enterprise was threatening to collapse under its own weight. Something had to be done.

Inspired by the punched railway tickets of the time, an inventor by the name of Herman Hollerith devised a system of punched manila cards to store information, and a machine, which he called the Hollerith Machine, to count and sort them. Hollerith was awarded a patent in 1889, and the government adopted the Hollerith Machine for the 1890 census. No one had ever seen anything like it. Wrote one awestruck observer, "The apparatus works as unerringly as the mills of the Gods, but beats them hollow as to speed." Another, however, reasoned that the invention was of limited use: "As no one will ever use it but governments, the inventor will not likely get very rich." This prediction, which Hollerith clipped and saved, would not prove entirely correct. Hollerith's firm merged with several others in 1911 to become the Computing-Tabulating-Recording Company. A few years later it was renamed—to International Business Machines, or IBM.

Sorting continued to spur the development of the computer through the next century. The first code ever written for a "stored program" computer was a program for efficient sorting. In fact, it was the computer's ability to outsort IBM's dedicated card-sorting machines that convinced the US government their enormous financial investment in a general-purpose machine was justified. By the 1960s, one study estimated that more than a quarter of the computing resources of the world were being spent on sorting. And no wonder—sorting is essential to working with almost any kind of information. Whether it's finding the largest or the smallest, the most common or the rarest, tallying, indexing, flagging duplicates, or just plain looking for the thing you want, they all generally begin under the hood with a sort.

But sorting is more pervasive, even, than this. After all, one of the main reasons things get sorted is to be shown in useful form to human eyes, which means that sorting is also key to the human experience of information. Sorted lists are so ubiquitous that—like the fish who asks, "What is water?"—we must consciously work to perceive them at all. And then we perceive them everywhere.

Our email inbox typically displays the top fifty messages of potentially thousands, sorted by time of receipt. When we look for restaurants on Yelp we're shown the top dozen or so of hundreds, sorted by proximity or by rating. A blog shows a cropped list of articles, sorted by date. The Facebook news feed, Twitter stream, and Reddit homepage all present themselves as lists, sorted by some proprietary measure. We refer to things like Google and Bing as "search engines," but that is something of a misnomer: they're really *sort* engines. What makes Google so dominant as a means of accessing the world's information is less that it *finds* our text within hundreds of millions of webpages—its 1990s competitors could generally do that part well enough—but that it *sorts* those webpages so well, and only shows us the most relevant ten.

The truncated top of an immense, sorted list is in many ways the universal user interface.

Computer science gives us a way to understand what's going on behind the scenes in all of these cases, which in turn can offer us some insight for those times when *we* are the one stuck making order—with our bills, our papers, our books, our socks, probably more times each day than we realize. By quantifying the vice (and the virtue) of mess, it also shows us the cases where we actually shouldn't make order at all.

What's more, when we begin looking, we see that sorting isn't just something we do with information. It's something we do with people. Perhaps the place where the computer science of establishing rank is most unexpectedly useful is on the sporting field and in the boxing ring—which is why knowing a little about sorting might help explain how human beings are able to live together while only occasionally coming to blows. That is to say, sorting offers some surprising clues about the nature of society— that other, larger, and more important kind of order that we make.

The Agony of Sorting

"To lower costs per unit of output, people usually increase the size of their operations," wrote J. C. Hosken in 1955, in the first scientific article published on sorting. This is the economy of scale familiar to any business student. But with sorting, size is a recipe for disaster: perversely, as a sort grows larger, "the unit cost of sorting, instead of falling, rises." Sorting involves steep *dis*economies of scale, violating our normal intuitions about the virtues of doing things in bulk. Cooking for two is typically no harder than cooking for one, and it's certainly easier than cooking for one person twice. But sorting, say, a shelf of a hundred books will take you longer than sorting two bookshelves of fifty apiece: you have twice as many things to organize, and there are twice as many places each of them could go. The more you take on, the worse it gets.

This is the first and most fundamental insight of sorting theory. Scale hurts.

From this we might infer that minimizing our pain and suffering when it comes to sorting is all about minimizing the number of things we have to sort. It's true: one of the best preventives against the computational difficulty of sock sorting is just doing your laundry more often. Doing laundry three times as frequently, say, could reduce your sorting overhead by a factor of nine. Indeed, if Hillis's roommate stuck with his peculiar procedure but went thirteen days between washes instead of fourteen, that alone would save him twenty-eight pulls from the hamper. (And going just a single day longer between washes would cost him thirty pulls more.)

Even at such a modest, fortnightly scope we can see the scale of sorting beginning to grow untenable. Computers, though, must routinely sort millions of items in a single go. For that, as the line from *Jaws* puts it, we're going to need a bigger boat—and a better algorithm.

But to answer the question of just how we ought to be sorting, and which methods come out on top, we need to figure out something else first: how we're going to keep score.

Big-O: A Yardstick for the Worst Case

The Guinness Book of World Records attributes the record for sorting a deck of cards to the Czech magician Zdeněk Bradáč. On May 15, 2008, Bradáč sorted a 52-card deck in just 36.16 seconds.* How did he do it? What sorting technique delivered him the title? Though the answer would shed interesting light on sorting theory, Bradáč declined to comment.

While we have nothing but respect for Bradáč's skill and dexterity, we are 100% certain of the following: we can personally break his record. In fact, we are 100% certain that we can attain an *unbreakable* record. All we need are about 80,658,175,170,943,878,571,660,636,856,403,766,975,289,505, 440,883,277,824,000,000,000,000 attempts at the title. This number, a bit over 80 unvigintillion, is 52 factorial, or "52!" in mathematical notation— the number of ways that a deck of 52 cards can possibly be ordered. By taking roughly that many attempts, sooner or later we are bound to start with a shuffled deck that is in fact completely sorted by chance. At that point we can proudly enter Christian-Griffiths into *The Guinness Book* alongside a not-too-shabby sort time of 0m00s.

To be fair, we'd almost certainly be trying until the heat death of the universe before we got our perfect record attempt. Nonetheless, this highlights the biggest fundamental difference between record keepers and computer scientists. The fine folks at Guinness care only about *best*-case performance (and beer). They're hardly blameworthy, of course: all records in sports reflect the single best performance. Computer science, however, almost never cares about the best case. Instead, computer scientists might want to know the *average* sort time of someone like Bradáč: get him to sort all of the 80 unvigintillion deck orders, or a reasonably sized sample, and score him on his average speed across all attempts. (You can see why they don't let computer scientists run these things.)

Moreover, a computer scientist would want to know the *worst* sort time. Worst-case analysis lets us make hard guarantees: that a critical process will finish in time, that deadlines won't be blown. So for the rest of this

---

*This is far from Bradáč's only record—he can escape from three pairs of handcuffs while underwater in roughly the same amount of time.

chapter—and the rest of this book, actually—we will be discussing only algorithms' worst-case performance, unless noted otherwise.

Computer science has developed a shorthand specifically for measuring algorithmic worst-case scenarios: it's called "Big-O" notation. Big-O notation has a particular quirk, which is that it's inexact by design. That is, rather than expressing an algorithm's performance in minutes and seconds, Big-O notation provides a way to talk about the kind of *relationship* that holds between the size of the problem and the program's running time. Because Big-O notation deliberately sheds fine details, what emerges is a schema for dividing problems into different broad classes.

Imagine you're hosting a dinner party with n guests. The time required to clean the house for their arrival doesn't depend on the number of guests at all. This is the rosiest class of problems there is: called "Big-O of one," written $O(1)$, it is also known as "constant time." Notably, Big-O notation doesn't care a whit how long the cleaning actually takes—just that it's always the same, totally invariant of the guest list. You've got the same work to do if you have one guest as if you have ten, a hundred, or any other n.

Now, the time required to pass the roast around the table will be "Big-O of n," written $O(n)$, also known as "linear time"—with twice the guests, you'll wait twice as long for the dish to come around. And again, Big-O notation couldn't care less about the number of courses that get served, or whether they go around for second helpings. In each case, the time still depends linearly on the guest list size—if you drew a graph of the number of guests vs. the time taken, it would be a straight line. What's more, the existence of *any* linear-time factors will, in Big-O notation, swamp *all* constant-time factors. That is to say, passing the roast once around the table, or remodeling your dining room for three months and *then* passing the roast once around the table, are both, to a computer scientist, effectively equivalent. If that seems crazy, remember that computers deal with values of n that could easily be in the thousands, millions, or billions. In other words, computer scientists are thinking about very, very big parties. With a guest list in the millions, passing the roast once around would indeed make the home remodel seem dwarfed to the point of insignificance.

What if, as the guests arrived, each one hugged the others in greeting? Your first guest hugs you; your second guest has two hugs to give; your third guest, three. How many hugs will there be in total? This turns out to be "Big-O of n-squared," written $O(n^2)$ and also known as "quadratic time." Here again, we only care about the basic contours of the relationship

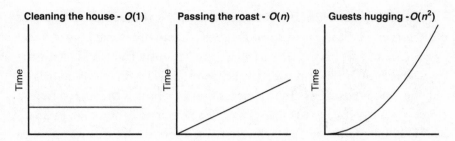

Constant time, written O(1); linear time, written O(n); and quadratic time, written O(n²).

between *n* and time. There's no $O(2n^2)$ for two hugs apiece, or $O(n^2 + n)$ for hugs plus passing the food around, or $O(n^2 + 1)$ for hugs plus home cleaning. It's all quadratic time, so $O(n^2)$ covers everything.

It gets worse from there. There's "exponential time," $O(2^n)$, where each additional guest *doubles* your work. Even worse is "factorial time," $O(n!)$, a class of problems so truly hellish that computer scientists only talk about it when they're joking—as we were in imagining shuffling a deck until it's sorted—or when they really, really wish they were.

The Squares: Bubble Sort and Insertion Sort

When then senator Obama visited Google in 2007, CEO Eric Schmidt jokingly began the Q&A like a job interview, asking him, "What's the best way to sort a million thirty-two-bit integers?" Without missing a beat, Obama cracked a wry smile and replied, "I think the **Bubble Sort** would be the wrong way to go." The crowd of Google engineers erupted in cheers. "He had me at Bubble Sort," one later recalled.

Obama was right to eschew Bubble Sort, an algorithm which has become something of a punching bag for computer science students: it's simple, it's intuitive, and it's extremely inefficient.

Imagine you want to alphabetize your unsorted collection of books. A natural approach would be just to scan across the shelf looking for out-of-order pairs—Wallace followed by Pynchon, for instance—and flipping them around. Put Pynchon ahead of Wallace, then continue your scan, looping around to the beginning of the shelf each time you reach the end. When you make a complete pass without finding any more out-of-order pairs on the entire shelf, then you know the job is done.

This is Bubble Sort, and it lands us in quadratic time. There are *n* books out of order, and each scan through the shelf can move each one at most one position. (We spot a tiny problem, make a tiny fix.) So in the worst case, where the shelf is perfectly backward, at least one book will need to be moved *n* positions. Thus a maximum of *n* passes through *n* books, which gives us $O(n^2)$ in the worst case.* It's not terrible—for one thing, it's worlds better than our $O(n!)$ shuffle-till-it's-sorted idea from earlier (in case you needed computer science to confirm that). But all the same, that squared term can get daunting quickly. For instance, it means that sorting five shelves of books will take not five times as long as sorting a single shelf, but *twenty-five* times as long.

You might take a different tack—pulling all the books off the shelf and putting them back in place one by one. You'd put the first book in the middle of the shelf, then take the second and compare it to the first, inserting it either to the right or to the left. Picking up the third book, you'd run through the books on the shelf from left to right until you found the right spot to tuck it in. Repeating this process, gradually all of the books would end up sorted on the shelf and you'd be done.

Computer scientists call this, appropriately enough, **Insertion Sort**. The good news is that it's arguably even more intuitive than Bubble Sort and doesn't have quite the bad reputation. The bad news is that it's not actually that much faster. You still have to do one insertion for each book. And each insertion still involves moving past about half the books on the shelf, on average, to find the correct place. Although in practice Insertion Sort does run a bit faster than Bubble Sort, again we land squarely, if you will, in quadratic time. Sorting anything more than a single book-shelf is still an unwieldy prospect.

Breaking the Quadratic Barrier: Divide and Conquer

At this point, having seen two entirely sensible approaches fall into unsustainable quadratic time, it's natural to wonder whether faster sorting is even possible.

The question sounds like it's about productivity. But talk to a computer scientist and it turns out to be closer to metaphysics—akin to thinking

---

*Actually, the *average* running time for Bubble Sort isn't any better, as books will, on average, be $n/2$ positions away from where they're supposed to end up. A computer scientist will still round $n/2$ passes of n books up to $O(n^2)$.

about the speed of light, time travel, superconductors, or thermodynamic entropy. What are the universe's fundamental rules and limits? What is possible? What is allowed? In this way computer scientists are glimpsing God's blueprints every bit as much as the particle physicists and cosmologists. What is the minimum effort requred to make order?

Could we find a constant-time sort, $O(1)$, one that (like cleaning the house before the bevy of guests arrive) can sort a list of any size in the same amount of time? Well, even just *confirming* that a shelf of n books is sorted cannot be done in constant time, since it requires checking all n of them. So actually sorting the books in constant time seems out of the question.

What about a linear-time sort, $O(n)$, as efficient as passing a dish around a table, where doubling the number of items to sort merely doubles the work? Thinking about the examples above, it's tough to imagine how that might work either. The n^2 in each case comes from the fact that you need to move n books, and the work required in each move scales with n as well. How would we get from n moves of size n down to just n by itself? In Bubble Sort, our $O(n^2)$ running time came from handling each of the n books and moving them as many as n places each. In Insertion Sort, quadratic running time came from handling each of the n books and comparing them to as many as n others before inserting them. A linear-time sort means handling each book for constant time regardless of how many others it needs to find its place among. Doesn't seem likely.

So we know that we can do at least as well as quadratic time, but probably not as well as linear time. Perhaps our limit lies somewhere *between* linear time and quadratic time. Are there any algorithms between linear and quadratic, between n and $n \times n$?

There are—and they were hiding in plain sight.

As we mentioned earlier, information processing began in the US censuses of the nineteenth century, with the development, by Herman Hollerith and later by IBM, of physical punch-card sorting devices. In 1936, IBM began producing a line of machines called "collators" that could merge two separately ordered stacks of cards into one. As long as the two stacks were themselves sorted, the procedure of merging them into a single sorted stack was incredibly straightforward and took linear time: simply compare the two top cards to each other, move the smaller of them to the new stack you're creating, and repeat until finished.

The program that John von Neumann wrote in 1945 to demonstrate the power of the stored-program computer took the idea of collating to its beautiful and ultimate conclusion. Sorting two cards is simple: just put the smaller one on top. And given a *pair* of two-card stacks, both of them sorted, you can easily collate them into an ordered stack of four. Repeating this trick a few times, you'd build bigger and bigger stacks, each one of them already sorted. Soon enough, you could collate yourself a perfectly sorted full deck—with a final climactic merge, like a riffle shuffle's order-creating twin, producing the desired result.

This approach is known today as **Mergesort**, one of the legendary algorithms in computer science. As a 1997 paper put it, "Mergesort is as important in the history of sorting as sorting in the history of computing."

The power of Mergesort comes from the fact that it indeed ends up with a complexity between linear and quadratic time—specifically, $O(n \log n)$, known as "linearithmic" time. Each pass through the cards doubles the size of the sorted stacks, so to completely sort n cards you'll need to make as many passes as it takes for the number 2, multiplied by itself, to equal n: the base-two logarithm, in other words. You can sort up to four cards in two collation passes, up to eight cards with a third pass, and up to sixteen cards with a fourth. Mergesort's divide-and-conquer approach inspired a host of other linearithmic sorting algorithms that quickly followed on its heels. And to say that linearithmic complexity is an improvement on quadratic complexity is a titanic understatement. In the case of sorting, say, a census-level number of items, it's the difference between making twenty-nine passes through your data set . . . and three hundred million. No wonder it's the method of choice for large-scale industrial sorting problems.

Mergesort also has real applications in small-scale domestic sorting problems. Part of the reason why it's so widely used is that it can easily be parallelized. If you're still strategizing about that bookshelf, the Mergesort solution would be to order a pizza and invite over a few friends. Divide the books evenly, and have each person sort their own stack. Then pair people up and have them merge their stacks. Repeat this process until there are just two stacks left, and merge them one last time onto the shelf. Just try to avoid getting pizza stains on the books.

Beyond Comparison: Outsmarting the Logarithm

In an inconspicuous industrial park near the town of Preston, Washington, tucked behind one nondescript gray entrance of many, lies the 2011

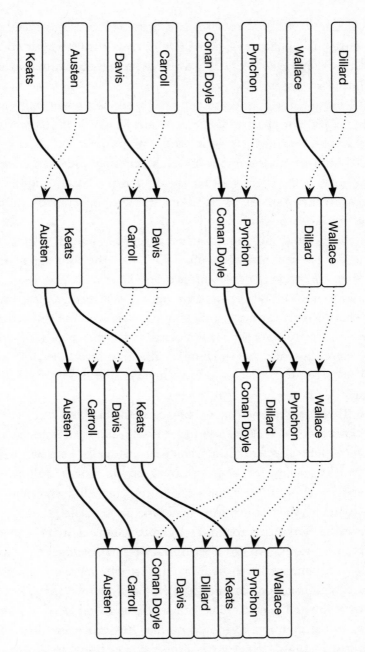

A Mergesort in action. Given a shelf of eight unsorted books, start by putting adjacent books into sorted pairs. Then collate the pairs into ordered sets of four, and finally collate those sets to get a fully sorted shelf.

and 2013 National Library Sorting Champion. A long, segmented conveyor belt moves 167 books a minute—85,000 a day—through a bar code scanner, where they are automatically diverted into bomb-bay doors that drop into one of 96 bins.

The Preston Sort Center is one of the biggest and most efficient book-sorting facilities in the world. It's run by the King County Library System, which has begun a healthy rivalry with the similarly equipped New York Public Library, with the title going back and forth over four closely contested years. "King County Library beating us this year?" said the NYPL's deputy director of BookOps, Salvatore Magaddino, before the 2014 showdown. "Fuhgeddaboutit."

There's something particularly impressive about the Preston Sort Center from a theoretician's point of view, too. The books going through its system are sorted in $O(n)$—linear time.

In an important sense, the $O(n \log n)$ linearithmic time offered by Mergesort is truly the best we can hope to achieve. It's been proven that if we want to fully sort n items via a series of head-to-head comparisons, there's just no way to compare them any fewer than $O(n \log n)$ times. It's a fundamental law of the universe, and there are no two ways around it.

But this doesn't, strictly speaking, close the book on sorting. Because sometimes you don't need a *fully* ordered set—and sometimes sorting can be done without any item-to-item comparisons at all. These two principles, taken together, allow for rough practical sorts in faster than linearithmic time. This is beautifully demonstrated by an algorithm known as **Bucket Sort**—of which the Preston Sort Center is a perfect illustration.

In Bucket Sort, items are grouped together into a number of sorted categories, with no regard for finer, intracategory sorting; that can come later. (In computer science the term "bucket" simply refers to a chunk of unsorted data, but some of the most powerful real-world uses of Bucket Sort, as at the KCLS, take the name entirely literally.) Here's the kicker: if you want to group n items into m buckets, the grouping can be done in $O(nm)$ time—that is, the time is simply proportional to the number of items times the number of buckets. And as long as the number of buckets is relatively small compared to the number of items, Big-O notation will round that to $O(n)$, or linear time.

The key to actually breaking the linearithmic barrier is knowing the distribution from which the items you're sorting are drawn. Poorly chosen buckets will leave you little better than when you started; if all the books end up in the same bin, for instance, you haven't made any progress at all. Well-chosen buckets, however, will divide your items into roughly equal-sized groups, which—given sorting's fundamental "scale hurts" nature—is a huge step toward a complete sort. At the Preston Sort Center, whose job is to sort books by their destination branch, rather than alphabetically, the choice of buckets is driven by circulation statistics. Some branches have a greater circulation volume than others, so they may have two bins allocated to them, or even three.

A similar knowledge of the material is useful to human sorters too. To see sorting experts in action, we took a field trip to UC Berkeley's Doe and Moffitt Libraries, where there are no less than fifty-two miles of book-shelves to keep in order—and it's all done by hand. Books returned to the library are first placed in a behind-the-scenes area, allocated to shelves designated by Library of Congress call numbers. For example, one set of shelves there contains a jumble of all the recently returned books with call numbers PS3000–PS9999. Then student assistants load those books onto carts, putting up to 150 books in proper order so they can be returned to the library shelves. The students get some basic training in sorting, but develop their own strategies over time. After a bit of experience, they can sort a full cart of 150 books in less than 40 minutes. And a big part of that experience involves knowing what to expect.

Berkeley undergraduate Jordan Ho, a chemistry major and star sorter, talked us through his process as he went through an impressive pile of books on the PS3000–PS9999 shelves:

> I know from experience that there's a lot of 3500s, so I want to look for any books that are below 3500 and rough-sort those out. And once I do that, then I sort those more finely. After I sort the ones under 3500, I know 3500 itself is a big section—3500–3599—so I want to make that a section itself. If there are a lot of those I might want to fine-tune it even more: 3510s, 3520s, 3530s.

Jordan aims to get a group of 25 or so books onto his cart before putting them in final order, which he does using an Insertion Sort. And his care-fully developed strategy is exactly the right way to get there: a Bucket Sort, with his well-informed forecast of how many books he'll have with various call numbers telling him what his buckets should be.

Sort Is Prophylaxis for Search

Knowing all these sorting algorithms should come in handy next time you decide to alphabetize your bookshelf. Like President Obama, you'll know not to use Bubble Sort. Instead, a good strategy—ratified by human and machine librarians alike—is to Bucket Sort until you get down to small enough piles that Insertion Sort is reasonable, or to have a Mergesort pizza party.

But if you actually asked a computer scientist to help implement this process, the first question they would ask is whether you should be sorting at all.

Computer science, as undergraduates are taught, is all about tradeoffs. We've already seen this in the tensions between looking and leaping, between exploring and exploiting. And one of the most central tradeoffs is between *sorting* and *searching*. The basic principle is this: the effort expended on sorting materials is just a preemptive strike against the effort it'll take to search through them later. What the precise balance should be depends on the exact parameters of the situation, but thinking about sorting as valuable *only* to support future search tells us something surprising:

Err on the side of messiness.

Sorting something that you will never search is a complete waste; searching something you never sorted is merely inefficient.

The question, of course, becomes how to estimate ahead of time what your future usage will be.

The poster child for the advantages of sorting would be an Internet search engine like Google. It seems staggering to think that Google can take the search phrase you typed in and scour the entire Internet for it in less than half a second. Well, it can't—but it doesn't need to. If you're Google, you are almost certain that (a) your data will be searched, (b) it will be searched not just once but repeatedly, and (c) the time needed to sort is somehow "less valuable" than the time needed to search. (Here, sorting is done by machines ahead of time, before the results are needed, and searching is done by users for whom time is of the essence.) All of these factors point in favor of tremendous up-front sorting, which is indeed what Google and its fellow search engines do.

So, should *you* alphabetize your bookshelves? For most domestic bookshelves, almost none of the conditions that make sorting worthwhile are true. It's fairly rare that we find ourselves searching for a particular title. The costs of an unsorted search are pretty low: for every book, if we know roughly where it is we can put our hands on it quickly. And the difference between the two seconds it would take to find the book on a sorted shelf and the ten seconds it would take to scan for it on an unsorted one is hardly a deal breaker. We rarely need to find a title so urgently that it's worth spending preparatory hours up front to shave off seconds later on. What's more, we search with our quick eyes and sort with slow hands.

The verdict is clear: ordering your bookshelf will take more time and energy than scanning through it ever will.

Your unsorted bookshelf might not be an everyday preoccupation, but your email inbox almost certainly is—and it's another domain where searching beats sorting handily. Filing electronic messages by hand into folders takes about the same amount of time as filing physical papers in the real world, but emails can be searched much more efficiently than their physical counterparts. As the cost of searching drops, sorting becomes less valuable.

Steve Whittaker is one of the world's experts on how people handle their email. A research scientist at IBM and professor at UC Santa Cruz, Whittaker, for almost two decades, has been studying how people manage personal information. (He wrote a paper on "email overload" in 1996, before many people even *had* email.) In 2011, Whittaker led a study of the searching and sorting habits of email users, resulting in a paper titled "Am I Wasting My Time Organizing Email?" Spoiler alert: the conclusion was an emphatic *Yes*. "It's empirical, but it's also experiential," Whittaker points out. "When I interview people about these kinds of organizational problems, that's something that they characteristically talk about, is that they sort of wasted a part of their life."

Computer science shows that the hazards of mess and the hazards of order are quantifiable and that their costs can be measured in the same currency: time. Leaving something unsorted might be thought of as an act of procrastination—passing the buck to one's future self, who'll have to pay off with interest what we chose not to pay up front. But the whole story is subtler than that. Sometimes mess is more than just the easy choice. It's the optimal choice.

Sorts and Sports

The search-sort tradeoff suggests that it's often more efficient to leave a mess. Saving time isn't the only reason we sort things, though: sometimes producing a final order is an end in itself. And nowhere is that clearer than on the sporting field.

In 1883, Charles Lutwidge Dodgson developed incredibly strong feelings about the state of British lawn tennis. As he explains:

> At a Lawn Tennis Tournament, where I chanced, some while ago, to be a spectator, the present method of assigning prizes was brought to my notice by the lamentations of one of the Players, who had been beaten (and had thus lost all chance of a prize) early in the contest, and who had had the mortification of seeing the 2nd prize carried off by a Player whom he knew to be quite inferior to himself.

Normal spectators might chalk up such "lamentations" to little more than the sting of defeat, but Dodgson was no ordinary sympathetic ear. He was an Oxford lecturer in mathematics, and the sportsman's complaints sent him on a deep investigation of the nature of sports tournaments.

Dodgson was more than just an Oxford mathematician—in fact, he's barely remembered as having been one. Today he's best known by his pen name, Lewis Carroll, under which he wrote *Alice's Adventures in Wonderland* and many other beloved works of nineteenth-century literature. Fusing his mathematical and literary talents, Dodgson produced one of his lesser-known works: "Lawn Tennis Tournaments: The True Method of Assigning Prizes with a Proof of the Fallacy of the Present Method."

Dodgson's complaint was directed at the structure of the **Single Elimination** tournament, where players are paired off with one another and eliminated from competition as soon as they lose a single match. As Dodgson forcefully argued, the true second-best player could be *any* of the players eliminated by the best—not just the last-eliminated one. Ironically, in the Olympics we do hold bronze medal matches, by which we appear to acknowledge that the Single Elimination format doesn't give us enough information to determine third place.* But in fact this format doesn't tell us enough to determine second place either—or, indeed, anything except the winner. As Dodgson put it, "The present method of assigning prizes is, except in the case of the first prize, entirely unmeaning." Said plainly, the silver medal is a lie.

---

*On rare occasions, as in boxing—where it is medically unsafe for a boxer to fight again after being recently knocked out—two bronzes are awarded instead.

"As a mathematical fact," he continued, "the chance that the 2nd best Player will get the prize he deserves is only 16/31sts; while the chance that the best 4 shall get their proper prizes is so small, that the odds are 12 to 1 against its happening!"

Despite the powers of his pen, it appears that Dodgson had little impact on the world of lawn tennis. His solution, an awkward take on triple elimination where the defeat of someone who had defeated you could also eliminate you, never caught on. But if Dodgson's solution was cumbersome, his critique of the problem was nevertheless spot on. (Alas, silver medals are still being handed out in Single Elimination tournaments to this day.)

But there's also a deeper insight in Dodgson's logic. We humans sort more than our data, more than our possessions. We sort *ourselves*.

The World Cup, the Olympics, the NCAA, NFL, NHL, NBA, and MLB—all of these implicitly implement sorting procedures. Their seasons, ladders, and playoffs are algorithms for producing rank order.

One of the most familiar algorithms in sports is the **Round-Robin** format, where each of n teams eventually plays every one of the other $n-1$ teams. While arguably the most comprehensive, it's also one of the most laborious. Having every team grapple with every other team is like having guests exchange hugs at our dinner party: the dreaded $O(n^2)$, quadratic time.

Ladder tournaments—popular in sports like badminton, squash, and racquetball—put players in a linear ranking, with each player allowed to issue a direct challenge to the player immediately above them, exchanging places if they prevail. Ladders are the Bubble Sorts of the athletic world and are thus also quadratic, requiring $O(n^2)$ games to reach a stable ranking.

Perhaps the most prevalent tournament format, however, is a bracket tournament—as in the famous NCAA basketball "March Madness," among many others. The March Madness tournament progresses from the "Round of 64" and the "Round of 32" to the "Sweet 16," "Elite Eight," "Final Four," and the finals. Each round divides the field in half: does that sound familiarly logarithmic? These tournaments are effectively Mergesort, beginning with unsorted pairs of teams and collating, collating, collating them.

We know that Mergesort operates in linearithmic time—$O(n \log n)$—and so, given that there are 64 teams, we can expect to only need something like 6 rounds (192 games), rather than the whopping 63 rounds (2,016 games) it would take to do a ladder or Round-Robin. That's a huge improvement: algorithm design at work.

Six rounds of March Madness sounds about right, but wait a second: 192 games? The NCAA tournament is only 63 games long.

In fact, March Madness is not a complete Mergesort—it doesn't produce a full ordering of all 64 teams. To truly rank the teams, we'd need an extra set of games to determine second place, another for third, and so on—taking a linearithmic number of games in sum. But March Madness doesn't do that. Instead, just like the lawn tennis tournament that Dodgson complained about, it uses a Single Elimination format where the eliminated teams are left unsorted. The advantage is that it runs in linear time: since every game eliminates exactly one team, in order to have one team left standing you need just $n - 1$ games—a linear number. The disadvantage is that, well, you never really figure out the standings aside from first place.

Ironically, in Single Elimination no tournament structure is actually necessary at all. *Any* 63 games will yield a single undefeated champion. For instance, you could simply have a single "king of the hill" team take on challengers one by one until it is dethroned, at which point whoever defeated it takes over its spot and continues. This format would have the drawback of needing 63 separate rounds, however, as games couldn't happen in parallel; also, one team could potentially have to play as many as 63 games in a row, which might not be ideal from a fatigue standpoint.

Though born well over a century after Dodgson, perhaps no one carries forward his mathematical take on sporting into the twenty-first century as strongly as Michael Trick. We met Trick back in our discussion of optimal stopping, but in the decades since his hapless application of the 37% Rule to his love life he's become not only a husband and a professor of operations research—he's now also one of the principal schedulers for Major League Baseball and for NCAA conferences like the Big Ten and the ACC, using computer science to decide the year's matchups.

As Trick points out, sports leagues aren't concerned with determining the rankings as quickly and expeditiously as possible. Instead, sports calendars are explicitly designed to maintain tension throughout the season, something that has rarely been a concern of sorting theory.

> For instance in Major League Baseball, you often have races to see who is going to win the division. Now, if we ignored the divisional setup, some of those races might get resolved fairly early in the season. But instead what we do is we make certain in the last five weeks, everybody plays everybody else within their division. The purpose of that is it doesn't matter who's in

a divisional race: they're going to have to play their next closest opponent at least six games in the final five weeks of the season. That allows for more interest in the schedule or interest in the season because in this case, uncertainty is delayed in its resolution.

What's more, sports are not, of course, always designed strictly to minimize the number of games. Without remembering this, some aspects of sports scheduling would otherwise seem mysterious to a computer scientist. As Trick says of baseball's regular season of 2,430 games, "We know that $n \log n$ is the right number of comparisons to do a full sort. That can get you *everybody*. Why do they do n^2 in order to just get, in some sense, the *top*, if that's all they care about?" In other words, why do a full $O(n^2)$ Round-Robin and then some, if we know we can do a full sort in linearithmic time, and can crown an undefeated Single Elimination champion in less than n games? Well, minimizing the number of games isn't actually in the league's interest. In computer science unnecessary comparisons are always bad, a waste of time and effort. But in sports that's far from the case. In many respects, after all, the games themselves are the point.

Griping Rights: Noise and Robustness

Another, perhaps even more important way of training an algorithmic lens on sports is to ask not what confidence we should have in the silver medal, but what confidence we should have in the *gold*.

As Michael Trick explains, in some sports, "for instance baseball, a team is going to lose 30% of their games and a team is going to win 30% of their games practically no matter who they are." This has disturbing implications for the Single Elimination format. If NCAA basketball games, say, are won by the stronger team 70% of the time, and winning the tournament involves prevailing in 6 straight games, then the best team has only a 0.70 to the 6th power—less than 12%—chance of winning the tournament! Put another way, the tournament would crown the league's truly best team just once a decade.

It may be that in some sports, having even 70% confidence in a game's outcome might be putting too much stock in the final score. UCSD physicist Tom Murphy applied numerical modeling techniques to soccer and concluded that soccer's low scores make game outcomes much closer to random than most fans would prefer to imagine. "A 3:2 score gives the winning team only a 5-in-8 chance of actually being a better team . . . Personally, I don't find this to be very impressive. Even a 6:1 blowout leaves a 7% chance that it was a statistical fluke."

Computer scientists call this phenomenon *noise*. All of the sorting algorithms that we've considered thus far assume perfect, flawless, fool-proof comparisons, ones that never mess up and mistakenly judge the lesser of two quantities to be the greater. Once you allow for a "noisy comparator," some of computer science's most hallowed algorithms go out the window—and some of its most maligned have their day of redemption.

Dave Ackley, professor of computer science at the University of New Mexico, works at the intersection of computer science and "artificial life"— he believes computers can stand to learn a few things from biology. For starters, organisms live in a world where few processes have anywhere near the level of reliability that computers depend on, so they are built from the ground up for what researchers call *robustness*. It's time, argues Ackley, that we started recognizing the virtues of robustness in algorithms too.

Thus, while the authoritative programming tome *Sorting and Searching* boldly declares that "bubble sort has no apparent redeeming features," the research of Ackley and his collaborators suggests that there may be a place for algorithms like Bubble Sort after all. Its very inefficiency—moving items only one position at a time—makes it fairly robust against noise, far more robust than faster algorithms like Mergesort, in which each comparison potentially moves an item a long way. Mergesort's very efficiency makes it brittle. An early error in a Mergesort is like a fluke loss in the first round of a Single Elimination tournament, which can not only dash a favored team's championship hopes but also permanently relegate them to the bottom half of the results.* In a Ladder tournament, on the other hand, as in a Bubble Sort, a fluke loss would only set a player back a single place in the standings.

But in fact it isn't Bubble Sort that emerges as the single best algorithm in the face of a noisy comparator. The winner of that particular honor is an algorithm called **Comparison Counting Sort**. In this algorithm, each

---

*It's interesting to note that NCAA's March Madness tournament is consciously designed to mit-igate this flaw in its algorithm. The biggest problem in Single Elimination, as we've said, would seem to be a scenario where the first team that gets eliminated by the winning team is actually the second-best team overall, yet lands in the (unsorted) bottom half. The NCAA works around this by seeding the teams, so that top-ranked teams cannot meet each other in the early rounds. The seeding process appears to be reliable at least in the most extreme case, as a sixteenth-seeded team has never defeated a first seed in the history of March Madness.

item is compared to all the others, generating a tally of how many items it is bigger than. This number can then be used directly as the item's rank. Since it compares all pairs, Comparison Counting Sort is a quadratic-time algorithm, like Bubble Sort. Thus it's not a popular choice in traditional computer science applications, but it's exceptionally fault-tolerant.

This algorithm's workings should sound familiar. Comparison Counting Sort operates *exactly* like a Round-Robin tournament. In other words, it strongly resembles a sports team's regular season—playing every other team in the division and building up a win-loss record by which they are ranked.

That Comparison Counting Sort is the single most robust sorting algorithm known, quadratic or better, should offer something very specific to sports fans: if your team doesn't make the playoffs, don't whine. The Mergesort postseason is chancy, but the Comparison Counting regular season is not; championship rings aren't robust, but divisional standings are literally as robust as it gets. Put differently, if your team is eliminated early in the postseason, it's tough luck. But if your team fails to *get* to the postseason, it's tough truth. You may get sports-bar sympathy from your fellow disappointed fans, but you won't get any from a computer scientist.

Blood Sort: Pecking Orders and Dominance Hierarchies

In all the examples we've considered so far, the sorting process in every case has been imposed from the top down: a librarian shelving books, the NCAA telling teams whom to play and when. But what if head-to-head comparisons happened only voluntarily? What does sorting look like when it emerges organically, from the bottom up?

It might look something like online poker.

Unlike most sports, which are governed by a ruling body of some kind, poker remains somewhat anarchic despite exploding in popularity over the past decade. Though some high-profile tournaments do explicitly sort their contestants (and remunerate them accordingly), a substantial portion of poker is still played in what are known as "cash games," where two or more players spontaneously agree to play with real money on the line with every hand.

Virtually no one knows this world more deeply than Isaac Haxton, one of the world's best cash-game poker players. In most sports it's sufficient to be as good as possible, and the less self-conscious one is about one's skills the better. But, Haxton explains, "In some ways the most important skill as a professional poker player is to be able to evaluate how good you are. If you're anything short of the very best poker player in the world, you can be pretty much assured of going broke if you are endlessly willing to play people better than you."

Haxton is a heads-up, no-limit specialist: "heads-up" meaning one-on-one poker, and "no-limit" meaning just that—the highest stakes, limited only by what they can bankroll and stomach. In multi-handed poker cash games, there will often be one weak player—a wealthy amateur, for instance—feeding a table full of professionals, who then don't much care who among them is better than whom. In the world of heads-up, it's different. "There has to be a disagreement between you and them about who's better—or somebody has to be willingly losing."

So what happens when there's a fairly established consensus and no one's willing to play anyone better than they are? You get something that looks a lot like players simply jockeying for seats. Most online poker sites have only a finite number of tables available. "So if you want to play heads-up no-limit, with blinds of fifty and one hundred dollars, there are only ten available tables for that," says Haxton, "and so only the consensus ten best players who are out right now . . . sit and wait for someone to show up who wants to play." And if a superior player arrives and sits down at one of these tables? If the person sitting isn't willing to ante up, they scram.

"Imagine two monkeys," says Christof Neumann. "One is sitting and feeding in its spot, very peacefully, and another one is coming up [to] where the other guy is sitting. And that guy would then stand up and leave."

Neumann isn't making a poker metaphor. He's a behavioral biologist at the University of Neuchâtel who studies dominance in macaques. What he's just described is known as *displacement*.

Displacement happens when an animal uses its knowledge of the hierarchy to determine that a particular confrontation simply isn't worth it. In many animal societies, resources and opportunities—food, mates, preferred spaces, and so forth—are scarce, and somehow it must be decided who gets what. Establishing an order ahead of time is less violent than coming to blows every time a mating opportunity or a prime spot of grass becomes available. Though we may cringe when we see creatures turning their claws and beaks on each other, biologists tend to think of pecking orders as the violence that preempts violence.

Sound familiar? It's the search-sort tradeoff.

The creation of a pecking order is a pugilistic solution to a fundamentally *computational* problem. For this reason, incidentally, debeaking chickens on farms may be a well-intentioned but counterproductive approach: it removes the authority of individual fights to resolve the order, and therefore makes it much harder for the flock to run any sorting procedure at all. So the amount of antagonism within the flock in many cases actually increases.

Looking at animal behavior from the perspective of computer science suggests several things. For one, it implies that the number of hostile confrontations encountered by each individual will grow substantially—at least logarithmically, and perhaps quadratically—as the group gets bigger. Indeed, studies of "agonistic behavior" in hens have found that "aggressive acts per hen increased as group size increased." Sorting theory thus suggests that the ethical raising of livestock may include limiting the size of the flock or herd. (In the wild, feral chickens roam in groups of ten to twenty, far smaller than flock sizes on commercial farms.) The studies also show that aggression appears to go away after a period of some weeks, unless new members are added to the flock—corroborating the idea that the group is sorting itself.

The key to thinking about decentralized sorting in nature, argues Jessica Flack, codirector of the Center for Complexity and Collective Computation at UW–Madison, is that dominance hierarchies are ultimately *information* hierarchies. There's a significant computational burden to these decentralized sorting systems, Flack points out. The number of fights in, say, a group of macaques is minimized only to the extent that every monkey has a detailed—and similar—understanding of the hierarchy. Otherwise violence will ensue.

If it comes down to how good the protagonists are at keeping track of the current order, we might expect to see fewer confrontations as animals become better able to reason and remember. And perhaps humans do come closest to optimally efficient sorting. As Haxton says of the poker world, "I'm one of the top heads-up, no-limit hold 'em players in the world, and in my head I have a fairly specific ranking of who I think the twenty or so best players are, and I think each of them has a similar ranking in their mind. I think there is a pretty high degree of consensus about what the list looks like." Only when these rankings differ will cash games ensue.

A Race Instead of a Fight

We've now seen two separate downsides to the desire of any group to sort itself. You have, at minimum, a linearithmic number of confrontations, making everyone's life more combative as the group grows—and you also oblige every competitor to keep track of the ever-shifting status of everyone else, otherwise they'll find themselves fighting battles they didn't need to. It taxes not only the body but the mind.

But it doesn't have to be that way. There are ways of making order without the costs.

There's one sporting contest, for instance, where tens of thousands of competitors are completely sorted within the time that it takes to hold just a single event. (A Round-Robin tournament with ten thousand players, on the other hand, would require a hundred million matchups.) The only caveat is that the time required for the event is determined by its slowest competitors. This sporting contest is the marathon, and it suggests something critical: a race is fundamentally different from a fight.

Consider the difference between boxers and skiers, between fencers and runners. An Olympic boxer must risk concussion $O(\log n)$ times, usually from 4 to 6, to make it to the podium; allowing a greater number of athletes into the games would imperil the health of all. But a skeleton racer or ski jumper or halfpipe specialist needs to make only a constant number of gambles with gravity, no matter the size of the field. A fencer puts herself at her opponent's mercy $O(\log n)$ times, but a marathoner must endure only one race. Being able to assign a simple numerical measure of performance results in a constant-time algorithm for status.

This move from "ordinal" numbers (which only express *rank*) to "cardinal" ones (which directly assign a *measure* to something's caliber) naturally orders a set without requiring pairwise comparisons. Accordingly, it makes possible dominance hierarchies that don't require direct head-to-head matchups. The *Fortune* 500 list, to the extent that it creates a kind of corporate hierarchy, is one of these. To find the most valuable company in the United States, analysts don't need to perform due diligence comparing Microsoft to General Motors, then General Motors to Chevron, Chevron to Walmart, and so forth. These seemingly apples-to-oranges contests (how many enterprise software installations equal how many oil futures?) become apples-to-apples in the medium of dollars. Having a benchmark—any benchmark—solves the computational problem of scaling up a sort.

In Silicon Valley, for instance, there's an adage about meetings: "You go to the money, the money doesn't come to you." Vendors go to founders, founders go to venture capitalists, venture capitalists go to their limited partners. It's possible for the individuals to resent the basis of this hierarchy, but not really to contest its verdict. As a result, individual pairwise interactions take place with a minimum of jockeying for status. By and large, any pair of people can tell, without needing to negotiate, who is supposed to show what level of respect to whom. Everyone knows where to meet.

Likewise, while maritime right-of-way is governed in theory by an extremely elaborate set of conventions, in practice one straightforward principle determines which ships give way to which: the "Law of Gross Tonnage." Quite simply, the smaller ship gets out of the way of the larger one. Some animals are also lucky enough to have such clear-cut dominance hierarchies. As Neumann observes, "Look at fish, for example: the bigger one is the dominant one. It's very simple." And because it's so simple, it's *peaceful*. Unlike chickens and primates, fish make order without shedding blood.

When we think about the factors that make large-scale human societies possible, it's easy to focus on technologies: agriculture, metals, machinery. But the cultural practice of measuring status with quantifiable metrics might be just as important. Money, of course, need not be the criterion; a rule like "respect your elders," for instance, likewise settles questions of people's status by reference to a common quantity. And the same principle is at work between nations as within them. It is often noted that a benchmark like national GDP—which underlies the invite lists to diplomatic summits such as the G20—is a crude, imperfect measurement. But the existence of any benchmark at all transforms the question of national status from one demanding at least a linearithmic number of tussles and resolutions into something with a single reference point that ranks all. Given that nation-to-nation status disputes often take military form, this saves not only time but lives.

Linearithmic numbers of fights might work fine for small-scale groups; they do in nature. But in a world where status is established through pairwise comparisons—whether they involve exchanging rhetoric or gunfire— the amount of confrontation quickly spirals out of control as society grows. Operating at industrial scale, with many thousands or millions of individuals sharing the same space, requires a leap beyond. A leap from ordinal to cardinal.

Much as we bemoan the daily rat race, the fact that it's a *race* rather than a *fight* is a key part of what sets us apart from the monkeys, the chickens—and, for that matter, the rats.

4 | Caching

Forget About It

In the practical use of our intellect, forgetting is as important a function as remembering.

—WILLIAM JAMES

You have a problem. Your closet is overflowing, spilling shoes, shirts, and underwear onto the floor. You think, "It's time to get organized." Now you have two problems.

Specifically, you first need to decide what to keep, and second, how to arrange it. Fortunately, there is a small industry of people who think about these twin problems for a living, and they are more than happy to offer their advice.

On what to keep, Martha Stewart says to ask yourself a few questions: "How long have I had it? Does it still function? Is it a duplicate of something I already own? When was the last time I wore it or used it?" On how to organize what you keep, she recommends "grouping like things together," and her fellow experts agree. Francine Jay, in *The Joy of Less*, stipulates, "Hang all your skirts together, pants together, dresses together, and coats together." Andrew Mellen, who bills himself as "The Most Organized Man in America," dictates, "Items will be sorted by type—all slacks together, shirts together, coats, etc. Within each type, they're further sorted by color and style—long-sleeved or short-sleeved, by neckline, etc." Other than the sorting problem this could entail, it looks like good advice; it certainly seems unanimous.

Except that there is another, larger industry of professionals who also think obsessively about storage—and they have their own ideas.

Your closet presents much the same challenge that a computer faces when managing its memory: space is limited, and the goal is to save both money and time. For as long as there have been computers, computer scientists have grappled with the dual problems of what to keep and how to arrange it. The results of these decades of effort reveal that in her four-sentence advice about what to toss, Martha Stewart actually makes several different, and not fully compatible, recommendations—one of which is much more critical than the others.

The computer science of memory management also reveals exactly how your closet (and your office) ought to be arranged. At first glance, computers appear to follow Martha Stewart's maxim of "grouping like things together." Operating systems encourage us to put our files into folders, like with like, forming hierarchies that branch as their contents become ever more specific. But just as the tidiness of a scholar's desk may hide the messiness of their mind, so does the apparent tidiness of a computer's file system obscure the highly engineered chaos of how data is actually being stored underneath the nested-folder veneer.

What's really happening is called *caching.*

Caching plays a critical role in the architecture of memory, and it underlies everything from the layout of processor chips at the millimeter scale to the geography of the global Internet. It offers a new perspective on all the various storage systems and memory banks of human life—not only our machines, but also our closets, our offices, our libraries. And our heads.

The Memory Hierarchy

A certain woman had a very sharp consciousness but almost no memory. . . .
She remembered enough to work, and she worked hard.

—LYDIA DAVIS

Starting roughly around 2008, anyone in the market for a new computer has encountered a particular conundrum when choosing their storage option. They must make a tradeoff between *size* and *speed.* The computer industry is currently in transition from hard disk drives to solid-state drives; at the same price point, a hard disk will offer dramatically greater capacity, but a solid-state drive will offer dramatically better performance—as most consumers now know, or soon discover when they begin to shop.

What casual consumers may not know is that this exact tradeoff is being made within the machine itself at a number of scales—to the point where it's considered one of the fundamental principles of computing.

In 1946, Arthur Burks, Herman Goldstine, and John von Neumann, working at the Institute for Advanced Study in Princeton, laid out a design proposal for what they called an electrical "memory organ." In an ideal world, they wrote, the machine would of course have limitless quantities of lightning-fast storage, but in practice this wasn't possible. (It still isn't.) Instead, the trio proposed what they believed to be the next best thing: "a hierarchy of memories, each of which has greater capacity than the preceding but which is less quickly accessible." By having effectively a pyramid of different forms of memory—a small, fast memory *and* a large, slow one—maybe we could somehow get the best of both.

The basic idea behind a memory hierarchy should be intuitive to anyone who has ever used a library. If you are researching a topic for a paper, let's say, there are some books you might need to refer to on multiple occasions. Rather than go back to the library each time, you of course check out the relevant books and take them home to your desk, where you can access them more easily.

In computing, this idea of a "memory hierarchy" remained just a theory until the development in 1962 of a supercomputer in Manchester, England, called Atlas. Its principal memory consisted of a large drum that could be rotated to read and write information, not unlike a wax phonograph cylinder. But Atlas also had a smaller, faster "working" memory built from polarized magnets. Data could be read from the drum to the magnets, manipulated there with ease, and the results then written back to the drum.

Shortly after the development of Atlas, Cambridge mathematician Maurice Wilkes realized that this smaller and faster memory wasn't just a convenient place to work with data before saving it off again. It could also be used to deliberately hold on to pieces of information likely to be needed later, *anticipating* similar future requests—and dramatically speeding up the operation of the machine. If what you needed was still in the working memory, you wouldn't have to load it from the drum at all. As Wilkes put it, the smaller memory "automatically accumulates to itself words that come from a slower main memory, and keeps them available for subsequent use without it being necessary for the penalty of main memory access to be incurred again."

The key, of course, would be managing that small, fast, precious memory so it had what you were looking for as often as possible. To continue the library analogy, if you're able to make just one trip to the stacks to get all the books you need, and then spend the rest of the week working at home, that's almost as good as if every book in the library had already been available at your desk. The more trips back to the library you make, the slower things go, and the less your desk is really doing for you.

Wilkes's proposal was implemented in the IBM 360/85 supercomputer later in the 1960s, where it acquired the name of the "cache." Since then, caches have appeared everywhere in computer science. The idea of keeping around pieces of information that you refer to frequently is so powerful that it is used in every aspect of computation. Processors have caches. Hard drives have caches. Operating systems have caches. Web browsers have caches. And the servers that deliver content to those browsers also have caches, making it possible to *instantly* show you the same video of a cat riding a vacuum cleaner that millions of . . . But we're getting ahead of ourselves a bit.

The story of the computer over the past fifty-plus years has been painted as one of exponential growth year after year—referencing, in part, the famously accurate "Moore's Law" prediction, made by Intel's Gordon Moore in 1975, that the number of transistors in CPUs would double every two years. What hasn't improved at that rate is the performance of memory, which means that relative to processing time, the cost of accessing memory is also increasing exponentially. The faster you can write your papers, for instance, the greater the loss of productivity from each trip to the library. Likewise, a factory that doubles its manufacturing speed each year—but has the same number of parts shipped to it from overseas at the same sluggish pace—will mean little more than a factory that's twice as idle. For a while it seemed that Moore's Law was yielding little except processors that twiddled their thumbs ever faster and ever more of the time. In the 1990s this began to be known as the "memory wall."

Computer science's best defense against hitting that wall has been an ever more elaborate hierarchy: caches for caches for caches, all the way down. Modern consumer laptops, tablets, and smartphones have on the order of a six-layer memory hierarchy, and managing memory smartly has never been as important to computer science as it is today.

So let's start with the first question that comes to mind about caches (or closets, for that matter). What do we do when they get full?

Eviction and Clairvoyance

Depend upon it there comes a time when for every addition of knowledge you forget something that you knew before. It is of the highest importance, therefore, not to have useless facts elbowing out the useful ones.

—SHERLOCK HOLMES

When a cache fills up, you are obviously going to need to make room if you want to store anything else, and in computer science this making of room is called "cache replacement" or "cache eviction." As Wilkes wrote, "Since the [cache] can only be a fraction of the size of the main memory, words cannot be preserved in it indefinitely, and there must be wired into the system an algorithm by which they are progressively overwritten." These algorithms are known as "replacement policies" or "eviction policies," or simply as caching algorithms.

IBM, as we've seen, played an early role in the deployment of caching systems in the 1960s. Unsurprisingly, it was also the home of seminal early research on caching algorithms—none, perhaps, as important as that of László "Les" Bélády. Bélády was born in 1928 in Hungary, where he studied as a mechanical engineer before fleeing to Germany during the 1956 Hungarian Revolution with nothing but a satchel containing "one change of underwear and my graduation paper." From Germany he went to France, and in 1961 immigrated to the United States, bringing his wife, "an infant son and $1,000 in my pocket, and that's it." It seems he had acquired a finely tuned sense of what to keep and what to leave behind by the time he found himself at IBM, working on cache eviction.

Bélády's 1966 paper on caching algorithms would become the most cited piece of computer science research for fifteen years. As it explains, the goal of cache management is to minimize the number of times you can't find what you're looking for in the cache and must go to the slower main memory to find it; these are known as "page faults" or "cache misses." The optimal cache eviction policy—essentially by definition, Bélády wrote—is, when the cache is full, to evict whichever item we'll need again *the longest from now.*

Of course, knowing exactly when you'll need something again is easier said than done.

The hypothetical all-knowing, prescient algorithm that would look ahead and execute the optimal policy is known today in tribute as **Bélády's Algorithm**. Bélády's Algorithm is an instance of what computer scientists call a "clairvoyant" algorithm: one informed by data from the future. It's not necessarily as crazy as it sounds—there are cases where a system might know what to expect—but in general clairvoyance is hard to come by, and software engineers joke about encountering "implementation difficulties" when they try to deploy Bélády's Algorithm in practice. So the challenge is to find an algorithm that comes as close to clairvoyance as we can get, for all those times when we're stuck firmly in the present and can only guess at what lies ahead.

We could just try **Random Eviction**, adding new data to the cache and overwriting old data at random. One of the startling early results in caching theory is that, while far from perfect, this approach is not half bad. As it happens, just having a cache at all makes a system more efficient, regardless of how you maintain it. Items you use often will end up back in the cache soon anyway. Another simple strategy is **First-In, First-Out** (**FIFO**), where you evict or overwrite whatever has been sitting in the cache the longest (as in Martha Stewart's question "How long have I had it?"). A third approach is **Least Recently Used** (**LRU**): evicting the item that's gone the longest untouched (Stewart's "When was the last time I wore it or used it?").

It turns out that not only do these two mantras of Stewart's suggest very different policies, one of her suggestions clearly outperforms the other. Bélády compared Random Eviction, FIFO, and variants of LRU in a number of scenarios and found that LRU consistently performed the closest to clairvoyance. The LRU principle is effective because of something computer scientists call "temporal locality": if a program has called for a particular piece of information once, it's likely to do so again in the near future. Temporal locality results in part from the way computers solve problems (for example, executing a loop that makes a rapid series of related reads and writes), but it emerges in the way people solve problems, too. If you are working on your computer, you might be switching among your email, a web browser, and a word processor. The fact that you accessed one of these recently is a clue that you're likely to do so again, and, all things being equal, the program that you haven't been using for the longest time is also probably the one that *won't* be used for some time to come.

In fact, this principle is even implicit in the interface that computers show to their users. The windows on your computer screen have what's called a "Z-order," a simulated depth that determines which programs are overlaid on top of which. The least recently used end up at the bottom. As former creative lead for Firefox, Aza Raskin, puts it, "Much of your time using a modern browser (computer) is spent in the digital equivalent of shuffling papers." This "shuffling" is also mirrored exactly in the Windows and Mac OS task switching interfaces: when you press Alt + Tab or Command + Tab, you see your applications listed in order from the most recently to the least recently used.

The literature on eviction policies goes about as deep as one can imagine—including algorithms that account for frequency as well as recency of use, algorithms that track the time of the *next*-to-last access rather than the last one, and so on. But despite an abundance of innovative caching schemes, some of which can beat LRU under the right conditions, LRU itself—and minor tweaks thereof—is the overwhelming favorite of computer scientists, and is used in a wide variety of deployed applications at a variety of scales. LRU teaches us that the next thing we can expect to need is the last one we needed, while the thing we'll need after that is probably the second-most-recent one. And the last thing we can expect to need is the one we've already gone longest without.

Unless we have good reason to think otherwise, it seems that our best guide to the future is a mirror image of the past. The nearest thing to clairvoyance is to assume that history repeats itself—backward.

Turning the Library Inside Out

Deep within the underground Gardner Stacks at the University of California, Berkeley, behind a locked door and a prominent "Staff Only" notice, totally off-limits to patrons, is one of the jewels of the UC library system. Cormac McCarthy, Thomas Pynchon, Elizabeth Bishop, and J. D. Salinger; Anaïs Nin, Susan Sontag, Junot Díaz, and Michael Chabon; Annie Proulx, Mark Strand, and Philip K. Dick; William Carlos Williams, Chuck Palahniuk, and Toni Morrison; Denis Johnson, Juliana Spahr, Jorie Graham, and David Sedaris; Sylvia Plath, David Mamet, David Foster

Wallace, and Neil Gaiman . . . It isn't the library's rare book collection; it's its cache.

As we have already discussed, libraries are a natural example of a memory hierarchy when used in concert with our own desk space. In fact, libraries in themselves, with their various sections and storage facilities, are a great example of a memory hierarchy with multiple levels. As a consequence, they face all sorts of caching problems. They have to decide which books to put in the limited display space at the front of the library, which to keep in their stacks, and which to consign to offsite storage. The policy for which books to shunt offsite varies from library to library, but almost all use a version of LRU. "For the Main Stacks, for example," says Beth Dupuis, who oversees the process in the UC Berkeley libraries, "if an item hasn't been used in twelve years, that's the cutoff."

At the other end of the spectrum from the books untouched in a dozen years is the library's "rough sorting" area, which we visited in the previous chapter. This is where books go just after they are returned, before they're fully sorted and shelved once again in the stacks. The irony is that the hardworking assistants putting them back on their shelves might, in some sense, be making them *less* ordered.

Here's why: if temporal locality holds, then the rough-sorting shelves contain the most important books in the whole building. These are the books that were most recently used, so they are the ones that patrons are most likely to be looking for. It seems a crime that arguably the juiciest and most browseworthy shelf of the libraries' miles of stacks is both hidden away and constantly eroded by earnest library staff just doing their jobs.

Meanwhile, the lobby of the Moffit Undergraduate Library—the location of the most prominent and accessible shelves—showcases the library's most recently *acquired* books. This is instantiating a kind of FIFO cache, privileging the items that were last added to the library, not last read.

The dominant performance of the LRU algorithm in most tests that computer scientists have thrown at it leads to a simple suggestion: *turn the library inside out*. Put acquisitions in the back, for those who want to find them. And put the most recently *returned* items in the lobby, where they are ripe for the browsing.

Humans are social creatures, and presumably the undergraduate body would find it interesting to peruse its own reading habits. It would nudge the campus toward a more organic and free-form version of what colleges strive for when they assign "common books": the facilitation of intellectual common points of reference. Here, the books being read on campus, whatever they happened to be, would become the books most likely to be serendipitously encountered by other students. A kind of grassroots, bottom-up analogue of the common book program.

But a system like this wouldn't only be more socially positive. Since the items most recently returned are the ones most likely to be next checked out, it would also be more efficient. It's true that students might be puzzled by the fact that popular books will sometimes be found in the stacks and sometimes in the lobby. However, recently returned books that await reshelving are missing from the stacks either way. It's just that currently they are off-limits during this brief limbo. Allowing the returned books to adorn the lobby instead would give students a chance to short-circuit the shelving process entirely. No employees would have to venture into the stacks to deposit the volumes, and no students would have to venture into the stacks to get them back out. That's exactly how caching is meant to work.

The Cloud at the End of the Street

> "We actually made a map of the country, on the scale of a mile to the mile!"
> "Have you used it much?" I enquired.
> "It has never been spread out, yet," said Mein Herr: "the farmers objected: they said it would cover the whole country, and shut out the sunlight! So we now use the country itself, as its own map, and I assure you it does nearly as well."
>
> —LEWIS CARROLL

We often think of the Internet as a flat, independent, and loosely connected network. In fact, it's none of those things. A quarter of all Internet traffic at present is handled by a single corporation, one that manages to stay almost entirely out of the headlines. This Massachusetts-based company is called Akamai, and they're in the caching business.

We also think of the Internet as abstract, dematerial, post-geographic. We're told our data is "in the cloud," which is meant to suggest a diffuse, distant place. Again, none of these are true. The reality is that the Internet is all about bundles of physical wires and racks of metal. And it's much more closely tied to geography than you might expect.

Engineers think about geography on a tiny scale when they design computer hardware: faster memory is usually placed closer to the processor, minimizing the length of the wires that information has to travel along. Today's processor cycles are measured in gigahertz, which is to say they are performing operations in fractions of nanoseconds. For reference, that's the time it takes light to travel a few *inches*—so the physical layout of a computer's internals is very much a concern. And applying the same principle at a dramatically larger scale, actual geography turns out to be critical for the functioning of the web, where the wires span not inches but potentially thousands of miles.

If you can create a cache of webpage content that is physically, geographically closer to the people who want it, you can serve those pages faster. Much of the traffic on the Internet is now handled by "content distribution networks" (CDNs), which have computers around the world that maintain copies of popular websites. This allows users requesting those pages to get their data from a computer that's nearby, without having to make the long haul across continents to the original server.

The largest of these CDNs is managed by Akamai: content providers pay for their websites to be "Akamaized" for better performance. An Australian who streams video from the BBC, for instance, is probably hitting local Akamai servers in Sydney; the request never makes it to London at all. It doesn't have to. Says Akamai's chief architect, Stephen Ludin, "It's our belief—and we build the company around the fact—that distance matters."

In our earlier discussion, we noted that certain types of computer memory have faster performance but cost more per unit of storage, leading to a "memory hierarchy" that tries to get the best of both. But it's not actually necessary to have memory made of different materials for caching to make sense. Caching is just as useful when it's proximity, rather than performance, that's the scarce resource.

This fundamental insight—that in-demand files should be stored near the location where they are used—also translates into purely physical environments. For example, Amazon's enormous fulfillment centers generally eschew any type of human-comprehensible organization, of the kind you'd find in a library or a department store. Instead, employees are told to place incoming items wherever they can find space in the warehouse—batteries cheek by jowl with pencil sharpeners, diapers, barbecue grills, and learn-the-dobro DVDs—and tag the location of each item in a central database using bar codes. But this deliberately disorganized-looking storage system still has one visible exception: high-demand items are placed in a different area, more quickly accessible than the rest. That area is Amazon's cache.

Recently, Amazon was granted a patent for an innovation that pushes this principle one step further. The patent talks about "anticipatory package shipping," which the press seized upon as though Amazon could somehow mail you something before you bought it. Amazon, like any technology company, would love to have that kind of Bélády-like clairvoyance—but for the next best thing, it turns to caching. Their patent is actually for shipping items that have been recently popular in a given region to a staging warehouse in that region—like having their own CDN for physical goods. Then, when somebody places an order, the item is just down the street. Anticipating the purchases of individuals is challenging, but when predicting the purchases of a few thousand people, the law of large numbers kicks in. *Somebody* in Berkeley is going to order, say, recycled toilet paper in a given day, and when they do it's already most of the way there.

When the things popular in an area are also *from* that area, an even more interesting geography of the cloud emerges. In 2011, film critic Micah Mertes created a map of the United States using each state's "Local Favorites" from Netflix—highlighting the movies uncommonly popular in each of those states. Overwhelmingly, it turned out, people love watching movies set where they live. Washingtonians favor *Singles*, set in Seattle; Louisianans watch *The Big Easy*, set in New Orleans; Angelinos unsurprisingly enjoy *L.A. Story*; Alaskans love *Braving Alaska*; and Montanans, *Montana Sky*.* And because nothing benefits quite so much from local caching as the enormous files that comprise full-length HD video, it's certain that Netflix has arranged it so the *files* for, say, *L.A. Story* live right in Los Angeles, just like its characters—and, more importantly, its fans.

Caching on the Home Front

While caching began as a scheme for organizing digital information inside computers, it's clear that it is just as applicable to organizing physical objects in human environments. When we spoke to John Hennessy—president of Stanford University, and a pioneering computer architect who helped develop modern caching systems—he immediately saw the link:

---

*For unknown reasons, *My Own Private Idaho* is best loved in Maine.

Caching is such an obvious thing because we do it all the time. I mean, the amount of information I get . . . certain things I have to keep track of right now, a bunch of things I have on my desk, and then other things are filed away, and then eventually filed away into the university archives system where it takes a whole *day* to get stuff out of it if I wanted. But we use that technique all the time to try to organize our lives.

The direct parallel between these problems means that there's the potential to consciously apply the solutions from computer science to the home.

First, when you are deciding what to keep and what to throw away, LRU is potentially a good principle to use—much better than FIFO. You shouldn't necessarily toss that T-shirt from college if you still wear it every now and then. But the plaid pants you haven't worn in ages? Those can be somebody else's thrift-store bonanza.

Second, exploit geography. Make sure things are in whatever cache is closest to the place where they're typically used. This isn't a concrete recommendation in most home-organization books, but it consistently turns up in the schemes that actual people describe as working well for them. "I keep running and exercise gear in a crate on the floor of my front coat closet," says one person quoted in Julie Morgenstern's *Organizing from the Inside Out*, for instance. "I like having it close to the front door."

A slightly more extreme example appears in the book *Keeping Found Things Found*, by William Jones:

> A doctor told me about her approach to keeping things. "My kids think I'm whacky, but I put things where I think I'll need them again later, even if it doesn't make much sense." As an example of her system, she told me that she keeps extra vacuum cleaner bags behind the couch in the living room. Behind the couch in the living room? Does that make any sense? . . . It turns out that when the vacuum cleaner is used, it is usually used for the carpet in the living room. . . . When a vacuum cleaner bag gets full and a new one is needed, it's usually in the living room. And that's just where the vacuum cleaner bags are.

A final insight, which hasn't yet made it into guides on closet organization, is that of the multi-level memory hierarchy. Having a cache is efficient, but having multiple levels of caches—from smallest and fastest to largest and slowest—can be even better. Where your belongings are concerned, your closet is one cache level, your basement another, and a self-storage locker a third. (These are in decreasing order of access speed, of course, so you should use the LRU principle as the basis for deciding what gets evicted from each level to the next.) But you might also be able to speed things up by adding yet another level of caching: an even smaller, faster, closer one than your closet.

Tom's otherwise extremely tolerant wife objects to a pile of clothes next to the bed, despite his insistence that it's in fact a highly efficient caching scheme. Fortunately, our conversations with computer scientists revealed a solution to this problem too. Rik Belew of UC San Diego, who studies search engines from a cognitive perspective, recommended the use of a valet stand. Though you don't see too many of them these days, a valet stand is essentially a one-outfit closet, a compound hanger for jacket, tie, and slacks—the perfect piece of hardware for your domestic caching needs. Which just goes to show that computer scientists won't only save you time; they might also save your marriage.

Filing and Piling

After deciding what to keep and where it should go, the final challenge is knowing how to organize it. We've talked about what goes in the closet and where the closet should be, but how should things be arranged inside?

One of the constants across all pieces of home-organization advice we've seen so far is the idea of grouping "like with like"—and perhaps no one so directly flies in the face of that advice as Yukio Noguchi. "I have to emphasize," says Noguchi, "that a very fundamental principle in my method is not to group files according to content." Noguchi is an economist at the University of Tokyo, and the author of a series of books that offer "super" tricks for sorting out your office and your life. Their titles translate roughly to *Super Persuasion Method*, *Super Work Method*, *Super Study Method*—and, most relevantly for us, *Super Organized Method*.

Early in his career as an economist, Noguchi found himself constantly inundated with information—correspondence, data, manuscripts—and losing a significant portion of each day just trying to organize it all. So he looked for an alternative. He began by simply putting each document into a file labeled with the document's title and date, and putting all the files into one big box. That saved time—he didn't have to think about the right place to put each document—but it didn't result in any form of organization. Then, sometime in the early 1990s, he had a breakthrough: he started to insert the files exclusively at the left-hand side of the box. And thus the "super" filing system was born.

The left-side insertion rule, Noguchi specifies, has to be followed for old files as well as new ones: every time you pull out a file to use its contents, you must put it back as the leftmost file when you return it to the box. And when you search for a file, you always start from the left-hand side as well. The most recently accessed files are thus the fastest to find.

This practice began, Noguchi explains, because returning every file to the left side was just easier than trying to reinsert it at the same spot it came from. Only gradually did he realize that this procedure was not only simple but also startlingly efficient.

The Noguchi Filing System clearly saves time when you're replacing something after you're done using it. There's still the question, however, of whether it's a good way to find the files you need in the first place. After all, it certainly goes against the recommendations of other efficiency gurus, who tell us that we should put similar things together. Indeed, even the etymology of the word "organized" evokes a body composed of *organs*—which are nothing if not cells grouped "like with like," marshalled together by similar form and function.

But computer science gives us something that most efficiency gurus don't: guarantees.

Though Noguchi didn't know it at the time, his filing system represents an extension of the LRU principle. LRU tells us that when we add something to our cache we should discard the oldest item—but it doesn't tell us where we should put the *new* item. The answer to that question comes from a line of research carried out by computer scientists in the 1970s and '80s. Their version of the problem is called "self-organizing lists," and its setup almost exactly mimics Noguchi's filing dilemma. Imagine that you have a set of items in a sequence, and you must periodically search through them to find specific items. The search itself is constrained to be linear—you must look through the items one by one, starting at the beginning—but once you find the item you're looking for, you can put it back anywhere in the sequence. Where should you replace the items to make searching as efficient as possible?

The definitive paper on self-organizing lists, published by Daniel Sleator and Robert Tarjan in 1985, examined (in classic computer science fashion) the worst-case performance of various ways to organize the list given all possible sequences of requests. Intuitively, since the search starts at the front, you want to arrange the sequence so that the items most likely to be searched for appear there. But which items will those be? We're back to wishing for clairvoyance again. "If you know the sequence ahead of time," says Tarjan, who splits his time between Princeton and Silicon Valley, "you can customize the data structure to minimize the total time for the entire sequence. That's the optimum offline algorithm: God's algorithm if you will, or the algorithm in the sky. Of course, nobody knows the future, so the question is, if you don't know the future, how *close* can you come to this optimum algorithm in the sky?" Sleator and Tarjan's results showed that some "very simple self-adjusting schemes, amazingly, come within a constant factor" of clairvoyance. Namely, if you follow the LRU principle—where you simply always put an item back at the very front of the list—then the total amount of time you spend searching will never be more than twice as long as if you'd known the future. That's not a guarantee any other algorithm can make.

Recognizing the Noguchi Filing System as an instance of the LRU principle in action tells us that it is not merely efficient. It's actually optimal.

Sleator and Tarjan's results also provide us with one further twist, and we get it by turning the Noguchi Filing System on its side. Quite simply, a box of files on its side becomes a pile. And it's the very nature of piles that you search them from top to bottom, and that each time you pull out a document it goes back not where you found it, but on top.*

In short, the mathematics of self-organizing lists suggests something radical: the big pile of papers on your desk, far from being a guilt-inducing fester of chaos, is actually one of the most well-designed and efficient structures available. What might appear to others to be an unorganized mess is, in fact, a self-organizing mess. Tossing things back on the top of the pile is the very best you can do, shy of knowing the future. In the previous

---

*You can force your computer to show your electronic documents in a pile, as well. Computers' default file-browsing interface makes you click through folders in alphabetical order—but the power of LRU suggests that you should override this, and display your files by "Last Opened" rather than "Name." What you're looking for will almost always be at or near the top.

chapter we examined cases where leaving something unsorted was more efficient than taking the time to sort everything; here, however, there's a very different reason why you don't need to organize it.

You already have.

The Forgetting Curve

Of course, no discussion of memory could be complete without mention of the "memory organ" closest to home: the human brain. Over the past few decades, the influence of computer science has brought about something of a revolution in how psychologists think about memory.

The science of human memory is said to have begun in 1879, with a young psychologist at the University of Berlin named Hermann Ebbinghaus. Ebbinghaus wanted to get to the bottom of how human memory worked, and to show that it was possible to study the mind with all the mathematical rigor of the physical sciences. So he began to experiment on himself.

Each day, Ebbinghaus would sit down and memorize a list of nonsense syllables. Then he would test himself on lists from previous days. Pursuing this habit over the course of a year, he established many of the most basic results in human memory research. He confirmed, for instance, that practicing a list multiple times makes it persist longer in memory, and that the number of items one can accurately recall goes down as time passes. His results mapped out a graph of how memory fades over time, known today by psychologists as "the forgetting curve."

Ebbinghaus's results established the credibility of a quantitative science of human memory, but they left open something of a mystery. Why this particular curve? Does it suggest that human memory is good or bad? What's the underlying story here? These questions have stimulated psychologists' speculation and research for more than a hundred years.

In 1987, Carnegie Mellon psychologist and computer scientist John Anderson found himself reading about the information retrieval systems of university libraries. Anderson's goal—or so he thought—was to write about how the design of those systems could be informed by the study of human memory. Instead, the opposite happened: he realized that information science could provide the missing piece in the study of the mind.

"For a long time," says Anderson, "I had felt that there was something missing in the existing theories of human memory, including my own. Basically, all of these theories characterize memory as an arbitrary and non-optimal configuration. . . . I had long felt that the basic memory processes were quite adaptive and perhaps even optimal; however, I had never been able to see a framework in which to make this point. In the computer science work on information retrieval, I saw that framework laid out before me."

A natural way to think about forgetting is that our minds simply run out of space. The key idea behind Anderson's new account of human memory is that the problem might be not one of *storage*, but of *organization*. According to his theory, the mind has essentially infinite capacity for memories, but we have only a finite amount of time in which to search for them. Anderson made the analogy to a library with a single, arbitrarily long shelf—the Noguchi Filing System at Library of Congress scale. You can fit as many items as you want on that shelf, but the closer something is to the front the faster it will be to find.

The key to a good human memory then becomes the same as the key to a good computer cache: predicting which items are most likely to be wanted in the future.

Barring clairvoyance, the best approach to making such predictions in the human world requires understanding the world itself. With his collaborator Lael Schooler, Anderson set out to perform Ebbinghaus-like studies not on human minds, but on human society. The question was straightforward: what patterns characterize the way the world itself "forgets"—the way that events and references fade over time? Anderson and Schooler analyzed three human environments: headlines from the *New York Times*, recordings of parents talking to their children, and Anderson's own email inbox. In all domains, they found that a word is most likely to appear again right after it had just been used, and that the likelihood of seeing it again falls off as time goes on.

In other words, reality itself has a statistical structure that mimics the Ebbinghaus curve.

This suggests something remarkable. If the pattern by which things fade from our minds is the very pattern by which things fade from use around us, then there may be a very good explanation indeed for the Ebbinghaus forgetting curve—namely, that it's a perfect tuning of

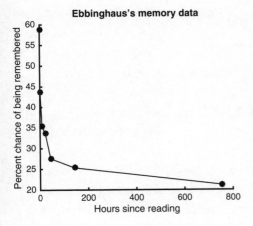

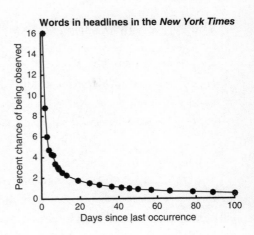

Human memory and human environments. The left panel shows the percentage of nonsense syllables Ebbinghaus correctly recalled from a list, as a function of the number of hours he waited after first memorizing the list. The right panel shows the chance that a word appears in the headlines of the New York Times *on a given day, as a function of the time since its previous appearance in print.*

the brain to the world, making available precisely the things most likely to be needed.

In putting the emphasis on time, caching shows us that memory involves unavoidable tradeoffs, and a certain zero-sumness. You can't have every library book at your desk, every product on display at the front of the store, every headline above the fold, every paper at the top of the pile. And in the same way, you can't have every fact or face or name at the front of your mind.

"Many people hold the bias that human memory is anything but optimal," wrote Anderson and Schooler. "They point to the many frustrating failures of memory. However, these criticisms fail to appreciate the task before human memory, which is to try to manage a huge stockpile of memories. In any system responsible for managing a vast data base there must be failures of retrieval. It is just too expensive to maintain access to an unbounded number of items."

This understanding has in turn led to a second revelation about human memory. If these tradeoffs really are unavoidable, and the brain appears to be optimally tuned to the world around it, then what we refer to as the inevitable "cognitive decline" that comes with age may in fact be something else.

The Tyranny of Experience

A big book is a big nuisance.
—CALLIMACHUS (305–410 BC), LIBRARIAN AT ALEXANDRIA

Why don't they make the whole plane out of that black box stuff?
—STEVEN WRIGHT

The need for a computer memory hierarchy, in the form of a cascade of caches, is in large part the result of our inability to afford making the entire memory out of the most expensive type of hardware. The fastest cache on current computers, for instance, is made with what's called SRAM, which costs roughly a thousand times as much per byte as the flash memory in solid-state drives. But the true motivation for caching goes deeper than that. In fact, even if we could get a bespoke machine that used exclusively the fastest form of memory possible, we'd still need caches.

As John Hennessy explains, size *alone* is enough to impair speed:

> When you make something bigger, it's inherently slower, right? If you make a city bigger, it takes longer to get from point A to point B. If you make a library bigger, it takes longer to find a book in the library. If you have a stack of papers on your desk that's bigger, it takes longer to find the paper you're looking for, right? Caches are actually a solution to that problem. . . . For example, right now, if you go to buy a processor, what you'll get is a Level 1 cache and a Level 2 cache on the chip. The reason that there are—even just on the *chip* there are two caches!—is that in order to keep up with the cycle rate of the processor, the first-level cache is limited in size.

Unavoidably, the larger a memory is, the more time it takes to search for and extract a piece of information from it.

Brian and Tom, in their thirties, already find themselves more frequently stalling a conversation as, for instance, they wait for the name of someone "on the tip of the tongue" to come to mind. Then again, Brian at age ten had two dozen schoolmates; twenty years later he has hundreds of contacts in his phone and thousands on Facebook, and has lived in four cities, each with its own community of friends, acquaintances, and colleagues. Tom, by this point in his academic career, has worked with hundreds of collaborators and taught thousands of students.

(In fact, this very book involved meeting with about a hundred people and citing a thousand.) Such effects are by no means limited to social connections, of course: a typical two-year-old knows two hundred words; a typical adult knows thirty thousand. And when it comes to episodic memory, well, every year adds a third of a million waking minutes to one's total lived experience.

Considered this way, it's a wonder that the two of us—or anyone—can mentally keep up at all. What's surprising is not memory's slowdown, but the fact that the mind can possibly stay afloat and responsive as so much data accumulates.

If the fundamental challenge of memory really is one of organization rather than storage, perhaps it should change how we think about the impact of aging on our mental abilities. Recent work by a team of psychologists and linguists led by Michael Ramscar at the University of Tübingen has suggested that what we call "cognitive decline"—lags and retrieval errors—may not be about the search process slowing or deteriorating, but (at least partly) an unavoidable consequence of the amount of information we have to navigate getting bigger and bigger. Regardless of whatever other challenges aging brings, older brains—which must manage a greater store of memories—are literally solving harder computational problems with every passing day. The old can mock the young for their speed: "It's because you don't know anything yet!"

Ramscar's group demonstrated the impact of extra information on human memory by focusing on the case of language. Through a series of simulations, the researchers showed that simply knowing more makes things harder when it comes to recognizing words, names, and even letters. No matter how good your organization scheme is, having to search through more things will inevitably take longer. It's not that we're forgetting; it's that we're remembering. We're becoming archives.

An understanding of the unavoidable computational demands of memory, Ramscar says, should help people come to terms with the effects of aging on cognition. "I think the most important tangible thing seniors can do is to try to get a handle on the idea that their minds are natural information processing devices," he writes. "Some things that might seem frustrating as we grow older (like remembering names!) are a function of the amount of stuff we have to sift through ... and are not necessarily a sign of a failing mind." As he puts it, "A lot of what is currently called decline is simply learning."

Caching gives us the language to understand what's happening. We say "brain fart" when we should really say "cache miss." The disproportionate occasional lags in information retrieval are a reminder of just how much we benefit the rest of the time by having what we need at the front of our minds.

So as you age, and begin to experience these sporadic latencies, take heart: the length of a delay is partly an indicator of the extent of your experience. The effort of retrieval is a testament to how much you know. And the rarity of those lags is a testament to how well you've arranged it: keeping the most important things closest to hand.

5 | Scheduling

First Things First

How we spend our days is, of course, how we spend our lives.

—ANNIE DILLARD

"Why don't we write a book on scheduling theory?" I asked. . . . "It shouldn't take much time!" Book-writing, like war-making, often entails grave mis-calculations. Fifteen years later, Scheduling *is still unfinished.*

—EUGENE LAWLER

It's Monday morning, and you have an as-yet blank schedule and a long list of tasks to complete. Some can be started only after others are finished (you can't load the dishwasher unless it's unloaded first), and some can be started only after a certain time (the neighbors will complain if you put the trash out on the curb before Tuesday night). Some have sharp deadlines, others can be done whenever, and many are fuzzily in between. Some are urgent, but not important. Some are important, but not urgent. "We are what we repeatedly do," you seem to recall Aristotle saying—whether it's mop the floor, spend more time with family, file taxes on time, learn French.

So what to do, and when, and in what order? Your life is waiting.

Though we always manage to find *some* way to order the things we do in our days, as a rule we don't consider ourselves particularly good at it—hence the perennial bestseller status of time-management guides. Unfortunately, the guidance we find in them is frequently divergent and inconsistent. *Getting Things Done* advocates a policy of immediately doing any task of two minutes or less as soon as it comes to mind. Rival bestseller *Eat That Frog!* advises beginning with the most difficult task and moving toward easier and easier things. *The Now Habit* suggests first scheduling one's social engagements and leisure time and then filling the gaps with work—rather than the other way around, as we so often do. William James, the "father of American psychology," asserts that "there's nothing so fatiguing as the eternal hanging on of an uncompleted task," but Frank Partnoy, in *Wait*, makes the case for deliberately *not* doing things right away.

Every guru has a different system, and it's hard to know who to listen to.

Spending Time Becomes a Science

Though time management seems a problem as old as time itself, the science of scheduling began in the machine shops of the industrial revolution. In 1874, Frederick Taylor, the son of a wealthy lawyer, turned down his acceptance at Harvard to become an apprentice machinist at Enterprise Hydraulic Works in Philadelphia. Four years later, he completed his apprenticeship and began working at the Midvale Steel Works, where he rose through the ranks from lathe operator to machine shop foreman and ultimately to chief engineer. In the process, he came to believe that the time of the machines (and people) he oversaw was not being used very well, leading him to develop a discipline he called "Scientific Management."

Taylor created a planning office, at the heart of which was a bulletin board displaying the shop's schedule for all to see. The board depicted every machine in the shop, showing the task currently being carried out by that machine and all the tasks waiting for it. This practice would be built upon by Taylor's colleague Henry Gantt, who in the 1910s developed the Gantt charts that would help organize many of the twentieth century's most ambitious construction projects, from the Hoover Dam to the Interstate Highway System. A century later, Gantt charts still adorn the walls and screens of project managers at firms like Amazon, IKEA, and SpaceX.

Taylor and Gantt made scheduling an object of study, and they gave it visual and conceptual form. But they didn't solve the fundamental problem of determining which schedules were best. The first hint that this problem even *could* be solved wouldn't appear until several decades later, in a 1954 paper published by RAND Corporation mathematician Selmer Johnson.

The scenario Johnson examined was bookbinding, where each book needs to be printed on one machine and then bound on another. But the most common instance of this two-machine setup is much closer to home: the laundry. When you wash your clothes, they have to pass through the washer and the dryer in sequence, and different loads will take different amounts of time in each. A heavily soiled load might take longer to wash but the usual time to dry; a large load might take longer to dry but the usual time to wash. So, Johnson asked, if you have several loads of laundry to do on the same day, what's the best way to do them?

His answer was that you should begin by finding the single step that takes the least amount of time—the load that will wash *or* dry the quickest. If that shortest step involves the washer, plan to do that load *first*. If it involves the dryer, plan to do it *last*. Repeat this process for the remaining loads, working from the two ends of the schedule toward the middle.

Intuitively, Johnson's algorithm works because regardless of how you sequence the loads, there's going to be some time at the start when the washer is running but not the dryer, and some time at the end when the dryer is running but not the washer. By having the shortest washing times at the start, and the shortest drying times at the end, you maximize the amount of overlap—when the washer and dryer are running simultaneously. Thus you can keep the total amount of time spent doing laundry to the absolute minimum. Johnson's analysis had yielded scheduling's first optimal algorithm: start with the lightest wash, end with the smallest hamper.

Beyond its immediate applications, Johnson's paper revealed two deeper points: first, that scheduling could be expressed algorithmically, and second, that optimal scheduling solutions existed. This kicked off what has become a sprawling literature, exploring strategies for a vast menagerie of hypothetical factories with every conceivable number and kind of machines.

We're going to focus on a tiny subset of this literature: the part that, unlike bookbinding or laundry, deals with scheduling for a *single* machine. Because the scheduling problem that matters the most involves just one machine: ourselves.

Handling Deadlines

With single-machine scheduling, we run into something of a problem right off the bat. Johnson's work on bookbinding was based on minimizing the total time required for the two machines to complete all of their jobs. In the case of single-machine scheduling, however, if we are going to do all the tasks assigned, then all schedules will take equally long to complete; the order is irrelevant.

This is a sufficiently fundamental and counterintuitive point that it's worth repeating. If you have only a single machine, and you're going to do all of your tasks, then any ordering of the tasks will take you the same amount of time.

Thus we encounter the first lesson in single-machine scheduling literally before we even begin: *make your goals explicit*. We can't declare some schedule a winner until we know how we're keeping score. This is something of a theme in computer science: before you can have a plan, you must first choose a metric. And as it turns out, which metric we pick here will directly affect which scheduling approaches fare best.

The first papers on single-machine scheduling followed quickly on the heels of Johnson's bookbinding work and offered several plausible metrics to consider. For each metric, they discovered a simple, optimal strategy.

It is of course common, for instance, for tasks to have a due date, with the lateness of a task being how far it has gone overdue. So we can think of the "maximum lateness" of a set of tasks as the lateness of whatever task has gone furthest past its due date—the kind of thing your employer might care about in a performance review. (Or what your customers might care about in a retail or service setting, where the "maximally late" task corresponds to the customer subjected to the longest wait time.)

If you're concerned with minimizing maximum lateness, then the best strategy is to start with the task due soonest and work your way toward the task due last. This strategy, known as **Earliest Due Date**, is fairly intuitive. (For instance, in a service-sector context, where each arriving patron's "due date" is effectively the instant they walk in the door, it just means serving customers in order of arrival.) But some of its implications are surprising. For example, how long each task will take to complete is entirely irrelevant: it doesn't change the plan, so in fact you don't even need to know. All that matters is when the tasks are due.

You might already be using Earliest Due Date to tackle your workload, in which case you probably don't need computer science to tell you that it's a sensible strategy. What you may not have known, though, is that it's the *optimal* strategy. More precisely, it is optimal assuming that you're only interested in one metric in particular: reducing your maximum lateness. If that's not your goal, however, then another strategy might be more applicable.

For instance, consider the refrigerator. If you're one of the many people who have a community-supported agriculture (CSA) subscription, then every week or two you've got a lot of fresh produce coming to your doorstep all at once. Each piece of produce is set to spoil on a different date—so eating them by Earliest Due Date, in order of their spoilage schedule, seems like a reasonable starting point. It's not, however, the end of the story. Earliest Due Date is optimal for reducing maximum lateness, which means it will minimize the rottenness of the *single most rotten thing* you'll have to eat; that may not be the most appetizing metric to eat by.

Maybe instead we want to minimize the *number* of foods that spoil. Here a strategy called **Moore's Algorithm** gives us our best plan. Moore's Algorithm says that we start out just like with Earliest Due Date—by scheduling out our produce in order of spoilage date, earliest first, one item at a time. However, as soon as it looks like we won't get to eating the next item in time, we pause, look back over the meals we've already planned, and *throw out* the biggest item (that is, the one that would take the most days to consume). For instance, that might mean forgoing the watermelon that would take a half dozen servings to eat; not even attempting it will mean getting to everything that follows a lot sooner. We then repeat this pattern, laying out the foods by spoilage date and tossing the largest already scheduled item any time we fall behind. Once everything that remains can be eaten in order of spoilage date without anything spoiling, we've got our plan.

Moore's Algorithm minimizes the number of items you'll need to throw away. Of course, you're also welcome to compost the food, donate it to the local food bank, or give it to your neighbor. In an industrial or bureaucratic context where you can't simply discard a project, but in which the *number*—rather than the severity—of late projects is still your biggest concern, Moore's Algorithm is just as indifferent about how those late tasks are handled. Anything booted from the main portion of your schedule can get done at the very end, in any order; it doesn't matter, as they're all already late.

Getting Things Done

> *Do the difficult things while they are easy and do the great things while they are small.*
>
> —LAO TZU

Sometimes due dates aren't our primary concern and we just want to get stuff done: as much stuff, as quickly as possible. It turns out that translating this seemingly simple desire into an explicit scheduling metric is harder than it sounds.

One approach is to take an outsider's perspective. We've noted that in single-machine scheduling, nothing we do can change how long it will take us to finish all of our tasks—but if each task, for instance, represents a waiting client, then there is a way to take up as little of *their* collective time as possible. Imagine starting on Monday morning with a four-day project and a one-day project on your agenda. If you deliver the bigger project on Thursday afternoon (4 days elapsed) and then the small one on Friday afternoon (5 days elapsed), the clients will have waited a total of $4 + 5 = 9$ days. If you reverse the order, however, you can finish the small project on Monday and the big one on Friday, with the clients waiting a total of only $1 + 5 = 6$ days. It's a full workweek for you either way, but now you've saved your clients three days of their combined time. Scheduling theorists call this metric the "sum of completion times."

Minimizing the sum of completion times leads to a very simple optimal algorithm called **Shortest Processing Time**: always do the quickest task you can.

Even if you don't have impatient clients hanging on every job, Shortest Processing Time gets things *done*. (Perhaps it's no surprise that it is compatible with the recommendation in *Getting Things Done* to immediately perform any task that takes less than two minutes.) Again, there's no way to change the total amount of time your work will take you, but Shortest Processing Time may ease your mind by shrinking the number of outstanding tasks as quickly as possible. Its sum-of-completion-times metric can be expressed another way: it's like focusing above all on reducing the length of your to-do list. If each piece of unfinished business is like a thorn in your side, then racing through the easiest items may bring some measure of relief.

Of course, not all unfinished business is created equal. Putting out an actual fire in the kitchen should probably be done before "putting out a fire" with a quick email to a client, even if the former takes a bit longer. In scheduling, this difference of importance is captured in a variable known as *weight*. When you're going through your to-do list, this weight might feel literal—the burden you get off your shoulders by finishing each task. A task's completion time shows how long you carry that burden, so minimizing the sum of *weighted* completion times (that is, each task's duration multiplied by its weight) means minimizing your total oppression as you work through your entire agenda.

The optimal strategy for that goal is a simple modification of Shortest Processing Time: divide the weight of each task by how long it will take to finish, and then work in order from the highest resulting importance-per-unit-time (call it "density" if you like, to continue the weight metaphor) to the lowest. And while it might be hard to assign a degree of importance to each one of your daily tasks, this strategy nonetheless offers a nice rule of thumb: only prioritize a task that takes twice as long if it's twice as important.

In business contexts, "weight" might easily be translated to the amount of money each task will bring in. The notion of dividing reward by duration translates, therefore, to assigning each task an hourly rate. (If you're a consultant or freelancer, that might in effect already be done for you: simply divide each project's fee by its size, and work your way from the highest hourly rate to the lowest.) Interestingly, this weighted strategy also shows up in studies of animal foraging, with nuts and berries taking the place of dollars and cents. Animals, seeking to maximize the rate at which they accumulate energy from food, should pursue foods in order of the ratio of their caloric energy to the time required to get and eat them—and indeed appear to do so.

When applied to debts rather than incomes, the same principle yields a strategy for getting in the black that's come to be called the "debt avalanche." This debt-reduction strategy says to ignore the number and size of your debts entirely, and simply funnel your money toward the debt with the single highest interest rate. This corresponds rather neatly to working through jobs in order of importance per unit time. And it's the strategy that will reduce the total burden of your debt as quickly as possible.

If, on the other hand, you're more concerned with reducing the *number* of debts than the *amount* of debt—if, for instance, the hassle of numerous bills and collection phone calls is a bigger deal than the difference in interest rates—then you're back to the unweighted, "just get stuff done" flavor of Shortest Processing Time, paying off the smallest debts first simply to get them out of the way. In debt-reduction circles, this approach is known as the "debt snowball." Whether people, in practice, ought to prioritize lowering the dollar amount of their debts or the quantity of them remains an active controversy, both in the popular press as well as in economics research.

Picking Our Problems

This brings us back to the note on which we began our discussion of single-machine scheduling. It's said that "a man with one watch knows what time it is; a man with two watches is never sure." Computer science can offer us optimal algorithms for various metrics available in single-machine scheduling, but choosing the metric we want to follow is up to us. In many cases, we get to decide what problem we want to be solving.

This offers a radical way to rethink procrastination, the classic pathology of time management. We typically think of it as a faulty algorithm. What if it's exactly the opposite? What if it's an optimal solution *to the wrong problem*?

There's an episode of *The X-Files* where the protagonist Mulder, bedridden and about to be consumed by an obsessive-compulsive vampire, spills a bag of sunflower seeds on the floor in self-defense. The vampire, powerless against his compulsion, stoops to pick them up one by one, and ultimately the sun rises before he can make a meal of Mulder. Computer scientists would call this a "ping attack" or a "denial of service" attack: give a system an overwhelming number of trivial things to do, and the important things get lost in the chaos.

We typically associate procrastination with laziness or avoidance behavior, but it can just as easily spring up in people (or computers, or vampires) who are trying earnestly and enthusiastically to get things done as quickly as possible. In a 2014 study led by Penn State's David Rosenbaum, for example, participants were asked to bring either one of two heavy buckets to the opposite end of a hallway. One of the buckets was right next to the participant; the other was partway down the hall. To the experimenters' surprise, people immediately picked up the bucket near them and lugged it all the way down—passing the other bucket on the way, which they could have carried a fraction of the distance. As the researchers write, "this seemingly irrational choice reflected a tendency to pre-crastinate, a term we introduce to refer to the hastening of subgoal completion, even at the expense of extra physical effort." Putting off work on a major project by attending instead to various trivial matters can likewise be seen as "the hastening of subgoal completion"—which is another way of saying that procrastinators are acting (optimally!) to reduce as quickly as possible the number of outstanding tasks on their minds. It's not that they have a bad strategy for getting things done; they have a great strategy for the wrong metric.

Working on a computer brings with it an additional hazard when it comes to being conscious and deliberate about our scheduling metrics: the user interface may subtly (or not so subtly) force its own metric upon us. A modern smartphone user, for instance, is accustomed to seeing "badges" hovering over application icons, ominous numbers in white-on-red signaling exactly how many tasks each particular app expects us to complete. If it's an email inbox blaring the figure of unread messages, then all messages are implicitly being given equal weight. Can we be blamed, then, for applying the unweighted Shortest Processing Time algorithm to the problem—dealing with all of the easiest emails first and deferring the hardest ones till last—to lower this numeral as quickly as possible?

Live by the metric, die by the metric. If all tasks are indeed of equal weight, then that's exactly what we should be doing. But if we don't want to become slaves to minutiae, then we need to take measures toward that end. This starts with making sure that the single-machine problem we're solving is the one we *want* to be solving. (In the case of app badges, if we can't get them to reflect our actual priorities, and can't overcome the impulse to optimally reduce any numerical figure thrown in our face, then perhaps the next best thing is simply to turn the badges off.)

Staying focused not just on getting things done but on getting *weighty* things done—doing the most important work you can at every moment—sounds like a surefire cure for procrastination. But as it turns out, even that is not enough. And a group of computer scheduling experts would encounter this lesson in the most dramatic way imaginable: on the surface of Mars, with the whole world watching.

Priority Inversion and Precedence Constraints

It was the summer of 1997, and humanity had a lot to celebrate. For the first time ever, a rover was navigating the surface of Mars. The $150 million Mars Pathfinder spacecraft had accelerated to a speed of 16,000 miles per hour, traveled across 309 million miles of empty space, and landed with space-grade airbags onto the rocky red Martian surface.

And now it was procrastinating.

Back on Earth, Jet Propulsion Laboratory (JPL) engineers were both worried and stumped. Pathfinder's highest priority task (to move data into and out of its "information bus") was mysteriously being neglected as the robot whiled away its time on tasks of middling importance. What was going on? Didn't the robot know any better?

Suddenly Pathfinder registered that the information bus hadn't been dealt with for an unacceptably long time, and, lacking a subtler recourse, initiated a complete restart, costing the mission the better part of a day's work. A day or so later, the same thing happened again.

Working feverishly, the JPL team finally managed to reproduce and then diagnose the behavior. The culprit was a classic scheduling hazard called *priority inversion*. What happens in a priority inversion is that a low-priority task takes possession of a system resource (access to a database, let's say) to do some work, but is then interrupted partway through that work by a timer, which pauses it and invokes the system scheduler. The scheduler tees up a high-priority task, but it can't run because the database is occupied. And so the scheduler moves down the priority list, running various unblocked medium-priority tasks instead—rather than the high-priority one (which is blocked), or the low-priority one that's blocking it (which is stuck in line behind all the medium-priority work). In these nightmarish scenarios, the system's highest priority can sometimes be neglected for arbitrarily long periods of time.*

Once JPL engineers had identified the Pathfinder problem as a case of priority inversion, they wrote up a fix and beamed the new code across millions of miles to Pathfinder. What was the solution they sent flying across the solar system? Priority *inheritance*. If a low-priority task is found to be blocking a high-priority resource, well, then all of a sudden that low-priority task should momentarily become the highest-priority thing on the system, "inheriting" the priority of the thing it's blocking.

The comedian Mitch Hedberg recounts a time when "I was at a casino, I was minding my own business, and this guy came up and said, 'You're gonna have to move, you're blocking the fire exit.' As though if there was a fire, I wasn't gonna run." The bouncer's argument was priority inversion; Hedberg's rebuttal was priority inheritance. Hedberg lounging casually in front of a fleeing mob puts his low-priority loitering ahead of their high-priority running for their lives—but not if he inherits their priority. And an onrushing mob has a way of making one inherit their priority rather

---

*Ironically enough, Pathfinder software team leader Glenn Reeves would blame the bug on "deadline pressures," and on the fact that fixing this particular issue during development had been deemed a "lower priority." So the root cause, in a sense, mirrored the problem itself.

quickly. As Hedberg explains, "If you're flammable and have legs, you are never blocking a fire exit."

The moral here is that a love of getting things done isn't enough to avoid scheduling pitfalls, and neither is a love of getting *important* things done. A commitment to fastidiously doing the most important thing you can, if pursued in a head-down, myopic fashion, can lead to what looks for all the world like procrastination. As with a car spinning its tires, the very desire to make immediate progress is how one gets stuck. "Things which matter most must never be at the mercy of things which matter least," Goethe allegedly proclaimed; but while that has the ring of wisdom about it, sometimes it's just not true. Sometimes that which matters most cannot be done until that which matters least is finished, so there's no choice but to treat that unimportant thing as being every bit as important as whatever it's blocking.

When a certain task can't be started until another one is finished, scheduling theorists call that a "precedence constraint." For operations research expert Laura Albert McLay, explicitly remembering this principle has made the difference on more than one occasion in her own household. "It can be really helpful if you can see these things. Of course, getting through the day with three kids, there's a lot of scheduling. . . . We can't get out the door unless the kids get breakfast first, and they can't get breakfast first if I don't remember to give them a spoon. Sometimes there's something very simple that you forget that just delays everything. In terms of scheduling algorithms, just knowing what [that] is, and keeping that moving, is incredibly helpful. That's how I get things done every day."

In 1978, scheduling researcher Jan Karel Lenstra was able to use the same principle while helping his friend Gene move into a new house in Berkeley. "Gene was postponing something that had to be finished before we could start something else which was urgent." As Lenstra recalls, they needed to return a van, but needed the van to return a piece of equipment, but needed the equipment to fix something in the apartment. The apartment fix didn't feel urgent (hence its postponement), but the van return did. Says Lenstra, "I explained to him that the former task should be considered even more urgent." While Lenstra is a central figure in scheduling theory, and thus was well positioned to give this advice to his friend, it came with a particularly delicious irony. This was a textbook case of priority inversion caused by precedence constraints. And arguably the twentieth century's single greatest expert on precedence constraints was none other than his friend, Eugene "Gene" Lawler.

The Speed Bump

Considering he spent much of his life thinking about how to most efficiently complete a sequence of tasks, Lawler took an intriguingly circuitous route to his own career. He studied mathematics at Florida State University before beginning graduate work at Harvard in 1954, though he left before finishing a doctorate. After time in law school, the army, and (thematically enough) working in a machine shop, he went back to Harvard in 1958, finishing his PhD and taking a position at the University of Michigan. Visiting Berkeley on sabbatical in 1969, he was arrested at a notorious Vietnam War protest. He became a member of the faculty at Berkeley the following year, and acquired a reputation there for being "the social conscience" of the computer science department. After his death in 1994, the Association for Computing Machinery established an award in Lawler's name, honoring people who demonstrate the humanitarian potential of computer science.

Lawler's first investigation into precedence constraints suggested that they could be handled quite easily. For instance, take the Earliest Due Date algorithm that minimizes the maximum lateness of a set of tasks. If your tasks have precedence constraints, that makes things trickier—you can't just plow forward in order of due date if some tasks can't be started until others are finished. But in 1968, Lawler proved that this is no trouble as long as you build the schedule back to front: look only at the tasks that no other tasks depend on, and put the one with the *latest* due date at the *end* of the schedule. Then simply repeat this process, again considering at each step only those tasks that no other (as-yet unscheduled) tasks depend upon as a prerequisite.

But as Lawler looked more deeply into precedence constraints, he found something curious. The Shortest Processing Time algorithm, as we saw, is the optimal policy if you want to cross off as many items as quickly as possible from your to-do list. But if some of your tasks have precedence constraints, there isn't a simple or obvious tweak to Shortest Processing Time to adjust for that. Although it looked like an elementary scheduling problem, neither Lawler nor any other researcher seemed to be able to find an efficient way to solve it.

In fact, it was much worse than this. Lawler himself would soon discover that this problem belongs to a class that most computer scientists believe *has* no efficient solution—it's what the field calls "intractable."* Scheduling theory's first speed bump turned out to be a brick wall.

---

*We will discuss "intractable" problems in more detail in chapter 8.

As we saw with the "triple or nothing" scenario for which optimal stopping theory has no sage words, not every problem that can be formally articulated has an answer. In scheduling, it's clear by definition that every set of tasks and constraints has *some* schedule that's the best, so scheduling problems aren't unanswerable, per se—but it may simply be the case that there's no straightforward algorithm that can find you the optimal schedule in a reasonable amount of time.

This led researchers like Lawler and Lenstra to an irresistible question. Just what proportion of scheduling problems was intractable, anyway? Twenty years after scheduling theory was kick-started by Selmer Johnson's bookbinding paper, the search for individual solutions was about to become something much grander and more ambitious by far: a quest to map the entire landscape of scheduling theory.

What the researchers found was that even the subtlest change to a scheduling problem often tips it over the fine and irregular line between tractable and intractable. For example, Moore's Algorithm minimizes the number of late tasks (or rotten fruits) when they're all of equal value—but if some are more important than others, the problem becomes intractable and no algorithm can readily provide the optimal schedule. Likewise, having to wait until a certain time to start some of your tasks makes nearly all of the scheduling problems for which we otherwise have efficient solutions into intractable problems. Not being able to put out the trash until the night before collection might be a reasonable municipal bylaw, but it will send your calendar headlong into intractability.

The drawing of the borders of scheduling theory continues to this day. A recent survey showed that the status of about 7% of all problems is still unknown, scheduling's terra incognita. Of the 93% of the problems that we do understand, however, the news isn't great: only 9% can be solved efficiently, and the other 84% have been proven intractable.* In other words, most scheduling problems admit no ready solution. If trying to perfectly manage your calendar feels overwhelming, maybe that's because it actually is. Nonetheless, the algorithms we have discussed are often the starting point for tackling those hard problems—if not perfectly, then at least as well as can be expected.

---

*Things aren't quite as bad as this number might make them seem, though, since it includes scheduling problems involving multiple machines—which is more like managing a group of employees than managing your calendar.

Drop Everything: Preemption and Uncertainty

The best time to plant a tree is twenty years ago. The second best time is now.
—PROVERB

So far we have considered only factors that make scheduling harder. But there is one twist that can make it easier: being able to stop one task partway through and switch to another. This property, "preemption," turns out to change the game dramatically.

Minimizing maximum lateness (for serving customers in a coffee shop) or the sum of completion times (for rapidly shortening your to-do list) both cross the line into intractability if some tasks can't be started until a particular time. But they return to having efficient solutions once preemption is allowed. In both cases, the classic strategies—Earliest Due Date and Shortest Processing Time, respectively—remain the best, with a fairly straightforward modification. When a task's starting time comes, compare that task to the one currently under way. If you're working by Earliest Due Date and the new task is due even sooner than the current one, switch gears; otherwise stay the course. Likewise, if you're working by Shortest Processing Time, and the new task can be finished faster than the current one, pause to take care of it first; otherwise, continue with what you were doing.

Now, on a good week a machine shop might know everything expected of them in the next few days, but most of us are usually flying blind, at least in part. We might not even be sure, for instance, when we'll be able to start a particular project (when will so-and-so give me a solid answer on the such-and-such?). And at any moment our phone can ring or an email can pop up with news of a whole new task to add to our agenda.

It turns out, though, that even if you don't know when tasks will begin, Earliest Due Date and Shortest Processing Time are *still* optimal strategies, able to guarantee you (on average) the best possible performance in the face of uncertainty. If assignments get tossed on your desk at unpredictable moments, the optimal strategy for minimizing maximum lateness is still the preemptive version of Earliest Due Date—switching to the job that just came up if it's due sooner than the one you're currently doing, and otherwise ignoring it. Similarly, the preemptive version of Shortest Processing Time—compare the time left to finish the current task to the time it would take to complete the new one—is still optimal for minimizing the sum of completion times.

In fact, the weighted version of Shortest Processing Time is a pretty good candidate for best general-purpose scheduling strategy in the face of uncertainty. It offers a simple prescription for time management: each time a new piece of work comes in, divide its importance by the amount of time it will take to complete. If that figure is higher than for the task you're currently doing, switch to the new one; otherwise stick with the current task. This algorithm is the closest thing that scheduling theory has to a skeleton key or Swiss Army knife, the optimal strategy not just for one flavor of problem but for many. Under certain assumptions it minimizes not just the sum of weighted completion times, as we might expect, but also the sum of the weights of the late jobs and the sum of the weighted lateness of those jobs.

Intriguingly, optimizing all these other metrics is intractable if we know the start times and durations of jobs ahead of time. So considering the impact of uncertainty in scheduling reveals something counterintuitive: there are cases where clairvoyance is a burden. Even with complete foreknowledge, finding the perfect schedule might be practically impossible. In contrast, thinking on your feet and reacting as jobs come in won't give you *as* perfect a schedule as if you'd seen into the future—but the best you *can* do is much easier to compute. That's some consolation. As business writer and coder Jason Fried says, "Feel like you can't proceed until you have a bulletproof plan in place? Replace 'plan' with 'guess' and take it easy." Scheduling theory bears this out.

When the future is foggy, it turns out you don't need a calendar—just a to-do list.

Preemption Isn't Free: The Context Switch

The hurrieder I go / The behinder I get
—NEEDLEPOINT SEEN IN BOONVILLE, CA

Programmers don't talk because they must not be interrupted. . . . To synchronize with other people (or their representation in telephones, buzzers and doorbells) can only mean interrupting the thought train. Interruptions mean certain bugs. You must not get off the train.
—ELLEN ULLMAN

Scheduling theory thus tells a reasonably encouraging story after all. There are simple, optimal algorithms for solving many scheduling problems, and those problems are tantalizingly close to situations we encounter daily in human lives. But when it comes to actually carrying out single-machine scheduling in the real world, things get complicated.

First of all, people and computer operating systems alike face a curious challenge: the machine that is doing the scheduling and the machine being scheduled are one and the same. Which makes straightening out your to-do list an item *on* your to-do list—needing, itself, to get prioritized and scheduled.

Second, preemption isn't free. Every time you switch tasks, you pay a price, known in computer science as a *context switch*. When a computer processor shifts its attention away from a given program, there's always a certain amount of necessary overhead. It needs to effectively bookmark its place and put aside all of its information related to that program. Then it needs to figure out which program to run next. Finally it must haul out all the relevant information for that program, find its place in the code, and get in gear.

None of this switching back and forth is "real work"—that is, none of it actually advances the state of any of the various programs the computer is switching between. It's *metawork*. Every context switch is wasted time.

Humans clearly have context-switching costs too. We feel them when we move papers on and off our desk, close and open documents on our computer, walk into a room without remembering what had sent us there, or simply say out loud, "Now, where was I?" or "What was I saying?" Psychologists have shown that for us, the effects of switching tasks can include both delays and errors—at the scale of minutes rather than microseconds. To put that figure in perspective, anyone you interrupt more than a few times an hour is in danger of doing no work at all.

Personally, we have found that both programming and writing require keeping in mind the state of the entire system, and thus carry inordinately large context-switching costs. A friend of ours who writes software says that the normal workweek isn't well suited to his workflow, since for him sixteen-hour days are more than twice as productive as eight-hour days. Brian, for his part, thinks of writing as a kind of blacksmithing, where it takes a while just to heat up the metal before it's malleable. He finds it somewhat useless to block out anything less than ninety minutes for writing, as nothing much happens in the first half hour except loading a giant block of "Now, where was I?" into his head. Scheduling expert Kirk Pruhs, of the University of Pittsburgh, has had the same experience. "If it's less than an hour I'll just do errands instead, because it'll take me the first thirty-five minutes to really figure out what I want to do and then I might not have time to do it."

Rudyard Kipling's celebrated 1910 poem "If—" ends with an exuberant call for time management: "If you can fill the unforgiving minute / With sixty seconds' worth of distance run . . ."

If only. The truth is, there's always overhead—time lost to metawork, to the logistics of bookkeeping and task management. This is one of the fundamental tradeoffs of scheduling. And the more you take on, the more overhead there is. At its nightmarish extreme, this turns into a phenomenon called *thrashing*.

Thrashing

> *Gage: Mr. Zuckerberg, do I have your full attention? . . .*
> *Zuckerberg: You have part of my attention—you have the minimum amount.*
> —THE SOCIAL NETWORK

Computers multitask through a process called "threading," which you can think of as being like juggling a set of balls. Just as a juggler only hurls one ball at a time into the air but keeps three aloft, a CPU only works on one program at a time, but by swapping between them quickly enough (on the scale of ten-thousandths of a second) it *appears* to be playing a movie, navigating the web, and alerting you to incoming email all at once.

In the 1960s, computer scientists began thinking about how to automate the process of sharing computer resources between different tasks and users. It was an exciting time, recounts Peter Denning, now one of the top experts on computer multitasking, who was then working on his doctorate at MIT. Exciting, and uncertain: "How do you partition a main memory among a bunch of jobs that are in there when some of them want to grow and some might want to shrink and they're going to interact with each other, trying to steal [memory] and all these kinds of things? . . . How do you manage that whole set of interactions? Nobody knew anything about that."

Not surprisingly, given that the researchers didn't really know yet what they were doing, the effort encountered difficulties. And there was one in particular that caught their attention. As Denning explains, under certain conditions a dramatic problem "shows up as you add more jobs to the multiprogramming mix. At some point you pass a critical threshold—

unpredictable exactly where it is, but you'll know it when you get there—and all of a sudden the system seems to die."

Think again about our image of a juggler. With one ball in the air, there's enough spare time while that ball is aloft for the juggler to toss some others upward as well. But what if the juggler takes on one more ball than he can handle? He doesn't drop *that* ball; he drops *everything*. The whole system, quite literally, goes down. As Denning puts it, "The presence of one additional program has caused a complete collapse of service. . . . The sharp difference between the two cases at first defies intuition, which might lead us to expect a gradual degradation of service as new programs are introduced into crowded main memory." Instead, catastrophe. And while we can understand a human juggler being overwhelmed, what could cause something like this to happen to a *machine*?

Here scheduling theory intersects caching theory. The whole idea of caches is to keep the "working set" of needed items available for quick access. One way this is done is by keeping the information the computer is currently using in fast memory rather than on the slow hard disk. But if a task requires keeping track of so many things that they won't all fit into memory, then you might well end up spending more time swapping information in and out of memory than doing the actual work. What's more, when you switch tasks, the newly active task might make space for its working set by evicting portions of *other* working sets from memory. The next task, upon reactivation, would then reacquire parts of *its* working set from the hard disk and muscle them back into memory, again displacing others. This problem—tasks stealing space from each other—can get even worse in systems with hierarchies of caches between the processor and the memory. As Peter Zijlstra, one of the head developers on the Linux operating system scheduler, puts it, "The caches are warm for the current workload, and when you context switch you pretty much invalidate all caches. And that hurts." At the extreme, a program may run *just* long enough to swap its needed items into memory, before giving way to another program that runs just long enough to overwrite them in turn.

This is thrashing: a system running full-tilt and accomplishing nothing at all. Denning first diagnosed this phenomenon in a memory-management context, but computer scientists now use the term "thrashing" to refer to pretty much any situation where the system grinds to a halt because it's entirely preoccupied with metawork. A thrashing computer's performance doesn't bog down gradually. It falls off a cliff. "Real work" has dropped to effectively zero, which also means it's going to be nearly impossible to get out.

Thrashing is a very recognizable human state. If you've ever had a moment where you wanted to stop doing everything just to have the chance to write down everything you were supposed to be doing, but couldn't spare the time, you've thrashed. And the cause is much the same for people as for computers: each task is a draw on our limited cognitive resources. When merely remembering everything we need to be doing occupies our full attention—or prioritizing every task consumes all the time we had to do them—or our train of thought is continually interrupted before those thoughts can translate to action—it feels like panic, like paralysis by way of hyperactivity. It's thrashing, and computers know it well.

If you've ever wrestled with a system in a state of thrashing—and if you've ever *been* in such a state—then you might be curious about the computer science of getting out. In his landmark 1960s paper on the subject, Denning noted that an ounce of prevention is worth a pound of cure. The easiest thing to do is simply to get more memory: enough RAM, for instance, to fit the working sets of all the running programs into memory at once and reduce the time taken by a context switch. But preventive advice for thrashing doesn't help you when you find yourself in the midst of it. Besides, when it comes to human attention, we're stuck with what we've got.

Another way to avert thrashing before it starts is to learn the art of saying no. Denning advocated, for instance, that a system should simply refuse to add a program to its workload if it didn't have enough free memory to hold its working set. This prevents thrashing in machines, and is sensible advice for anyone with a full plate. But this, too, might seem like an unattainable luxury to those of us who find ourselves *already* overloaded—or otherwise unable to throttle the demands being placed on us.

In these cases there's clearly no way to work any harder, but you can work . . . *dumber*. Along with considerations of memory, one of the biggest sources of metawork in switching contexts is the very act of choosing what to do next. This, too, can at times swamp the actual doing of the work. Faced with, say, an overflowing inbox of n messages, we know from sorting theory that repeatedly scanning it for the most important one to answer next will take $O(n^2)$ operations—n scans of n messages apiece. This means that waking up to an inbox that's three times as full as usual could take you *nine* times as long to process. What's more, scanning through those emails means swapping every message into your mind, one after another, before you respond to any of them: a surefire recipe for memory thrashing.

In a thrashing state, you're making essentially no progress, so even doing tasks in the wrong order is better than doing nothing at all. Instead of answering the most important emails first—which requires an assessment of the whole picture that may take longer than the work itself—maybe you should sidestep that quadratic-time quicksand by just answering the emails in *random* order, or in whatever order they happen to appear on-screen. Thinking along the same lines, the Linux core team, several years ago, replaced their scheduler with one that was less "smart" about calculating process priorities but more than made up for it by taking less time to calculate them.

If you still want to maintain your priorities, though, there's a different and even more interesting bargain you can strike to get your productivity back.

Interrupt Coalescing

Part of what makes real-time scheduling so complex and interesting is that it is fundamentally a negotiation between two principles that aren't fully compatible. These two principles are called *responsiveness* and *throughput*: how quickly you can respond to things, and how much you can get done overall. Anyone who's ever worked in an office environment can readily appreciate the tension between these two metrics. It's part of the reason there are people whose job it is to answer the phone: they are responsive so that others may have throughput.

Again, life is harder when—like a computer—you must make the responsiveness/throughput tradeoff yourself. And the best strategy for getting things done might be, paradoxically, to slow down.

Operating system schedulers typically define a "period" in which every program is guaranteed to run at least a little bit, with the system giving a "slice" of that period to each program. The more programs are running, the smaller those slices become, and the more context switches are happening every period, maintaining responsiveness at the cost of throughput. Left unchecked, however, this policy of guaranteeing each process at least *some* attention every period could lead to catastrophe. With enough programs running, a task's slice would shrink to the point that the system was spending the *entire* slice on context switching, before immediately context-switching again to the next task.

The culprit is the hard responsiveness guarantee. So modern operating systems in fact set a minimum length for their slices and will refuse to subdivide the period any more finely. (In Linux, for instance, this minimum useful slice turns out to be about three-quarters of a millisecond, but in humans it might realistically be at least several minutes.) If more processes are added beyond that point, the period will simply get longer. This means that processes will have to wait longer to get their turn, but the turns they get will at least be long enough to do something.

Establishing a minimum amount of time to spend on any one task helps to prevent a commitment to responsiveness from obliterating throughput entirely: if the minimum slice is longer than the time it takes to context-switch, then the system can never get into a state where context switching is the *only* thing it's doing. It's also a principle that is easy to translate into a recommendation for human lives. Methods such as "timeboxing" or "pomodoros," where you literally set a kitchen timer and commit to doing a single task until it runs out, are one embodiment of this idea.

But what slice size should you aim for? Faced with the question of how long to wait between intervals of performing a recurring task, like checking your email, the answer from the perspective of throughput is simple: as long as possible. But that's not the end of the story; higher throughput, after all, also means lower responsiveness.

For your computer, the annoying interruption that it has to check on regularly isn't email—it's you. You might not move the mouse for minutes or hours, but when you do, you expect to see the pointer on the screen move immediately, which means the machine expends a lot of effort simply checking in on you. The more frequently it checks on the mouse and keyboard, the quicker it can react when there is input, but the more context switches it has to do. So the rule that computer operating systems follow when deciding how long they can afford to dedicate themselves to some task is simple: as long as possible without seeming jittery or slow to the user.

When we humans leave the house to run a quick errand, we might say something like, "You won't even notice I'm gone." When our machines context-switch into a computation, they must literally return to us before we notice they're gone. To find this balancing point, operating systems programmers have turned to psychology, mining papers in psychophysics for the exact number of milliseconds of delay it takes for a human brain to register lag or flicker. There is no point in attending to the user any more often than that.

Thanks to these efforts, when operating systems are working right you don't even notice how hard your computer is exerting itself. You continue to be able to move your mouse around the screen fluidly even when your processor is hauling full-tilt. The fluidity *is* costing you some throughput, but that's a design tradeoff that has been explicitly made by the system engineers: your system spends as much time as it possibly can away from interacting with you, then gets around to redrawing the mouse just in time.

And again, this is a principle that can be transferred to human lives. The moral is that you should try to stay on a single task as long as possible without decreasing your responsiveness below the minimum acceptable limit. Decide how responsive you need to be—and then, if you want to get things done, be no more responsive than that.

If you find yourself doing a lot of context switching because you're tackling a heterogeneous collection of short tasks, you can also employ another idea from computer science: "interrupt coalescing." If you have five credit card bills, for instance, don't pay them as they arrive; take care of them all in one go when the fifth bill comes. As long as your bills are never due less than thirty-one days after they arrive, you can designate, say, the first of each month as "bill-paying day," and sit down at that point to process every bill on your desk, no matter whether it came three weeks or three hours ago. Likewise, if none of your email correspondents require you to respond in less than twenty-four hours, you can limit yourself to checking your messages once a day. Computers themselves do something like this: they wait until some fixed interval and check everything, instead of context-switching to handle separate, uncoordinated interrupts from their various subcomponents.*

On occasion, computer scientists notice the absence of interrupt coalescing in their own lives. Says Google director of research Peter Norvig: "I had to go downtown three times today to run errands, and I said, 'Oh, well, that's just a one-line bug in your algorithm. You should have just waited, or added it to the to-do queue, rather than executing them sequentially as they got added one at a time.'"

---

*Given that many computers tend to brashly pop up error messages and cursor-stealing dialogue boxes whenever they want something from us, their behavior is somewhat hypocritical. The user interface demands the user's attention in a way that the CPU itself would rarely tolerate.

At human scale, we get interrupt coalescing for free from the postal system, just as a consequence of their delivery cycle. Because mail gets delivered only once a day, something mailed only a few minutes late might take an extra twenty-four hours to reach you. Considering the costs of context switching, the silver lining to this should by now be obvious: you can only get interrupted by bills and letters at most once a day. What's more, the twenty-four-hour postal rhythm demands minimal responsiveness from you: it doesn't make any difference whether you mail your reply five minutes or five hours after receiving a letter.

In academia, holding office hours is a way of coalescing interruptions from students. And in the private sector, interrupt coalescing offers a redemptive view of one of the most maligned office rituals: the weekly meeting. Whatever their drawbacks, regularly scheduled meetings are one of our best defenses against the spontaneous interruption and the unplanned context switch.

Perhaps the patron saint of the minimal-context-switching lifestyle is the legendary programmer Donald Knuth. "I do one thing at a time," he says. "This is what computer scientists call batch processing—the alternative is swapping in and out. I don't swap in and out." Knuth isn't kidding. On January 1, 2014, he embarked on "The TeX Tuneup of 2014," in which he fixed all of the bugs that had been reported in his TeX typesetting software over the previous *six years*. His report ends with the cheery sign-off "Stay tuned for The TeX Tuneup of 2021!" Likewise, Knuth has not had an email address since 1990. "Email is a wonderful thing for people whose role in life is to be on top of things. But not for me; my role is to be on the bottom of things. What I do takes long hours of studying and uninterruptible concentration." He reviews all his postal mail every three months, and all his faxes every six.

But one does not need to take things to Knuth's extreme to wish that more of our lives used interrupt coalescing as a design principle. The post office gives it to us almost by accident; elsewhere, we need to build it, or demand it, for ourselves. Our beeping and buzzing devices have "Do Not Disturb" modes, which we could manually toggle on and off throughout the day, but that is too blunt an instrument. Instead, we might agitate for settings that would provide an explicit option for interrupt coalescing—the same thing at a human timescale that the devices are doing internally. Alert me only once every ten minutes, say; then tell me everything.

6 | Bayes's Rule

Predicting the Future

All human knowledge is uncertain, inexact, and partial.

—BERTRAND RUSSELL

The sun'll come out tomorrow. You can bet your bottom dollar there'll be sun.

—ANNIE

In 1969, before embarking on a doctorate in astrophysics at Princeton, J. Richard Gott III took a trip to Europe. There he saw the Berlin Wall, which had been built eight years earlier. Standing in the shadow of the wall, a stark symbol of the Cold War, he began to wonder how much longer it would continue to divide the East and West.

On the face of it, there's something absurd about trying to make this kind of prediction. Even setting aside the impossibility of forecasting geopolitics, the question seems mathematically laughable: it's trying to make a prediction from *a single data point*.

But as ridiculous as this might seem on its face, we make such predictions all the time, by necessity. You arrive at a bus stop in a foreign city and learn, perhaps, that the other tourist standing there has been waiting seven minutes. When is the next bus likely to arrive? Is it worthwhile to wait—and if so, how long should you do so before giving up?

Or perhaps a friend of yours has been dating somebody for a month and wants your advice: is it too soon to invite them along to an upcoming family wedding? The relationship is off to a good start—but how far ahead is it safe to make plans?

A famous presentation made by Peter Norvig, Google's director of research, carried the title "The Unreasonable Effectiveness of Data" and enthused about "how billions of trivial data points can lead to understanding." The media constantly tell us that we're living in an "age of big data," when computers can sift through those billions of data points and find patterns invisible to the naked eye. But often the problems most germane to daily human life are at the opposite extreme. Our days are full of "small data." In fact, like Gott standing at the Berlin Wall, we often have to make an inference from the smallest amount of data we could possibly have: a single observation.

So how do we do it? And how *should* we?

The story begins in eighteenth-century England, in a domain of inquiry irresistible to great mathematical minds of the time, even those of the clergy: gambling.

Reasoning Backward with the Reverend Bayes

> *If we be, therefore, engaged by arguments to put trust in past experience, and make it the standard of our future judgement, these arguments must be probable only.*
>
> —DAVID HUME

More than 250 years ago, the question of making predictions from small data weighed heavily on the mind of the Reverend Thomas Bayes, a Presbyterian minister in the charming spa town of Tunbridge Wells, England.

If we buy ten tickets for a new and unfamiliar raffle, Bayes imagined, and five of them win prizes, then it seems relatively easy to estimate the raffle's chances of a win: 5/10, or 50%. But what if instead we buy a single ticket and it wins a prize? Do we really imagine the probability of winning to be 1/1, or 100%? That seems too optimistic. Is it? And if so, by how much? What should we actually guess?

For somebody who has had such an impact on the history of reasoning under uncertainty, Bayes's own history remains ironically uncertain. He was born in 1701, or perhaps 1702, in the English county of Hertfordshire, or maybe it was London. And in either 1746, '47, '48, or '49 he would write one of the most influential papers in all of mathematics, abandon it unpublished, and move on to other things.

Between those two events we have a bit more certainty. The son of a minister, Bayes went to the University of Edinburgh to study theology, and was ordained like his father. He had mathematical as well as theological interests, and in 1736 he wrote an impassioned defense of Newton's newfangled "calculus" in response to an attack by Bishop George Berkeley. This work resulted in his election in 1742 as a Fellow of the Royal Society, to whom he was recommended as "a Gentleman . . . well skilled in Geometry and all parts of Mathematical and Philosophical Learning."

After Bayes died in 1761, his friend Richard Price was asked to review his mathematical papers to see if they contained any publishable material. Price came upon one essay in particular that excited him—one he said "has great merit, and deserves to be preserved." The essay concerned exactly the kind of raffle problem under discussion:

> Let us then imagine a person present at the drawing of a lottery, who knows nothing of its scheme or of the proportion of *Blanks* to *Prizes* in it. Let it further be supposed, that he is obliged to infer this from the number of *blanks* he hears drawn compared with the number of *prizes*; and that it is enquired what conclusions in these circumstances he may reasonably make.

Bayes's critical insight was that trying to use the winning and losing tickets we see to figure out the overall ticket pool that they came from is essentially reasoning backward. And to do that, he argued, we need to first reason *forward* from hypotheticals. In other words, we need to first determine how probable it is that we would have drawn the tickets we did *if* various scenarios were true. This probability—known to modern statisticians as the "likelihood"—gives us the information we need to solve the problem.

For instance, imagine we bought three tickets and all three were winners. Now, if the raffle was of the particularly generous sort where *all* the tickets are winners, then our three-for-three experience would of course happen all of the time; it has a 100% chance in that scenario. If, instead, only half of the raffle's tickets were winners, our three-for-three experience would happen $\frac{1}{2} \times \frac{1}{2} \times \frac{1}{2}$ of the time, or in other words $\frac{1}{8}$ of the time. And if the raffle rewarded only one ticket in a thousand, our outcome would have been incredibly unlikely: $\frac{1}{1,000} \times \frac{1}{1,000} \times \frac{1}{1,000}$, or one in a billion.

Bayes argued that we should accordingly judge it to be more probable that all the raffle tickets are winners than that half of them are, and in turn more probable that half of them are than that only one in a thousand is. Perhaps we had already intuited as much, but Bayes's logic offers us the ability to quantify that intuition. All things being equal, we should imagine it to be exactly eight times likelier that all the tickets are winners than that half of them are—because the tickets we drew are exactly *eight times likelier* (100% versus one-in-eight) in that scenario. Likewise, it's exactly 125 million times likelier that half the raffle tickets are winners than that there's only one winning ticket per thousand, which we know by comparing one-in-eight to one-in-a-billion.

This is the crux of Bayes's argument. Reasoning forward from hypothetical pasts lays the foundation for us to then work backward to the most probable one.

It was an ingenious and innovative approach, but it did not manage to provide a full answer to the raffle problem. In presenting Bayes's results to the Royal Society, Price was able to establish that if you buy a single raffle ticket and it's a winner, then there's a 75% chance that at least half the tickets are winners. But thinking about the probabilities *of* probabilities can get a bit head-spinning. What's more, if someone pressed us, "Well, fine, but what do you think the raffle odds actually *are*?" we still wouldn't know what to say.

The answer to this question—how to distill all the various possible hypotheses into a single specific expectation—would be discovered only a few years later, by the French mathematician Pierre-Simon Laplace.

Laplace's Law

Laplace was born in Normandy in 1749, and his father sent him to a Catholic school with the intent that he join the clergy. Laplace went on to study theology at the University of Caen, but unlike Bayes—who balanced spiritual and scientific devotions his whole life—he ultimately abandoned the cloth entirely for mathematics.

In 1774, completely unaware of the previous work by Bayes, Laplace published an ambitious paper called "Treatise on the Probability of the Causes of Events." In it, Laplace finally solved the problem of how to make inferences backward from observed effects to their probable causes.

Bayes, as we saw, had found a way to compare the relative probability of one hypothesis to another. But in the case of a raffle, there is literally an infinite number of hypotheses: one for every conceivable proportion of winning tickets. Using calculus, the once-controversial mathematics of which Bayes had been an important defender, Laplace was able to prove that this vast spectrum of possibilities could be distilled down to a single estimate, and a stunningly concise one at that. If we really know nothing about our raffle ahead of time, he showed, then after drawing a winning ticket on our first try we should expect that the proportion of winning tickets in the whole pool is exactly 2/3. If we buy three tickets and all of them are winners, the expected proportion of winning tickets is exactly 4/5. In fact, for any possible drawing of w winning tickets in n attempts, the expectation is simply the number of wins plus one, divided by the number of attempts plus two: $\frac{w+1}{n+2}$.

This incredibly simple scheme for estimating probabilities is known as **Laplace's Law**, and it is easy to apply in any situation where you need to assess the chances of an event based on its history. If you make ten attempts at something and five of them succeed, Laplace's Law estimates your overall chances to be 6/12 or 50%, consistent with our intuitions. If you try only once and it works out, Laplace's estimate of 2/3 is both more reasonable than assuming you'll win every time, and more actionable than Price's guidance (which would tell us that there is a 75% metaprobability of a 50% or greater chance of success).

Laplace went on to apply his statistical approach to a wide range of problems of his time, including assessing whether babies are truly equally likely to be born male or female. (He established, to a virtual certainty, that male infants are in fact slightly more likely than female ones.) He also wrote the *Philosophical Essay on Probabilities*, arguably the first book about probability for a general audience and still one of the best, laying out his theory and considering its applications to law, the sciences, and everyday life.

Laplace's Law offers us the first simple rule of thumb for confronting small data in the real world. Even when we've made only a few observations—or only one—it offers us practical guidance. Want to calculate the chance your bus is late? The chance your softball team will win? Count the number of times it has happened in the past plus one, then divide by the number of opportunities plus two. And the beauty of Laplace's Law is that it works equally well whether we have a single data point or millions of them. Little Annie's faith that the sun will rise tomorrow is justified, it tells us: with an Earth that's seen the sun rise for about 1.6 trillion days in a row, the chance of another sunrise on the next "attempt" is all but indistinguishable from 100%.

Bayes's Rule and Prior Beliefs

All these suppositions are consistent and conceivable. Why should we give the preference to one, which is no more consistent or conceivable than the rest?

—DAVID HUME

Laplace also considered another modification to Bayes's argument that would prove crucial: how to handle hypotheses that are simply *more probable* than others. For instance, while it's possible that a lottery might give away prizes to 99% of the people who buy tickets, it's more likely—we'd assume—that they would give away prizes to only 1%. That assumption should be reflected in our estimates.

To make things concrete, let's say a friend shows you two different coins. One is a normal, "fair" coin with a 50–50 chance of heads and tails; the other is a two-headed coin. He drops them into a bag and then pulls one out at random. He flips it once: heads. Which coin do you think your friend flipped?

Bayes's scheme of working backward makes short work of this question. A flip coming up heads happens 50% of the time with a fair coin and 100% of the time with a two-headed coin. Thus we can assert confidently that it's $\frac{100\%}{50\%}$, or exactly twice as probable, that the friend had pulled out the two-headed coin.

Now consider the following twist. This time, the friend shows you *nine* fair coins and one two-headed coin, puts all ten into a bag, draws one at random, and flips it: heads. Now what do you suppose? Is it a fair coin or the two-headed one?

Laplace's work anticipated this wrinkle, and here again the answer is impressively simple. As before, a fair coin is exactly half as likely to come up heads as a two-headed coin. But now, a fair coin is also nine times as likely to have been drawn in the first place. It turns out that we can just take these two different considerations and multiply them together: it is exactly four and a half times more likely that your friend is holding a fair coin than the two-headed one.

The mathematical formula that describes this relationship, tying together our previously held ideas and the evidence before our eyes, has come to be known—ironically, as the real heavy lifting was done by Laplace—as **Bayes's Rule**. And it gives a remarkably straightforward solution to the problem of how to combine preexisting beliefs with observed evidence: multiply their probabilities together.

Notably, having *some* preexisting beliefs is crucial for this formula to work. If your friend simply approached you and said, "I flipped one coin from this bag and it came up heads. How likely do you think it is that this is a fair coin?," you would be totally unable to answer that question unless you had at least some sense of what coins were in the bag to begin with. (You can't multiply the two probabilities together when you don't *have* one of them.) This sense of what was "in the bag" before the coin flip—the chances for each hypothesis to have been true before you saw any data—is known as the prior probabilities, or "prior" for short. And Bayes's Rule always needs some prior from you, even if it's only a guess. How many two-headed coins exist? How easy are they to get? How much of a trickster is your friend, anyway?

The fact that Bayes's Rule is dependent on the use of priors has at certain points in history been considered controversial, biased, even unscientific. But in reality, it is quite rare to go into a situation so totally unfamiliar that our mind is effectively a blank slate—a point we'll return to momentarily.

When you do have some estimate of prior probabilities, meanwhile, Bayes's Rule applies to a wide range of prediction problems, be they of the big-data variety or the more common small-data sort. Computing the probability of winning a raffle or tossing heads is only the beginning. The methods developed by Bayes and Laplace can offer help any time you have uncertainty and a bit of data to work with. And that's exactly the situation we face when we try to predict the future.

The Copernican Principle

> It's difficult to make predictions, especially about the future.
>
> —DANISH PROVERB

When J. Richard Gott arrived at the Berlin Wall, he asked himself a very simple question: *Where am I?* That is to say, where in the total life span of this artifact have I happened to arrive? In a way, he was asking the temporal version of the spatial question that had obsessed the astronomer Nicolaus Copernicus four hundred years earlier: Where are we? Where in the universe is the Earth? Copernicus would make the radical paradigm shift of imagining that the Earth was not the bull's-eye center of the universe—that it was, in fact, nowhere special in particular. Gott decided to take the same step with regard to time.

He made the assumption that the moment when he encountered the Berlin Wall wasn't special—that it was equally likely to be any moment in the wall's total lifetime. And if any moment was equally likely, then on average his arrival should have come precisely at the halfway point (since it was 50% likely to fall before halfway and 50% likely to fall after). More generally, unless we know better we can expect to have shown up precisely halfway into the duration of *any* given phenomenon.* And if we assume that we're arriving precisely halfway into something's duration, the best guess we can make for how long it will last into the future becomes obvious: *exactly as long as it's lasted already.* Gott saw the Berlin Wall eight years after it was built, so his best guess was that it would stand for eight years more. (It ended up being twenty.)

This straightforward reasoning, which Gott named the **Copernican Principle**, results in a simple algorithm that can be used to make predictions about all sorts of topics. Without any preconceived expectations, we might use it to obtain predictions for the end of not only the Berlin Wall but any number of other short- and long-lived phenomena. The Copernican Principle predicts that the United States of America will last as a nation until approximately the year 2255, that Google will last until roughly 2032, and that the relationship your friend began a month ago will probably last about another month (maybe tell him not to RSVP to that wedding invitation just yet). Likewise, it tells us to be skeptical when, for instance, a recent *New Yorker* cover depicts a man holding a six-inch smartphone with a familiar grid of square app icons, and the caption reads "2525." Doubtful. The smartphone as we know it is barely a decade old, and the Copernican Principle tells us that it isn't likely to be around in 2025, let alone five centuries later. By 2525 it'd be mildly surprising if there were even a New York City.

More practically, if we're considering employment at a construction site

---

*There's a certain irony here: when it comes to time, assuming that there's nothing special about our arrival does result in us imagining ourselves at the very center after all.

whose signage indicates that it's been "7 days since the last industrial accident," we might want to stay away, unless it's a particularly short job we plan to do. And if a municipal transit system cannot afford the incredibly useful but expensive real-time signs that tell riders when the next bus is going to arrive, the Copernican Principle suggests that there might be a dramatically simpler and cheaper alternative. Simply displaying how long it's been since the *previous* bus arrived at that stop offers a substantial hint about when the next one will.

But is the Copernican Principle right? After Gott published his conjecture in *Nature*, the journal received a flurry of critical correspondence. And it's easy to see why when we try to apply the rule to some more familiar examples. If you meet a 90-year-old man, the Copernican Principle predicts he will live to 180. Every 6-year-old boy, meanwhile, is predicted to face an early death at the tender age of 12.

To understand why the Copernican Principle works, and why it sometimes doesn't, we need to return to Bayes. Because despite its apparent simplicity, the Copernican Principle is really an instance of Bayes's Rule.

Bayes Meets Copernicus

When predicting the future, such as the longevity of the Berlin Wall, the hypotheses we need to evaluate are all the possible durations of the phenomenon at hand: will it last a week, a month, a year, a decade? To apply Bayes's Rule, as we have seen, we first need to assign a prior probability to each of these durations. And it turns out that the Copernican Principle is exactly what results from applying Bayes's Rule using what is known as an *uninformative* prior.

At first this may seem like a contradiction in terms. If Bayes's Rule always requires us to specify our prior expectations and beliefs, how could we tell it that we don't have any? In the case of a raffle, one way to plead ignorance would be to assume what's called the "uniform prior," which considers every proportion of winning tickets to be equally likely.* In the case of the Berlin Wall, an uninformative prior means saying that we don't know

---

*This is precisely what Laplace's Law does in its simplest form: it assumes that having 1% or 10% of the tickets be winners is just as likely as 50% or 100%. The $\frac{w+1}{n+2}$ formula might seem naive in its suggestion that after buying a single losing Powerball ticket you have a 1/3 chance of winning on your next one—but that result faithfully reflects the odds in a raffle where you come in knowing nothing at all.

anything about the time span we're trying to predict: the wall could equally well come down in the next five minutes or last for five millennia.

Aside from that uninformative prior, the only piece of data we supply to Bayes's Rule, as we've seen, is the fact that we've encountered the Berlin Wall when it is eight years old. Any hypothesis that would have predicted a less than eight-year life span for the wall is thereby ruled out immediately, since those hypotheses can't account for our situation at all. (Similarly, a two-headed coin is ruled out by the first appearance of tails.) Anything longer than eight years is within the realm of possibility—but if the wall were going to be around for a million years, it would be a big coincidence that we happened to bump into it so very close to the start of its existence. Therefore, even though enormously long life spans cannot be ruled out, neither are they very likely.

When Bayes's Rule combines all these probabilities—the more-probable short time spans pushing down the average forecast, the less-probable yet still possible long ones pushing it up—the Copernican Principle emerges: if we want to predict how long something will last, and have no other knowledge about it whatsoever, the best guess we can make is that it will continue just as long as it's gone on so far.

In fact, Gott wasn't even the first to propose something like the Copernican Principle. In the mid-twentieth century, the Bayesian statistician Harold Jeffreys had looked into determining the number of tramcars in a city given the serial number on just one tramcar, and came up with the same answer: double the serial number. And a similar problem had arisen even earlier, during World War II, when the Allies sought to estimate the number of tanks being produced by Germany. Purely mathematical estimates based on captured tanks' serial numbers predicted that the Germans were producing 246 tanks every month, while estimates obtained by extensive (and highly risky) aerial reconnaissance suggested the figure was more like 1,400. After the war, German records revealed the true figure: 245.

Recognizing that the Copernican Principle is just Bayes's Rule with an uninformative prior answers a lot of questions about its validity. The Copernican Principle seems reasonable exactly in those situations where we know nothing at all—such as looking at the Berlin Wall in 1969, when we're not even sure what timescale is appropriate. And it feels completely wrong in those cases where we *do* know something about the subject matter. Predicting that a 90-year-old man will live to 180 years seems unreasonable precisely because we go into the problem already knowing a lot about human life spans—and so we can do better. The richer the prior information we bring to Bayes's Rule, the more useful the predictions we can get out of it.

Real-World Priors . . .

In the broadest sense, there are two types of things in the world: things that tend toward (or cluster around) some kind of "natural" value, and things that don't.

Human life spans are clearly in the former category. They roughly follow what's termed a "normal" distribution—also known as the "Gaussian" distribution, after the German mathematician Carl Friedrich Gauss, and informally called the "bell curve" for its characteristic shape. This shape does a good job of characterizing human life spans; the average life span for men in the United States, for instance, is centered at about 76 years, and the probabilities fall off fairly sharply to either side. Normal distributions tend to have a single appropriate scale: a one-digit life span is considered tragic, a three-digit one extraordinary. Many other things in the natural world are normally distributed as well, from human height, weight, and blood pressure to the noontime temperature in a city and the diameter of fruits in an orchard.

There are a number of things in the world that *don't* look normally distributed, however—not by a long shot. The average population of a town in the United States, for instance, is 8,226. But if you were to make a graph of the number of towns by population, you wouldn't see anything remotely like a bell curve. There would be *way* more towns smaller than 8,226 than larger. At the same time, the larger ones would be *way* bigger than the average. This kind of pattern typifies what are called "power-law distributions." These are also known as "scale-free distributions" because they characterize quantities that can plausibly range over many scales: a town can have tens, hundreds, thousands, tens of thousands, hundreds of thousands, or millions of residents, so we can't pin down a single value for how big a "normal" town should be.

The power-law distribution characterizes a host of phenomena in everyday life that have the same basic quality as town populations: most things below the mean, and a few enormous ones above it. Movie box-office grosses, which can range from four to ten figures, are another example.

Most movies don't make much money at all, but the occasional *Titanic* makes . . . well, titanic amounts.

In fact, money in general is a domain full of power laws. Power-law distributions characterize both people's wealth and people's incomes. The mean income in America, for instance, is $55,688—but because income is roughly power-law distributed, we know, again, that many more people will be below this mean than above it, while those who *are* above might be practically off the charts. So it is: two-thirds of the US population make less than the mean income, but the top 1% make almost ten times the mean. And the top 1% *of* the 1% make ten times more than that.

It's often lamented that "the rich get richer," and indeed the process of "preferential attachment" is one of the surest ways to produce a power-law distribution. The most popular websites are the most likely to get incoming links; the most followed online celebrities are the ones most likely to gain new fans; the most prestigious firms are the ones most likely to attract new clients; the biggest cities are the ones most likely to draw new residents. In every case, a power-law distribution will result.

Bayes's Rule tells us that when it comes to making predictions based on limited evidence, few things are as important as having good priors—that is, a sense of the distribution from which we expect that evidence to have come. Good predictions thus begin with having good instincts about when we're dealing with a normal distribution and when with a power-law distribution. As it turns out, Bayes's Rule offers us a simple but dramatically different predictive rule of thumb for each.

. . . and Their Prediction Rules

Did you mean "this could go on forever" in a good way?
—BEN LERNER

Examining the Copernican Principle, we saw that when Bayes's Rule is given an uninformative prior, it always predicts that the total life span of an object will be exactly double its current age. In fact, the uninformative prior, with its wildly varying possible scales—the wall that might last for months or for millennia—*is* a power-law distribution. And for any power-law distribution, Bayes's Rule indicates that the appropriate prediction strategy is a **Multiplicative Rule**: multiply the quantity observed so far by some constant factor. For an uninformative prior, that constant factor happens to be 2, hence the Copernican prediction; in other power-law cases, the multiplier will depend on the exact distribution you're working with. For the grosses of movies, for instance, it happens to be about 1.4. So if you hear a movie has made $6 million so far, you can guess it will make about $8.4 million overall; if it's made $90 million, guess it will top out at $126 million.

This multiplicative rule is a direct consequence of the fact that power-law distributions do not specify a natural scale for the phenomenon they're describing. The only thing that gives us a sense of scale for our prediction, therefore, is the single data point we have—such as the fact that the Berlin Wall has stood for eight years. The larger the value of that single data point, the larger the scale we're probably dealing with, and vice versa. It's *possible* that a movie that's grossed $6 million is actually a blockbuster in its first hour of release, but it's far more likely to be just a single-digit-millions kind of movie.

When we apply Bayes's Rule with a normal distribution as a prior, on the other hand, we obtain a very different kind of guidance. Instead of a multiplicative rule, we get an **Average Rule**: use the distribution's "natural" average—its single, specific scale—as your guide. For instance, if somebody is younger than the average life span, then simply predict the average; as their age gets close to and then exceeds the average, predict that they'll live a few years more. Following this rule gives reasonable predictions for the 90-year-old and the 6-year-old: 94 and 77, respectively. (The 6-year-old gets a tiny edge over the population average of 76 by virtue of having made it through infancy: we know he's not in the distribution's left tail.)

Movie running times, like human lifetimes, also follow a normal distribution: most films cluster right around a hundred minutes or so, with diminishing numbers of exceptions tailing off to either side. But not all human activities are so well behaved. The poet Dean Young once remarked that whenever he's listening to a poem in numbered sections, his heart sinks if the reader announces the start of section four: if there are more than three parts, all bets are off, and Young needs to hunker down for an earful. It turns out that Young's dismay is, in fact, perfectly Bayesian. An analysis of poems shows that, unlike movie running times, poems follow something closer to a power-law than a normal distribution: most poems are short, but some are epics. So when it comes to poetry, make sure you've got a comfortable seat. Something normally distributed that's gone on seemingly too long is bound to end shortly; but the longer something in a power-law distribution has gone on, the *longer* you can expect it to keep going.

Between those two extremes, there's actually a third category of things in life: those that are *neither* more nor less likely to end just because they've gone on for a while. Sometimes things are simply . . . invariant. The Danish mathematician Agner Krarup Erlang, who studied such phenomena, formalized the spread of intervals between independent events into the function that now carries his name: the Erlang distribution. The shape of this curve differs from both the normal and the power-law: it has a wing-like contour, rising to a gentle hump, with a tail that falls off faster than a power-law but more slowly than a normal distribution. Erlang himself, working for the Copenhagen Telephone Company in the early twentieth century, used it to model how much time could be expected to pass between successive calls on a phone network. Since then, the Erlang distribution has also been used by urban planners and architects to model car and pedestrian traffic, and by networking engineers designing infrastructure for the Internet. There are a number of domains in the natural world, too, where events are completely independent from one another and the intervals between them thus fall on an Erlang curve. Radioactive decay is one example, which means that the Erlang distribution perfectly models when to expect the next ticks of a Geiger counter. It also turns out to do a pretty good job of describing certain human endeavors—such as the amount of time politicians stay in the House of Representatives.

The Erlang distribution gives us a third kind of prediction rule, the **Additive Rule**: always predict that things will go on just a constant amount longer. The familiar refrain of "Just five more minutes! . . . [*five minutes later*] Five more minutes!" that so often characterizes human claims regarding, say, one's readiness to leave the house or office, or the time until the completion of some task, may seem indicative of some chronic failure to make realistic estimates. Well, in the cases where one's up against an Erlang distribution, anyway, that refrain happens to be correct.

If a casino card-playing enthusiast tells his impatient spouse, for example, that he'll quit for the day after hitting one more blackjack (the odds of which are about 20 to 1), he might cheerily predict, "I'll be done in about twenty more hands!" If, an unlucky twenty hands later, she returns, asking how long he's going to make her wait *now*, his answer will be unchanged: "I'll be done in about twenty more hands!" It sounds like our indefatigable card shark has suffered a short-term memory loss—but, in fact, his prediction is entirely correct. Indeed, distributions that yield the same prediction, no matter their history or current state, are known to statisticians as "memoryless."

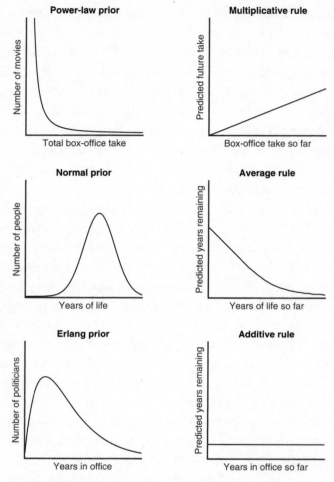

Different prior distributions and their prediction rules.

These three very different patterns of optimal prediction—the Multiplicative, Average, and Additive Rules—all result directly from applying Bayes's Rule to the power-law, normal, and Erlang distributions, respectively. And given the way those predictions come out, the three distributions offer us different guidance, too, on how *surprised* we should be by certain events.

In a power-law distribution, the longer something has gone on, the longer we expect it to *continue* going on. So a power-law event is more surprising the longer we've been waiting for it—and maximally surprising right before it happens. A nation, corporation, or institution only grows more venerable with each passing year, so it's always stunning when it collapses.

In a normal distribution, events are surprising when they're early—since we expected them to reach the average—but not when they're late. Indeed, by that point they seem overdue to happen, so the longer we wait, the more we expect them.

And in an Erlang distribution, events by definition are never any more or less surprising no matter *when* they occur. Any state of affairs is always equally likely to end regardless of how long it's lasted. No wonder politicians are always thinking about their next election.

Gambling is characterized by a similar kind of steady-state expectancy. If your wait for, say, a win at the roulette wheel were characterized by a normal distribution, then the Average Rule would apply: after a run of bad luck, it'd tell you that your number should be coming any second, probably followed by more losing spins. (In that case, it'd make sense to press on to the next win and then quit.) If, instead, the wait for a win obeyed a power-law distribution, then the Multiplicative Rule would tell you that winning spins follow quickly after one another, but the longer a drought had gone on the longer it would probably continue. (In that scenario, you'd be right to keep playing for a while after any win, but give up after a losing streak.) Up against a memoryless distribution, however, you're stuck. The Additive Rule tells you the chance of a win now is the same as it was an hour ago, and the same as it will be an hour from now. Nothing ever changes. You're not rewarded for sticking it out and ending on a high note; neither is there a tipping point when you should just cut your losses. In "The Gambler," Kenny Rogers famously advised that you've got to "Know when to walk away / Know when to run"—but for a memoryless distribution, there *is* no right time to quit. This may in part explain these games' addictiveness.

Knowing what distribution you're up against can make all the difference. When the Harvard biologist and prolific popularizer of science Stephen Jay Gould discovered that he had cancer, his immediate impulse was to read the relevant medical literature. Then he found out why his doctors had discouraged him from doing so: half of all patients with his form of cancer died within eight months of discovery.

But that one statistic—eight months—didn't tell him anything about the *distribution* of survivors. If it were a normal distribution, then the Average Rule would give a pretty clear forecast of how long he could expect to live: about eight months. But if it were a power-law, with a tail that stretches far out to the right, then the situation would be quite different: the Multiplicative Rule would tell him that the longer he lived, the more evidence it would provide that he would live longer. Reading further, Gould discovered that "the distribution was indeed, strongly right skewed, with a long tail (however small) that extended for several years above the eight month median. I saw no reason why I shouldn't be in that small tail, and I breathed a very long sigh of relief." Gould would go on to live for twenty more years after his diagnosis.

Small Data and the Mind

The three prediction rules—Multiplicative, Average, and Additive—are applicable in a wide range of everyday situations. And in those situations, people in general turn out to be remarkably good at using the right prediction rule. When he was in graduate school, Tom, along with MIT's Josh Tenenbaum, ran an experiment asking people to make predictions for a variety of everyday quantities—such as human life spans, the grosses of movies, and the time that US representatives would spend in office—based on just one piece of information in each case: current age, money earned so far, and years served to date. Then they compared the predictions people made to the predictions given by applying Bayes's Rule to the actual real-world data across each of those domains.

As it turned out, the predictions that people had made were extremely close to those produced by Bayes's Rule. Intuitively, people made different types of predictions for quantities that followed different distributions—power-law, normal, and Erlang—in the real world. In other words, while you might not know or consciously remember which situation calls for the Multiplicative, Average, or Additive Rule, the predictions you make every day tend to *implicitly* reflect the different cases where these distributions appear in everyday life, and the different ways they behave.

In light of what we know about Bayes's Rule, this remarkably good human performance suggests something critical that helps to understand how people make predictions. *Small data is big data in disguise.* The reason we can often make good predictions from a small number of observations—or just a single one—is that our priors are so rich. Whether we know it or not, we appear to carry around in our heads surprisingly accurate priors about movie grosses and running times, poem lengths, and political terms of office, not to mention human life spans. We don't need to gather them explicitly; we absorb them from the world.

The fact that, on the whole, people's hunches seem to closely match the predictions of Bayes's Rule also makes it possible to *reverse-engineer* all kinds of prior distributions, even ones about which it's harder to get authoritative real-world data. For instance, being kept on hold by customer service is a lamentably common facet of human experience, but there aren't publicly available data sets on hold times the way there are for Hollywood box-office grosses. But if people's predictions are informed by their experiences, we can use Bayes's Rule to conduct indirect reconnaissance about the world by mining people's expectations. When Tom and Josh asked people to predict hold times from a single data point, the results suggested that their subjects were using the Multiplicative Rule: the total wait people expect is one and a third times as long as they've waited so far. This is consistent with having a power-law distribution as a prior, where a wide range of scales is possible. Just hope you don't end up on the *Titanic* of hold times. Over the past decade, approaches like these have enabled cognitive scientists to identify people's prior distributions across a broad swath of domains, from vision to language.

There's a crucial caveat here, however. In cases where we don't have good priors, our predictions *aren't* good. In Tom and Josh's study, for instance, there was one subject where people's predictions systematically diverged from Bayes's Rule: predicting the length of the reign of Egyptian pharaohs. (As it happens, pharaohs' reigns follow an Erlang distribution.) People simply didn't have enough everyday exposure to have an intuitive feel for the range of those values, so their predictions, of course, faltered. Good predictions require good priors.

This has a number of important implications. Our judgments betray our expectations, and our expectations betray our experience. What we project about the future reveals a lot—about the world we live in, and about our own past.

What Our Predictions Tell Us About Ourselves

When Walter Mischel ran his famous "marshmallow test" in the early 1970s, he was trying to understand how the ability to delay gratification develops with age. At a nursery school on the Stanford campus, a series of three-, four-, and five-year-olds had their willpower tested. Each child would be shown a delicious treat, such as a marshmallow, and told that the adult running the experiment was about to leave the room for a while. If they wanted to, they could eat the treat right away. But if they waited until the experimenter came back, they would get *two* treats.

Unable to resist, some of the children ate the treat immediately. And some of them stuck it out for the full fifteen minutes or so until the experimenter returned, and got two treats as promised. But perhaps the most interesting group comprised the ones in between—the ones who managed to wait a little while, but then surrendered and ate the treat.

These cases, where children struggled mightily and suffered valiantly, only to give in and lose the extra marshmallow anyway, have been interpreted as suggesting a kind of irrationality. If you're going to cave, why not just cave immediately, and skip the torture? But it all depends on what kind of situation the children think they are in. As the University of Pennsylvania's Joe McGuire and Joe Kable have pointed out, if the amount of time it takes for adults to come back is governed by a power-law distribution—with long absences suggesting even longer waits lie ahead—then cutting one's losses at some point can make perfect sense.

In other words, the ability to resist temptation may be, at least in part, a matter of expectations rather than willpower. If you predict that adults tend to come back after short delays—something like a normal distribution—you should be able to hold out. The Average Rule suggests that after a painful wait, the thing to do is hang in there: the experimenter should be returning any minute now. But if you have no idea of the timescale of the disappearance—consistent with a power-law distribution—then it's an uphill battle. The Multiplicative Rule then suggests that a protracted wait is just a small fraction of what's to come.

Decades after the original marshmallow experiments, Walter Mischel and his colleagues went back and looked at how the participants were faring in life. Astonishingly, they found that children who had waited for two treats grew into young adults who were more successful than the others, even measured by quantitative metrics like their SAT scores. If the marshmallow test is about willpower, this is a powerful testament to the impact that learning self-control can have on one's life. But if the test is less about will than about expectations, then this tells a different, perhaps more poignant story.

A team of researchers at the University of Rochester recently explored how prior experiences might affect behavior in the marshmallow test.

Before marshmallows were even mentioned, the kids in the experiment embarked on an art project. The experimenter gave them some mediocre supplies, and promised to be back with better options soon. But, unbeknownst to them, the children were divided into two groups. In one group, the experimenter was reliable, and came back with the better art supplies as promised. In the other, she was unreliable, coming back with nothing but apologies.

The art project completed, the children went on to the standard marshmallow test. And here, the children who had learned that the experimenter was unreliable were more likely to eat the marshmallow before she came back, losing the opportunity to earn a second treat.

Failing the marshmallow test—and being less successful in later life—may not be about lacking willpower. It could be a result of believing that adults are not dependable: that they can't be trusted to keep their word, that they disappear for intervals of arbitrary length. Learning self-control is important, but it's equally important to grow up in an environment where adults are consistently present and trustworthy.

Priors in the Age of Mechanical Reproduction

As if someone were to buy several copies of the morning paper to assure himself that what it said was true.

—LUDWIG WITTGENSTEIN

He is careful of what he reads, for that is what he will write. He is careful of what he learns, for that is what he will know.

—ANNIE DILLARD

The best way to make good predictions, as Bayes's Rule shows us, is to be accurately informed about the things you're predicting. That's why we can do a good job of projecting human life spans, but perform poorly when asked to estimate the reigns of pharaohs.

Being a good Bayesian means representing the world in the correct proportions—having good priors, appropriately calibrated. By and large, for humans and other animals this happens naturally; as a rule, when something surprises us, it *ought* to surprise us, and when it doesn't, it ought not to. Even when we accumulate biases that aren't objectively correct, they still usually do a reasonable job of reflecting the specific part of the world we live in. For instance, someone living in a desert climate might overestimate the amount of sand in the world, and someone living at the poles might overestimate the amount of snow. Both are well tuned to their own ecological niche.

Everything starts to break down, however, when a species gains language. What we talk about isn't what we experience—we speak chiefly of interesting things, and those tend to be things that are uncommon. More or less by definition, events are always *experienced* at their proper frequencies, but this isn't at all true of language. Anyone who has experienced a snake bite or a lightning strike will tend to retell those singular stories for the rest of their lives. And those stories will be so salient that they will be picked up and retold by others.

There's a curious tension, then, between communicating with others and maintaining accurate priors about the world. When people talk about what interests them—and offer stories they think their listeners will find interesting—it skews the statistics of our experience. That makes it hard to maintain appropriate prior distributions. And the challenge has only increased with the development of the printing press, the nightly news, and social media—innovations that allow our species to spread language *mechanically.*

Consider how many times you've seen either a crashed plane or a crashed car. It's entirely possible you've seen roughly as many of each—yet many of those cars were on the road next to you, whereas the planes were probably on another continent, transmitted to you via the Internet or television. In the United States, for instance, the total number of people who have lost their lives in commercial plane crashes since the year 2000 would not be enough to fill Carnegie Hall even half full. In contrast, the number of people in the United States killed in car accidents over that same time is greater than the entire population of Wyoming.

Simply put, the representation of events in the media does not track their frequency in the world. As sociologist Barry Glassner notes, the murder rate in the United States declined by 20% over the course of the 1990s, yet during that time period the presence of gun violence on American news *increased* by 600%.

If you want to be a good intuitive Bayesian—if you want to naturally make good predictions, without having to think about what kind of prediction rule is appropriate—you need to protect your priors. Counterintuitively, that might mean turning off the news.

7 | Overfitting

When to Think Less

When Charles Darwin was trying to decide whether he should propose to his cousin Emma Wedgwood, he got out a pencil and paper and weighed every possible consequence. In favor of marriage he listed children, companionship, and the "charms of music & female chit-chat." Against marriage he listed the "terrible loss of time," lack of freedom to go where he wished, the burden of visiting relatives, the expense and anxiety provoked by children, the concern that "perhaps my wife won't like London," and having less money to spend on books. Weighing one column against the other produced a narrow margin of victory, and at the bottom Darwin scrawled, "Marry—Marry—Marry Q.E.D." *Quod erat demonstrandum*, the mathematical sign-off that Darwin himself then restated in English: "It being proved necessary to Marry."

The pro-and-con list was already a time-honored algorithm by Darwin's time, being endorsed by Benjamin Franklin a century before. To get over "the Uncertainty that perplexes us," Franklin wrote,

> my Way is, divide half a Sheet of Paper by a Line into two Columns, writing over the one Pro, and over the other Con. Then during three or four Days Consideration I put down under the different Heads short Hints of the different Motives that at different Times occur to me for or against the Measure. When I have thus got them all together in one View, I endeavour to estimate their respective Weights; and where I find two, one on each side, that seem equal, I strike them both out: If I find a Reason pro equal to

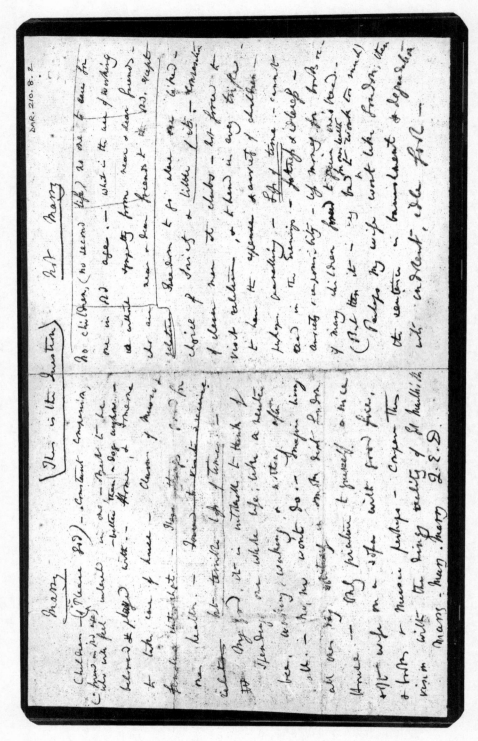

Darwin's journal, July 1838. Reprinted with permission of Cambridge University Library.

some two Reasons con, I strike out the three. If I judge some two Reasons con equal to some three Reasons pro, I strike out the five; and thus proceeding I find at length where the Ballance lies; and if after a Day or two of farther Consideration nothing new that is of Importance occurs on either side, I come to a Determination accordingly.

Franklin even thought about this as something like a computation, saying, "I have found great Advantage from this kind of Equation, in what may be called Moral or Prudential Algebra."

When we think about thinking, it's easy to assume that more is better: that you will make a better decision the more pros and cons you list, make a better prediction about the price of a stock the more relevant factors you identify, and write a better report the more time you spend working on it. This is certainly the premise behind Franklin's system. In this sense, Darwin's "algebraic" approach to matrimony, despite its obvious eccentricity, seems remarkably and maybe even laudably rational.

However, if Franklin or Darwin had lived into the era of machine-learning research—the science of teaching computers to make good judgments from experience—they'd have seen Moral Algebra shaken to its foundations. The question of how hard to think, and how many factors to consider, is at the heart of a knotty problem that statisticians and machine-learning researchers call "overfitting." And dealing with that problem reveals that there's a wisdom to deliberately thinking *less*. Being aware of overfitting changes how we should approach the market, the dining table, the gym . . . and the altar.

The Case Against Complexity

Anything you can do I can do better; I can do anything better than you.
—*ANNIE GET YOUR GUN*

Every decision is a kind of prediction: about how much you'll like something you haven't tried yet, about where a certain trend is heading, about how the road less traveled (or more so) is likely to pan out. And every prediction, crucially, involves thinking about two distinct things: what you know and what you don't. That is, it's an attempt to formulate a theory that will account for the experiences you've had to date *and* say something about the future ones you're guessing at. A good theory, of course, will do

Life satisfaction as a function of time since marriage.

both. But the fact that every prediction must in effect pull double duty creates a certain unavoidable tension.

As one illustration of this tension, let's look at a data set that might have been relevant to Darwin: people's life satisfaction over their first ten years of marriage, from a recent study conducted in Germany. Each point on that chart is taken from the study itself; our job is to figure out the formula for a line that would fit those points and extend into the future, allowing us to make predictions past the ten-year mark.

One possible formula would use just a single factor to predict life satisfaction: the time since marriage. This would create a straight line on the chart. Another possibility is to use two factors, *time* and *time squared*; the resulting line would have a parabolic U-shape, letting it capture a potentially more complex relationship between time and happiness. And if we expand the formula to include yet more factors (time cubed and so on), the line will acquire ever more inflection points, getting more and more "bendy" and flexible. By the time we get to a nine-factor formula, we can capture very complex relationships indeed.

Mathematically speaking, our two-factor model incorporates all the information that goes into the one-factor model, and has another term it could use as well. Likewise, the nine-factor model leverages all of the information at the disposal of the two-factor model, plus potentially lots more. By this logic, it seems like the nine-factor model ought to always give us the best predictions.

As it turns out, things are not quite so simple.

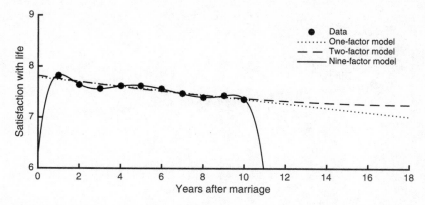

Predictions of life satisfaction using models with different numbers of factors.

The result of applying these models to the data is shown above. The one-factor model, unsurprisingly, misses a lot of the exact data points, though it captures the basic trend—a comedown after the honeymoon bliss. However, its straight-line prediction forecasts that this decrease will continue forever, ultimately resulting in infinite misery. Something about that trajectory doesn't sound quite right. The two-factor model comes closer to fitting the survey data, and its curved shape makes a different long-term prediction, suggesting that after the initial decline life satisfaction more or less levels out over time. Finally, the nine-factor model passes through each and every point on the chart; it is essentially a perfect fit for all the data from the study.

In that sense it seems like the nine-factor formula is indeed our best model. But if you look at the predictions it makes for the years *not* included in the study, you might wonder about just how useful it really is: it predicts misery at the altar, a giddily abrupt rise in satisfaction after several months of marriage, a bumpy roller-coaster ride thereafter, and a sheer drop after year ten. By contrast, the leveling off predicted by the two-factor model is the forecast most consistent with what psychologists and economists say about marriage and happiness. (They believe, incidentally, that it simply reflects a return to normalcy—to people's baseline level of satisfaction with their lives—rather than any displeasure with marriage itself.)

The lesson is this: it is indeed true that including more factors in a model will always, by definition, make it a better fit for the data we have already. But a better fit for the available data does not necessarily mean a better prediction.

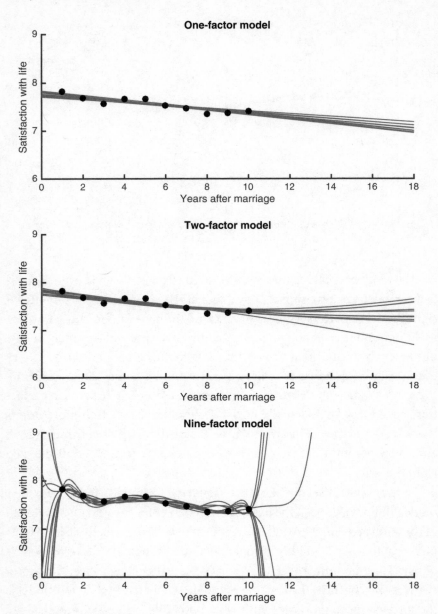

Adding small amounts of random "noise" to the data (simulating the effects of repeating the survey with a different group of participants) produces wild undulations in the nine-factor model, while the one- and two-factor models in comparison are much more stable and consistent in their predictions.

Granted, a model that's too simple—for instance, the straight line of the one-factor formula—can fail to capture the essential pattern in the data. If the truth looks like a curve, no straight line can ever get it right. On the other hand, a model that's too complicated, such as our nine-factor model here, becomes oversensitive to the particular data points that we happened to observe. As a consequence, precisely because it is tuned so finely to that specific data set, the solutions it produces are highly variable. If the study were repeated with different people, producing slight variations on the same essential pattern, the one- and two-factor models would remain more or less steady—but the nine-factor model would gyrate wildly from one instance of the study to the next. This is what statisticians call *overfitting*.

So one of the deepest truths of machine learning is that, in fact, it's not always better to use a more complex model, one that takes a greater number of factors into account. And the issue is not just that the extra factors might offer diminishing returns—performing better than a simpler model, but not enough to justify the added complexity. Rather, they might make our predictions dramatically worse.

The Idolatry of Data

If we had copious data, drawn from a perfectly representative sample, completely mistake-free, and representing exactly what we're trying to evaluate, then using the most complex model available would indeed be the best approach. But if we try to perfectly fit our model to the data when any of these factors fails to hold, we risk overfitting.

In other words, overfitting poses a danger any time we're dealing with noise or mismeasurement—and we almost always are. There can be errors in how the data were collected, or in how they were reported. Sometimes the phenomena being investigated, such as human happiness, are hard to even define, let alone measure. Thanks to their flexibility, the most complex models available to us can fit any patterns that appear in the data, but this means that they will also do so even when those patterns are mere phantoms and mirages in the noise.

Throughout history, religious texts have warned their followers against idolatry: the worshipping of statues, paintings, relics, and other tangible artifacts in lieu of the intangible deities those artifacts represent. The First Commandment, for instance, warns against bowing down to "any graven image, or any likeness of any thing that is in heaven." And in the Book of Kings, a bronze snake made at God's orders becomes an object of worship and incense-burning, instead of God himself. (God is not amused.) Fundamentally, overfitting is a kind of idolatry of data, a consequence of focusing on what we've been able to measure rather than what matters.

This gap between the data we have and the predictions we want is virtually everywhere. When making a big decision, we can only guess at what will please us later by thinking about the factors important to us right now. (As Harvard's Daniel Gilbert puts it, our future selves often "pay good money to remove the tattoos that we paid good money to get.") When making a financial forecast, we can only look at what correlated with the price of a stock in the past, not what might in the future. Even in our small daily acts this pattern holds: writing an email, we use our own read-through of the text to predict that of the recipient. No less than in public surveys, the data in our own lives are thus also always noisy, at best a proxy metric for the things we really care about.

As a consequence, considering more and more factors and expending more effort to model them can lead us into the error of optimizing for the wrong thing—offering prayers to the bronze snake of data rather than the larger force behind it.

Overfitting Everywhere

Once you know about overfitting, you see it everywhere.

Overfitting, for instance, explains the irony of our palates. How can it be that the foods that taste best to us are broadly considered to be bad for our health, when the entire function of taste buds, evolutionarily speaking, is to prevent us from eating things that are bad?

The answer is that taste is our body's proxy metric for health. Fat, sugar, and salt are important nutrients, and for a couple hundred thousand years, being drawn to foods containing them was a reasonable measure for a sustaining diet.

But being able to modify the foods available to us broke that relationship. We can now add fat and sugar to foods beyond amounts that are good for us, and then eat those foods exclusively rather than the mix of plants, grains, and meats that historically made up the human diet. In other words, we can overfit taste. And the more skillfully we can manipulate food (and the more our lifestyles diverge from those of our ancestors), the more imperfect a metric taste becomes. Our human agency thus turns into a curse, making us dangerously able to have exactly what we want even when we don't quite want exactly the right thing.

Beware: when you go to the gym to work off the extra weight from all that sugar, you can also risk overfitting fitness. Certain visible signs of fitness—low body fat and high muscle mass, for example—are easy to measure, and they are *related* to, say, minimizing the risk of heart disease and other ailments. But they, too, are an imperfect proxy measure. Overfitting the signals—adopting an extreme diet to lower body fat and taking steroids to build muscle, perhaps—can make you the picture of good health, but only the picture.

Overfitting also shows up in sports. For instance, Tom has been a fencer, on and off, since he was a teenager. The original goal of fencing was to teach people how to defend themselves in a duel, hence the name: "defencing." And the weapons used in modern fencing are similar to those that were used to train for such encounters. (This is particularly true of the épée, which was still used in formal duels less than fifty years ago.) But the introduction of electronic scoring equipment—a button on the tip of the blade that registers a hit—has changed the nature of the sport, and techniques that would serve you poorly in a serious duel have become critical skills in competition. Modern fencers use flexible blades that allow them to "flick" the button at their opponent, grazing just hard enough to register and score. As a result, they can look more like they're cracking thin metal whips at each other than cutting or thrusting. It's as exciting a sport as ever, but as athletes overfit their tactics to the quirks of scorekeeping, it becomes less useful in instilling the skills of real-world swordsmanship.

Perhaps nowhere, however, is overfitting as powerful and troublesome as in the world of business. "Incentive structures work," as Steve Jobs put it. "So you have to be very careful of what you incent people to do, because various incentive structures create all sorts of consequences that you can't anticipate." Sam Altman, president of the startup incubator Y Combinator, echoes Jobs's words of caution: "It really is true that the company will build whatever the CEO decides to measure."

In fact, it's incredibly difficult to come up with incentives or measurements that do not have some kind of perverse effect. In the 1950s, Cornell management professor V. F. Ridgway cataloged a host of such "Dysfunctional Consequences of Performance Measurements." At a job-placement firm, staffers were evaluated on the number of interviews they conducted, which motivated them to run through the meetings as quickly as possible, without spending much time actually helping their clients find jobs. At a federal law enforcement agency, investigators given monthly performance quotas were found to pick easy cases at the end of the month rather than the most urgent ones. And at a factory, focusing on production metrics led supervisors to neglect maintenance and repairs, setting up future catastrophe. Such problems can't simply be dismissed as a failure to achieve management goals. Rather, they are the opposite: the ruthless and clever optimization of the wrong thing.

The twenty-first-century shift into real-time analytics has only made the danger of metrics more intense. Avinash Kaushik, digital marketing evangelist at Google, warns that trying to get website users to see as many ads as possible naturally devolves into trying to cram sites with ads: "When you are paid on a [cost per thousand impressions] basis the incentive is to figure out how to show the most possible ads on every page [and] ensure the visitor sees the most possible pages on the site. . . . That incentive removes a focus from the important entity, your customer, and places it on the secondary entity, your advertiser." The website might gain a little more money in the short term, but ad-crammed articles, slow-loading multi-page slide shows, and sensationalist clickbait headlines will drive away readers in the long run. Kaushik's conclusion: "Friends don't let friends measure Page Views. Ever."

In some cases, the difference between a model and the real world is literally a matter of life and death. In the military and in law enforcement, for example, repetitive, rote training is considered a key means for instilling line-of-fire skills. The goal is to drill certain motions and tactics to the point that they become totally automatic. But when overfitting creeps in, it can prove disastrous. There are stories of police officers who find themselves, for instance, taking time out during a gunfight to put their spent casings in their pockets—good etiquette on a firing range. As former Army Ranger and West Point psychology professor Dave Grossman writes, "After the smoke had settled in many real gunfights, officers were shocked to discover empty brass in their pockets with no memory of how it got there. On several occasions, dead cops were found with brass in their hands, dying in the middle of an administrative procedure that had been drilled into them." Similarly, the FBI was forced to change its training after agents were found reflexively firing two shots and then holstering their weapon—a standard cadence in training—regardless of whether their shots had hit the target and whether there was still a threat. Mistakes like these are known in law enforcement and the military as "training scars," and they reflect the fact that it's possible to overfit one's own preparation. In one particularly dramatic case, an officer instinctively grabbed the gun out of the hands of an assailant and then instinctively handed it right back—just as he had done time and time again with his trainers in practice.

Detecting Overfitting: Cross-Validation

Because overfitting presents itself initially as a theory that perfectly fits the available data, it may seem insidiously hard to detect. How can we expect to tell the difference between a genuinely good model and one that's overfitting? In an educational setting, how can we distinguish between a class of students excelling at the subject matter and a class merely being "taught to the test"? In the business world, how can we tell a genuine star performer from an employee who has just cannily overfit their work to the company's key performance indicators—or to the boss's perception?

Teasing apart those scenarios is indeed challenging, but it is not impossible. Research in machine learning has yielded several concrete strategies for detecting overfitting, and one of the most important is what's known as **Cross-Validation**.

Simply put, Cross-Validation means assessing not only how well a model fits the data it's given, but how well it generalizes to data it hasn't seen. Paradoxically, this may involve using *less* data. In the marriage example, we might "hold back," say, two points at random, and fit our models only to the other eight. We'd then take those two test points and use them to gauge how well our various functions generalize beyond the eight "training" points they've been given. The two held-back points function as canaries in the coal mine: if a complex model nails the eight training points but wildly misses the two test points, it's a good bet that overfitting is at work.

Aside from withholding some of the available data points, it is also useful to consider testing the model with data derived from some other form of evaluation entirely. As we have seen, the use of proxy metrics—taste as a proxy for nutrition, number of cases solved as a proxy for investigator diligence—can also lead to overfitting. In these cases, we'll need to cross-validate the primary performance measure we're using against other possible measures.

In schools, for example, standardized tests offer a number of benefits, including a distinct economy of scale: they can be graded cheaply and rapidly by the thousands. Alongside such tests, however, schools could randomly assess some small fraction of the students—one per class, say, or one in a hundred—using a different evaluation method, perhaps something like an essay or an oral exam. (Since only a few students would be tested this way, having this secondary method scale well is not a big concern.) The standardized tests would provide immediate feedback—you could have students take a short computerized exam every week and chart the class's progress almost in real time, for instance—while the secondary data points would serve to cross-validate: to make sure that the students were actually acquiring the knowledge that the standardized test is meant to measure, and not simply getting better at test-taking. If a school's standardized scores rose while its "nonstandardized" performance moved in the opposite direction, administrators would have a clear warning sign that "teaching to the test" had set in, and the pupils' skills were beginning to overfit the mechanics of the test itself.

Cross-Validation also offers a suggestion for law enforcement and military personnel looking to instill good reflexes without hammering in habits from the training process itself. Just as essays and oral exams can cross-validate standardized tests, so occasional unfamiliar "cross-training" assessments might be used to measure whether reaction time and shooting accuracy are generalizing to unfamiliar tasks. If they aren't, then that's a strong signal to change the training regimen. While nothing may truly prepare one for actual combat, exercises like this may at least warn in advance where "training scars" are likely to have formed.

How to Combat Overfitting: Penalizing Complexity

> *If you can't explain it simply, you don't understand it well enough.*
> —ANONYMOUS

We've seen some of the ways that overfitting can rear its head, and we've looked at some of the methods to detect and measure it. But what can we actually do to alleviate it?

From a statistics viewpoint, overfitting is a symptom of being too sensitive to the actual data we've seen. The solution, then, is straightforward: we must balance our desire to find a good fit against the complexity of the models we use to do so.

One way to choose among several competing models is the Occam's razor principle, which suggests that, all things being equal, the simplest possible hypothesis is probably the correct one. Of course, things are rarely completely equal, so it's not immediately obvious how to apply something like Occam's razor in a mathematical context. Grappling with this challenge in the 1960s, Russian mathematician Andrey Tikhonov proposed one answer: introduce an additional term to your calculations that penalizes more complex solutions. If we introduce a complexity penalty, then more complex models need to do not merely a better job but a *significantly* better job of explaining the data to justify their greater complexity. Computer scientists refer to this principle—using constraints that penalize models for their complexity—as **Regularization**.

So what do these complexity penalties look like? One algorithm, discovered in 1996 by biostatistician Robert Tibshirani, is called the **Lasso** and uses as its penalty the total weight of the different factors in the model.* By putting this downward pressure on the weights of the factors, the Lasso drives as many of them as possible completely to zero. Only the factors that have a big impact on the results remain in the equation—thus potentially transforming, say, an overfitted nine-factor model into a simpler, more robust formula with just a couple of the most critical factors.

Techniques like the Lasso are now ubiquitous in machine learning, but the same kind of principle—a penalty for complexity—also appears in nature. Living organisms get a certain push toward simplicity almost automatically, thanks to the constraints of time, memory, energy, and attention. The burden of metabolism, for instance, acts as a brake on the complexity of organisms, introducing a caloric penalty for overly elaborate machinery. The fact that the human brain burns about a fifth of humans' total daily caloric intake is a testament to the evolutionary advantages that our intellectual abilities provide us with: the brain's contributions must somehow more than pay for that sizable fuel bill. On the other hand, we can also infer that a substantially *more* complex brain probably didn't provide sufficient dividends, evolutionarily speaking. We're as brainy as we have needed to be, but not extravagantly more so.

The same kind of process is also believed to play a role at the neural level. In computer science, software models based on the brain, known as "artificial neural networks," can learn arbitrarily complex functions— they're even more flexible than our nine-factor model above—but precisely

---

*For the mathematically inclined, that's the sum of the absolute values of the variables' coefficients.

because of this very flexibility they are notoriously vulnerable to overfitting. Actual, biological neural networks sidestep some of this problem because they need to trade off their performance against the costs of maintaining it. Neuroscientists have suggested, for instance, that brains try to minimize the number of neurons that are firing at any given moment—implementing the same kind of downward pressure on complexity as the Lasso.

Language forms yet another natural Lasso: complexity is punished by the labor of speaking at greater length and the taxing of our listener's attention span. Business plans get compressed to an elevator pitch; life advice becomes proverbial wisdom only if it is sufficiently concise and catchy. And anything that needs to be remembered has to pass through the inherent Lasso of memory.

The Upside of Heuristics

The economist Harry Markowitz won the 1990 Nobel Prize in Economics for developing modern portfolio theory: his groundbreaking "mean-variance portfolio optimization" showed how an investor could make an optimal allocation among various funds and assets to maximize returns at a given level of risk. So when it came time to invest his own retirement savings, it seems like Markowitz should have been the one person perfectly equipped for the job. What did he decide to do?

> I should have computed the historical covariances of the asset classes and drawn an efficient frontier. Instead, I visualized my grief if the stock market went way up and I wasn't in it—or if it went way down and I was completely in it. My intention was to minimize my future regret. So I split my contributions fifty-fifty between bonds and equities.

Why in the world would he do that? The story of the Nobel Prize winner and his investment strategy could be presented as an example of human irrationality: faced with the complexity of real life, he abandoned the rational model and followed a simple heuristic. But it's precisely because of the complexity of real life that a simple heuristic might in fact be the rational solution.

When it comes to portfolio management, it turns out that unless you're highly confident in the information you have about the markets, you may actually be better off ignoring that information altogether. Applying Markowitz's optimal portfolio allocation scheme requires having good estimates of the statistical properties of different investments. An error in those estimates can result in very different asset allocations, potentially increasing risk. In contrast, splitting your money evenly across stocks and bonds is not affected at all by what data you've observed. This strategy doesn't even try to fit itself to the historical performance of those investment types—so there's no way it can *overfit*.

Of course, just using a fifty-fifty split is not necessarily the complexity sweet spot, but there's something to be said for it. If you happen to know the expected mean and expected variance of a set of investments, then use mean-variance portfolio optimization—the optimal algorithm is optimal for a reason. But when the odds of estimating them all correctly are low, and the weight that the model puts on those untrustworthy quantities is high, then an alarm should be going off in the decision-making process: it's time to regularize.

Inspired by examples like Markowitz's retirement savings, psychologists Gerd Gigerenzer and Henry Brighton have argued that the decision-making shortcuts people use in the real world are in many cases exactly the kind of thinking that makes for good decisions. "In contrast to the widely held view that less processing reduces accuracy," they write, "the study of heuristics shows that less information, computation, and time can in fact improve accuracy." A heuristic that favors simpler answers—with fewer factors, or less computation—offers precisely these "less is more" effects.

Imposing penalties on the ultimate complexity of a model is not the only way to alleviate overfitting, however. You can also nudge a model toward simplicity by controlling the speed with which you allow it to adapt to incoming data. This makes the study of overfitting an illuminating guide to our history—both as a society and as a species.

The Weight of History

> *Every food a living rat has eaten has, necessarily, not killed it.*
> —SAMUEL REVUSKY AND ERWIN BEDARF, "ASSOCIATION OF
> ILLNESS WITH PRIOR INGESTION OF NOVEL FOODS"

The soy milk market in the United States more than quadrupled from the mid-1990s to 2013. But by the end of 2013, according to news headlines, it already seemed to be a thing of the past, a distant second place to almond milk. As food and beverage researcher Larry Finkel told *Bloomberg Businessweek*: "Nuts are trendy now. Soy sounds more like old-fashioned health food." The Silk company, famous for popularizing soy milk (as the name implies), reported in late 2013 that its almond milk products had grown by more than 50% in the previous quarter alone. Meanwhile, in other beverage news, the leading coconut water brand, Vita Coco, reported in 2014 that its sales had doubled since 2011—and had increased an astounding three-hundred-fold since 2004. As the *New York Times* put it, "coconut water seems to have jumped from invisible to unavoidable without a pause in the realm of the vaguely familiar." Meanwhile, the kale market grew by 40% in 2013 alone. The biggest purchaser of kale the year before had been Pizza Hut, which put it in their salad bars—as decoration.

Some of the most fundamental domains of human life, such as the question of what we should put in our bodies, seem curiously to be the ones most dominated by short-lived fads. Part of what enables these fads to take the world by storm is how quickly our culture can change. Information now flows through society faster than ever before, while global supply chains enable consumers to rapidly change their buying habits en masse (and marketing encourages them to do so). If some particular study happens to suggest a health benefit from, say, star anise, it can be all over the blogosphere within the week, on television the week after that, and in seemingly every supermarket in six months, with dedicated star anise cookbooks soon rolling off the presses. This breathtaking speed is both a blessing and a curse.

In contrast, if we look at the way organisms—including humans—evolve, we notice something intriguing: change happens slowly. This means that the properties of modern-day organisms are shaped not only by their present environments, but also by their history. For example, the oddly cross-wired arrangement of our nervous system (the left side of our body controlled by the right side of our brain and vice versa) reflects the evolutionary history of vertebrates. This phenomenon, called "decussation," is theorized to have arisen at a point in evolution when early vertebrates' bodies twisted 180 degrees with respect to their heads; whereas the nerve cords of invertebrates such as lobsters and earthworms run on the "belly" side of the animal, vertebrates have their nerve cords along the spine instead.

The human ear offers another example. Viewed from a functional perspective, it is a system for translating sound waves into electrical signals by way of amplification via three bones: the malleus, incus, and stapes. This amplification system is impressive—but the specifics of how it works have a lot to do with historical constraints. Reptiles, it turns out, have just a single bone in their ear, but additional bones in the jaw that mammals lack. Those jawbones were apparently repurposed in the mammalian ear. So the exact form and configuration of our ear anatomy reflects our evolutionary history at least as much as it does the auditory problem being solved.

The concept of overfitting gives us a way of seeing the virtue in such evolutionary baggage. Though crossed-over nerve fibers and repurposed jawbones may seem like suboptimal arrangements, we don't necessarily want evolution to fully optimize an organism to every shift in its environmental niche—or, at least, we should recognize that doing so would make it extremely sensitive to further environmental changes. Having to make use of existing materials, on the other hand, imposes a kind of useful restraint. It makes it harder to induce drastic changes in the structure of organisms, harder to overfit. As a species, being constrained by the past makes us less perfectly adjusted to the present we know but helps keep us robust for the future we don't.

A similar insight might help us resist the quick-moving fads of human society. When it comes to culture, tradition plays the role of the evolutionary constraints. A bit of conservatism, a certain bias in favor of history, can buffer us against the boom-and-bust cycle of fads. That doesn't mean we ought to ignore the latest data either, of course. Jump toward the bandwagon, by all means—but not necessarily on it.

In machine learning, the advantages of moving slowly emerge most concretely in a regularization technique known as **Early Stopping**. When we looked at the German marriage survey data at the beginning of the chapter, we went straight to examining the best-fitted one-, two-, and nine-factor models. In many situations, however, tuning the parameters to find the best possible fit for given data is a process in and of itself. What happens if we stop that process early and simply don't allow a model the *time* to become too complex? Again, what might seem at first blush like being halfhearted or unthorough emerges, instead, as an important strategy in its own right.

Many prediction algorithms, for instance, start out by searching for the single most important factor rather than jumping to a multi-factor model.

Only after finding that first factor do they look for the next most important factor to add to the model, then the next, and so on. Their models can therefore be kept from becoming overly complex simply by stopping the process short, before overfitting has had a chance to creep in. A related approach to calculating predictions considers one data point at a time, with the model tweaked to account for each new point before more points are added; there, too, the complexity of the model increases gradually, so stopping the process short can help keep it from overfitting.

This kind of setup—where more time means more complexity—characterizes a lot of human endeavors. Giving yourself more time to decide about something does not necessarily mean that you'll make a better decision. But it does guarantee that you'll end up considering more factors, more hypotheticals, more pros and cons, and thus risk overfitting.

Tom had exactly this experience when he became a professor. His first semester, teaching his first class ever, he spent a huge amount of time perfecting his lectures—more than ten hours of preparation for every hour of class. His second semester, teaching a different class, he wasn't able to put in as much time, and worried that it would be a disaster. But a strange thing happened: the students liked the second class. In fact, they liked it more than the first one. Those extra hours, it turned out, had been spent nailing down nitty-gritty details that only confused the students, and wound up getting cut from the lectures the next time Tom taught the class. The underlying issue, Tom eventually realized, was that he'd been using his own taste and judgment as a kind of proxy metric for his students'. This proxy metric worked reasonably well as an approximation, but it wasn't worth overfitting—which explained why spending extra hours painstakingly "perfecting" all the slides had been counterproductive.

The effectiveness of regularization in all kinds of machine-learning tasks suggests that we can make better decisions by deliberately thinking and doing less. If the factors we come up with first are likely to be the most important ones, then beyond a certain point thinking more about a problem is not only going to be a waste of time and effort—it will lead us to worse solutions. Early Stopping provides the foundation for a reasoned argument against reasoning, the thinking person's case against thought. But turning this into practical advice requires answering one more question: *when* should we stop thinking?

When to Think Less

As with all issues involving overfitting, how early to stop depends on the gap between what you can measure and what really matters. If you have all the facts, they're free of all error and uncertainty, and you can directly assess whatever is important to you, then don't stop early. Think long and hard: the complexity and effort are appropriate.

But that's almost never the case. If you have high uncertainty and limited data, then do stop early by all means. If you don't have a clear read on how your work will be evaluated, and by whom, then it's not worth the extra time to make it perfect with respect to your own (or anyone else's) idiosyncratic guess at what perfection might be. The greater the uncertainty, the bigger the gap between what you can measure and what matters, the more you should watch out for overfitting—that is, the more you should prefer simplicity, and the earlier you should stop.

When you're truly in the dark, the best-laid plans will be the simplest. When our expectations are uncertain and the data are noisy, the best bet is to paint with a broad brush, to think in broad strokes. Sometimes literally. As entrepreneurs Jason Fried and David Heinemeier Hansson explain, the further ahead they need to brainstorm, the thicker the pen they use—a clever form of simplification by stroke size:

> When we start designing something, we sketch out ideas with a big, thick Sharpie marker, instead of a ball-point pen. Why? Pen points are too fine. They're too high-resolution. They encourage you to worry about things that you shouldn't worry about yet, like perfecting the shading or whether to use a dotted or dashed line. You end up focusing on things that should still be out of focus.
>
> A Sharpie makes it impossible to drill down that deep. You can only draw shapes, lines, and boxes. That's good. The big picture is all you should be worrying about in the beginning.

As McGill's Henry Mintzberg puts it, "What would happen if we started from the premise that we can't measure what matters and go from there? Then instead of measurement we'd have to use something very scary: it's called judgment."

The upshot of Early Stopping is that sometimes it's not a matter of choosing between being rational and going with our first instinct. Going with our first instinct can *be* the rational solution. The more complex, unstable, and uncertain the decision, the more rational an approach that is.

To return to Darwin, his problem of deciding whether to propose could probably have been resolved based on just the first few pros and cons he identified, with the subsequent ones adding to the time and anxiety expended on the decision without necessarily aiding its resolution (and in all likelihood impeding it). What seemed to make up his mind was the thought that "it is intolerable to think of spending one's whole life like a neuter bee, working, working, & nothing after all." Children and companionship—the very first points he mentioned—were precisely those that ultimately swayed him in favor of marriage. His book budget was a distraction.

Before we get too critical of Darwin, however, painting him as an inveterate overthinker, it's worth taking a second look at this page from his diary. Seeing it in facsimile shows something fascinating. Darwin was no Franklin, adding assorted considerations for days. Despite the seriousness with which he approached this life-changing choice, Darwin made up his mind exactly when his notes reached the bottom of the diary sheet. *He was regularizing to the page.* This is reminiscent of both Early Stopping and the Lasso: anything that doesn't make the page doesn't make the decision.

His mind made up to marry, Darwin immediately went on to overthink the timing. "When? Soon or Late," he wrote above another list of pros and cons, considering everything from happiness to expenses to "awkwardness" to his long-standing desire to travel in a hot air balloon and/or to Wales. But by the end of the page he resolved to "Never mind, trust to chance"— and the result, within several months' time, was a proposal to Emma Wedgwood, the start of a fulfilling partnership and a happy family life.

8 | Relaxation

Let It Slide

In 2010 Meghan Bellows was working on her PhD in chemical engineering at Princeton by day and planning her wedding by night. Her research revolved around finding the right places to put amino acids in a protein chain to yield a molecule with particular characteristics. ("If you maximize the binding energy of two proteins then you can successfully design a peptidic inhibitor of some biological function so you can actually inhibit a disease's progress.") On the nuptial front, she was stuck on the problem of seating.

There was a group of nine college friends, and Bellows agonized over who else to throw into the midst of such a mini-reunion to make a table of ten. Even worse, she'd counted up eleven close relatives. Who would get the boot from the honored parents' table, and how could she explain it to them? And what about folks like her childhood neighbors and babysitter, or her parents' work colleagues, who didn't really know anyone at the wedding at all?

The problem seemed every bit as hard as the protein problem she was working on at the lab. Then it hit her. It *was* the problem she was working on at the lab. One evening, as she stared at her seating charts, "I realized that there was literally a one-to-one correlation between the amino acids and proteins in my PhD thesis and people sitting at tables at my wedding." Bellows called out to her fiancé for a piece of paper and began scribbling equations. Amino acids became guests, binding energies became relationships, and the molecules' so-called nearest-neighbor interactions became—well—nearest-neighbor interactions. She could use the algorithms from her research to solve her own wedding.

Bellows worked out a way to numerically define the strength of the relationships among all the guests. If a particular pair of people didn't know one another they got a 0, if they did they got a 1, and if they were a couple they got a 50. (The sister of the bride got to give a score of 10 to all the people she wanted to sit with, as a special prerogative.) Bellows then specified a few constraints: the maximum table capacity, and a minimum score necessary for each table, so that no one table became the awkward "miscellaneous" group full of strangers. She also codified the program's goal: to maximize the relationship scores between the guests and their tablemates.

There were 107 people at the wedding and 11 tables, which could accommodate ten people each. This means there were about 11^{107} possible seating plans: that's a 112-digit number, more than 200 billion googols, a figure that dwarfs the (merely 80-digit) number of atoms in the observable universe. Bellows submitted the job to her lab computer on Saturday evening and let it churn. When she came in on Monday morning, it was still running; she had it spit out the best assignment it had found so far and put it back onto protein design.

Even with a high-powered lab computer cluster and a full thirty-six hours of processing time, there was no way for the program to evaluate more than a tiny fraction of the potential seating arrangements. The odds are that the truly optimal solution, the one that would have earned the very highest score, never came up in its permutations. Still, Bellows was pleased with the computer's results. "It identified relationships that we were forgetting about," she says, offering delightful, unconventional possibilities that the human planners hadn't even considered. For instance, it proposed removing her *parents* from the family table, putting them instead with old friends they hadn't seen for years. Its final recommendation was an arrangement that all parties agreed was a hit—although the mother of the bride couldn't resist making just a few manual tweaks.

The fact that all the computing power of a lab at Princeton couldn't find the perfect seating plan might seem surprising. In most of the domains we've discussed so far, straightforward algorithms could guarantee optimal solutions. But as computer scientists have discovered over the past few decades, there are entire classes of problems where a perfect solution is essentially unreachable, no matter how fast we make our computers or how cleverly we program them. In fact, no one understands as well as a computer scientist that in the face of a seemingly unmanageable challenge, you should neither toil forever nor give up, but—as we'll see—try a third thing entirely.

The Difficulty of Optimization

Before leading the country through the American Civil War, before drafting the Emancipation Proclamation or delivering the Gettysburg Address, Abraham Lincoln worked as a "prairie lawyer" in Springfield, Illinois, traveling the Eighth Judicial Circuit twice a year for sixteen years. Being a circuit lawyer meant literally making a circuit—moving through towns in fourteen different counties to try cases, riding hundreds of miles over many weeks. Planning these circuits raised a natural challenge: how to visit all the necessary towns while covering as few miles as possible and without going to any town twice.

This is an instance of what's known to mathematicians and computer scientists as a "constrained optimization" problem: how to find the single best arrangement of a set of variables, given particular rules and a scorekeeping measure. In fact, it's the most famous optimization problem of them all. If it had been studied in the nineteenth century it might have become forever known as "the prairie lawyer problem," and if it had first come up in the twenty-first century it might have been nicknamed "the delivery drone problem." But like the secretary problem, it emerged in the mid-twentieth century, a period unmistakably evoked by its canonical name: "the traveling salesman problem."

The problem of route planning didn't get the attention of the mathematics community until the 1930s, but then it did so with a vengeance. Mathematician Karl Menger spoke of "the postal messenger problem" in 1930, noting that no easier solution was known than simply trying out every possibility in turn. Hassler Whitney posed the problem in a 1934 talk at Princeton, where it lodged firmly in the brain of fellow mathematician Merrill Flood (who, you might recall from chapter 1, is also credited with circulating the first solution to the secretary problem). When Flood moved to California in the 1940s he spread it in turn to his colleagues at the RAND Institute, and the problem's iconic name first appeared in print in a 1949 paper by mathematician Julia Robinson. As the problem swept through mathematical circles, it grew in notoriety. Many of the greatest minds of the time obsessed over it, and no one seemed able to make real headway.

In the traveling salesman problem, the question isn't whether a computer (or a mathematician) *could* find the shortest route: theoretically, one can simply crank out a list of all the possibilities and measure each one. Rather, the issue is that as the number of towns grows, the list of possible routes connecting them explodes. A route is just an ordering of the towns, so trying them all by brute force is the dreaded $O(n!)$ "factorial time"—the computational equivalent of sorting a deck of cards by throwing them in the air until they happen to land in order.

The question is: is there any hope of doing better?

Decades of work did little to tame the traveling salesman problem. Flood, for instance, wrote in 1956, more than twenty years after first encountering it: "It seems very likely that quite a different approach from any yet used may be required for successful treatment of the problem. In fact, there may well be no general method for treating the problem and impossibility results would also be valuable." Another decade later, the mood was only more grim. "I conjecture," wrote Jack Edmonds, "that there is no good algorithm for the traveling salesman problem."

These words would prove prophetic.

Defining Difficulty

In the mid-1960s, Edmonds, at the National Institute of Standards and Technology, along with Alan Cobham of IBM, developed a working definition for what makes a problem feasible to solve or not. They asserted what's now known as the Cobham–Edmonds thesis: an algorithm should be considered "efficient" if it runs in what's called "polynomial time"—that is, $O(n^2)$, $O(n^3)$, or in fact n to the power of any number at all. A problem, in turn, is considered "tractable" if we know how to solve it using an efficient algorithm. A problem we *don't* know how to solve in polynomial time, on the other hand, is considered "intractable." And at anything but the smallest scales, intractable problems are beyond the reach of solution by computers, no matter how powerful.*

This amounts to what is arguably the central insight of computer science.

---

*It may look strange, given that $O(n^2)$ seemed so odious in the sorting context, to call it "efficient" here. But the truth is that even exponential time with an unassumingly small base number, like $O(2^n)$, quickly gets hellish even when compared to a polynomial with a large base, like n^{10}. The exponent will always overtake the polynomial at some problem size—in this case, if you're sorting more than several dozen items, n^{10} starts to look like a walk in the park compared to 2^n. Ever since Cobham and Edmonds's work, this chasm between "polynomials" (n-to-the-something) and "exponentials" (something-to-the-n) has served as the field's de facto out-of-bounds marker.

It's possible to quantify the difficulty of a problem. And some problems are just . . . *hard*.

Where does this leave the traveling salesman problem? Curiously enough, we are still not quite sure. In 1972, Berkeley's Richard Karp demonstrated that the traveling salesman problem is linked to a controversially borderline class of problems that have not yet been definitively proven to be either efficiently solvable or not. But so far there have been no efficient solutions found for any of those problems—making them effectively intractable—and most computer scientists believe that there *aren't* any to be found. So the "impossibility result" for the traveling salesman problem that Flood imagined in the 1950s is likely to be its ultimate fate. What's more, many other optimization problems—with implications for everything from political strategy to public health to fire safety—are similarly intractable.

But for the computer scientists who wrestle with such problems, this verdict isn't the end of the story. Instead, it's more like a call to arms. Having determined a problem to be intractable, you can't just throw up your hands. As scheduling expert Jan Karel Lenstra told us, "When the problem is hard, it doesn't mean that you can forget about it, it means that it's just in a different status. It's a serious enemy, but you still have to fight it." And this is where the field figured out something invaluable, something we can all learn from: how to best approach problems whose optimal answers are out of reach. How to relax.

Just Relax

The perfect is the enemy of the good.

—VOLTAIRE

When somebody tells you to relax, it's probably because you're uptight—making a bigger deal of things than you should. When computer scientists are up against a formidable challenge, their minds also turn to relaxation, as they pass around books like *An Introduction to Relaxation Methods* or *Discrete Relaxation Techniques*. But they don't relax themselves; they relax the problem.

One of the simplest forms of relaxation in computer science is known as **Constraint Relaxation**. In this technique, researchers remove some of the problem's constraints and set about solving the problem they *wish* they had. Then, after they've made a certain amount of headway, they try to add the constraints back in. That is, they make the problem temporarily easier to handle before bringing it back to reality.

For instance, you can relax the traveling salesman problem by letting the salesman visit the same town more than once, and letting him retrace his steps for free. Finding the shortest route under these looser rules produces what's called the "minimum spanning tree." (If you prefer, you can also think of the minimum spanning tree as the fewest miles of road needed to connect every town to at least one other town. The shortest traveling salesman route and the minimum spanning tree for Lincoln's judicial circuit are shown below.) As it turns out, solving this looser problem takes a computer essentially no time at all. And while the minimum spanning tree doesn't necessarily lead straight to the solution of the real problem, it is quite useful all the same. For one thing, the spanning tree, with its free backtracking, will never be any longer than the real solu-

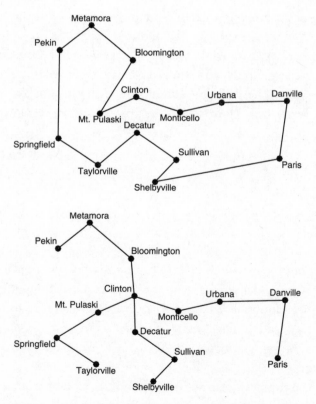

Figure 8.1 The shortest traveling salesman route (top) and minimum spanning tree (bottom) for Lincoln's 1855 judicial circuit.

tion, which has to follow all the rules. Therefore, we can use the relaxed problem—the fantasy—as a lower bound on the reality. If we calculate that the spanning tree distance for a particular set of towns is 100 miles, we can be sure the traveling salesman distance will be no less than that. And if we find, say, a 110-mile route, we can be certain it is at most 10% longer than the best solution. Thus we can get a grasp of how close we are to the real answer even without knowing what it is.

Even better, in the traveling salesman problem it turns out that the minimum spanning tree is actually one of the best starting points from which to begin a search for the real solution. Approaches like these have allowed even one of the largest traveling salesman problems imaginable—finding the shortest route that visits every single town on Earth—to be solved to within less than 0.05% of the (unknown) optimal solution.

Though most of us haven't encountered the formal algorithmic version of Constraint Relaxation, its basic message is familiar to almost anyone who's dreamed big about life questions. *What would you do if you weren't afraid?* reads a mantra you might have seen in a guidance counselor's office or heard at a motivational seminar. *What would you do if you could not fail?* Similarly, when considering questions of profession or career, we ask questions like *What would you do if you won the lottery?* or, taking a different tack, *What would you do if all jobs paid the same?* The idea behind such thought exercises is exactly that of Constraint Relaxation: to make the intractable tractable, to make progress in an idealized world that can be ported back to the real one. If you can't solve the problem in front of you, solve an easier version of it—and then see if that solution offers you a starting point, or a beacon, in the full-blown problem. Maybe it does.

What relaxation cannot do is offer you a guaranteed shortcut to the perfect answer. But computer science can also quantify the tradeoff that relaxation offers between time and solution quality. In many cases, the ratio is dramatic, a no-brainer—for instance, an answer at least half as good as the perfect solution in a quadrillionth of the time. The message is simple but profound: if we're willing to accept solutions that are close enough, then even some of the hairiest problems around can be tamed with the right techniques.

Temporarily removing constraints, as in the minimum spanning tree and the "what if you won the lottery?" examples, is the most straightforward form of algorithmic relaxation. But there are also two other, subtler types of relaxation that repeatedly show up in optimization research. They have proven instrumental in solving some of the field's most important intractable problems, with direct real-world implications for everything from city planning and disease control to the cultivation of athletic rivalries.

Uncountably Many Shades of Gray: Continuous Relaxation

The traveling salesman problem, like Meghan Bellows's search for the best seating arrangement, is a particular kind of optimization problem known as "discrete optimization"—that is, there's no smooth continuum among its solutions. The salesman goes either to this town or to that one; you're either at table five or at table six. There are no shades of gray in between.

Such discrete optimization problems are all around us. In cities, for example, planners try to place fire trucks so that every house can be reached within a fixed amount of time—say, five minutes. Mathematically, each fire truck "covers" whatever houses can be reached within five minutes from its location. The challenge is finding the minimal set of locations such that all houses are covered. "The whole [fire and emergency] profession has just adopted this coverage model, and it's great," says University of Wisconsin–Madison's Laura Albert McLay. "It's a nice, clear thing we can model." But since a fire truck either exists at a location or it doesn't, calculating that minimal set involves discrete optimization. And as McLay notes, "that's where a lot of problems become computationally hard, when you can't do half of this and half of that."

The challenge of discrete optimization shows up in social settings, too. Imagine you wanted to throw a party for all your friends and acquaintances, but didn't want to pay for all the envelopes and stamps that so many invitations would entail. You could instead decide to mail invitations to a few well-connected friends, and tell them to "bring everyone we know." What you'd ideally want to find, then, is the smallest subgroup of your friends that knows all the rest of your social circle—which would let you lick the fewest envelopes and still get everyone to attend. Granted, this might sound like a lot of work just to save a few bucks on stamps, but it's exactly the kind of problem that political campaign managers and corporate marketers want to solve to spread their message most effectively. And it's also the problem that epidemiologists study in thinking about, say, the minimum number of people in a population—and *which* people—to vaccinate to protect a society from communicable diseases.

As we noted, discrete optimization's commitment to whole numbers—a fire department can have one engine in the garage, or two, or three, but not two and a half fire trucks, or π of them—is what makes discrete optimization problems so hard to solve. In fact, both the fire truck problem and the party invitation problem are intractable: no general efficient solution for them exists. But, as it turns out, there do exist a number of efficient strategies for solving the *continuous* versions of these problems, where any fraction or decimal is a possible solution. Researchers confronted with a discrete optimization problem might gaze at those strategies enviously—but they also can do more than that. They can try to relax their discrete problem into a continuous one and see what happens.

In the case of the invitation problem, relaxing it from discrete to continuous optimization means that a solution may tell us to send someone a quarter of an invitation, and someone else two-thirds of one. What does that even mean? It obviously can't be the answer to the original question, but, like the minimum spanning tree, it does give us a place to start. With the relaxed solution in hand, we can decide how to translate those fractions back into reality. We could, for example, choose to simply round them as necessary, sending invitations to everyone who got "half an invitation" or more in the relaxed scenario. We could also interpret these fractions as probabilities—for instance, flipping a coin for every location where the relaxed solution tells us to put half a fire truck, and actually placing a truck there only if it lands heads. In either case, with these fractions turned back to whole numbers, we'll have a solution that makes sense in the context of our original, discrete problem.

The final step, as with any relaxation, is to ask how good this solution is compared to the actual best solution we might have come up with by exhaustively checking every single possible answer to the original problem. It turns out that for the invitations problem, Continuous Relaxation with rounding will give us an easily computed solution that's not half bad: it's mathematically guaranteed to get everyone you want to the party while sending out at most twice as many invitations as the best solution obtainable by brute force. Similarly, in the fire truck problem, Continuous Relaxation with probabilities can quickly get us within a comfortable bound of the optimal answer.

Continuous Relaxation is not a magic bullet: it still doesn't give us an efficient way to get to the truly optimal answers, only to their approximations. But delivering twice as many mailings or inoculations as optimal is still far better than the unoptimized alternatives.

Just a Speeding Ticket: Lagrangian Relaxation

> *Vizzini: Inconceivable!*
> *Inigo Montoya: You keep using that word. I do not think it means what you*
> *think it means.*
>
> —THE PRINCESS BRIDE

One day as a child, Brian was complaining to his mother about all the things he had to do: his homework, his chores.... "Technically, you don't *have* to do anything," his mother replied. "You don't *have* to do what your teachers tell you. You don't *have* to do what I tell you. You don't even *have* to obey the law. There are consequences to everything, and you get to decide whether you want to face those consequences."

Brian's kid-mind was blown. It was a powerful message, an awakening of a sense of agency, responsibility, moral judgment. It was something else, too: a powerful computational technique called **Lagrangian Relaxation**. The idea behind Lagrangian Relaxation is simple. An optimization problem has two parts: the rules and the scorekeeping. In Lagrangian Relaxation, we take some of the problem's constraints and bake them into the scoring system instead. That is, we take the impossible and downgrade it to costly. (In a wedding seating optimization, for instance, we might relax the constraint that tables each hold ten people max, allowing overfull tables but with some kind of elbow-room penalty.) When an optimization problem's constraints say "Do it, or else!," Lagrangian Relaxation replies, "Or else what?" Once we can color outside the lines—even just a little bit, and even at a steep cost—problems become tractable that weren't tractable before.

Lagrangian Relaxations are a huge part of the theoretical literature on the traveling salesman problem and other hard problems in computer science. They're also a critical tool for a number of practical applications. For instance, recall Carnegie Mellon's Michael Trick, who, as we mentioned in chapter 3, is in charge of scheduling for Major League Baseball and a number of NCAA conferences. What we hadn't mentioned is how he does it. The composition of each year's schedule is a giant discrete optimization problem, much too complex for any computer to solve by brute force. So each year Trick and his colleagues at the Sports Scheduling Group turn to Lagrangian Relaxation to get the job done. Every time you turn on the television or take a seat in a stadium, know that the meeting of those teams on that court on that particular night ... well, it's not necessarily the optimum matchup. But it's close. And for that we have not only Michael Trick but eighteenth-century French mathematician Joseph-Louis Lagrange to thank.

In scheduling a sports season, Trick finds that the Continuous Relaxation we described above doesn't necessarily make his life any easier. "If you end up with fractional games, you just don't get anything useful." It's one thing to end up with fractional allocations of party invitations or fire trucks, where the numbers can always be rounded up if necessary. But in sports, the integer constraints—on how many teams play a game, how many games are played in sum, and how many times each team plays every other team—are just too strong. "And so we cannot relax in that direction. We really have got to keep the fundamental [discrete] part of the model."

Nonetheless, something has to be done to reckon with the sheer complexity of the problem. So "we have to work with the leagues to relax some of the constraints they might like to have," Trick explains. The number of such constraints that go into scheduling a sports season is immense, and it includes not only the requirements arising from the league's basic structure but also all sorts of idiosyncratic requests and qualms. Some leagues are fine with the second half of the season mirroring the first, just with home and away games reversed; other leagues don't want that structure, but nonetheless demand that no teams meet for a second time until they've already met every other team once. Some leagues insist on making the most famous rivalries happen in the final game of the season. Some teams cannot play home games on certain dates due to conflicting events at their arenas. In the case of NCAA basketball, Trick also has to consider further constraints coming from the television networks that broadcast the games. Television channels define a year in advance what they anticipate "A games" and "B games" to be—the games that will attract the biggest audience. (Duke vs. UNC is a perennial A game, for instance.) The channels then expect one A game and one B game each week—but never two A games at the same time, lest it split the viewership.

Unsurprisingly, given all these demands, Trick has found that computing a sports schedule often becomes possible only by softening some of these hard constraints.

> Generally, when people first come to us with a sports schedule, they will claim . . . "We never do x and we never do y." Then we look at their schedules and we say, "Well, twice you did x and three times you did y last year." Then "Oh, yeah, well, okay. Then other than that we never do it." And then

we go back the year before. . . . We generally realize that there are some things they think they never do that people *do* do. People in baseball believe that the Yankees and the Mets are never at home at the same time. And it's not true. It's never been true. They are at home perhaps three games, perhaps six games in a year at the same day. But in the broad season, eighty-one games at home for each of the teams, it's relatively rare, people forget about them.

Occasionally it takes a bit of diplomatic finesse, but a Lagrangian Relaxation—where some impossibilities are downgraded to penalties, the inconceivable to the undesirable—enables progress to be made. As Trick says, rather than spending eons searching for an unattainable perfect answer, using Lagrangian Relaxation allows him to ask questions like, "How close can you get?" Close enough, it turns out, to make everyone happy—the league, the schools, the networks—and to stoke the flames of March Madness, year after year.

Learning to Relax

Of the various ways that computational questions present themselves to us, optimization problems—one part goals, one part rules—are arguably the most common. And *discrete* optimization problems, where our options are stark either/or choices, with no middle ground, are the most typical of those. Here, computer science hands down a disheartening verdict. Many discrete optimization problems are truly hard. The field's brightest minds have come up empty in every attempt to find an easy path to perfect answers, and in fact are increasingly more devoted to proving that such paths don't exist than to searching for them.

If nothing else, this should offer us some consolation. If we're up against a problem that seems gnarly, thorny, impassable—well, we might be right. And having a computer won't necessarily help.

At least, not unless we can learn to relax.

There are many ways to relax a problem, and we've seen three of the most important. The first, Constraint Relaxation, simply removes some constraints altogether and makes progress on a looser form of the problem before coming back to reality. The second, Continuous Relaxation, turns discrete or binary choices into continua: when deciding between iced tea and lemonade, first imagine a 50–50 "Arnold Palmer" blend and then round it up or down. The third, Lagrangian Relaxation, turns impossibilities into mere penalties, teaching the art of bending the rules (or breaking them and accepting the consequences). A rock band deciding which songs to cram into a limited set, for instance, is up against what computer scientists call the "knapsack problem"—a puzzle that asks one to decide which of a set of items of different bulk and importance to pack into a confined volume. In its strict formulation the knapsack problem is famously intractable, but that needn't discourage our relaxed rock stars. As demonstrated in several celebrated examples, sometimes it's better to simply play a bit past the city curfew and incur the related fines than to limit the show to the available slot. In fact, even when you don't commit the infraction, simply *imagining* it can be illuminating.

The conservative British columnist Christopher Booker says that "when we embark on a course of action which is unconsciously driven by wishful thinking, all may seem to go well for a time"—but that because "this make-believe can never be reconciled with reality," it will inevitably lead to what he describes as a multi-stage breakdown: "dream," "frustration," "nightmare," "explosion." Computer science paints a dramatically rosier view. Then again, as an optimization technique, relaxation is all about being *consciously* driven by wishful thinking. Perhaps that's partly what makes the difference.

Relaxations offer us a number of advantages. For one, they offer a bound on the quality of the true solution. If we're trying to pack our calendar, imagining that we can magically teleport across town will instantaneously make it clear that eight one-hour meetings is the most we could possibly expect to fit into a day; such a bound might be useful in setting expectations before we face the full problem. Second, relaxations are designed so that they can indeed be reconciled with reality—and this gives us bounds on the solution from the other direction. When Continuous Relaxation tells us to give out fractional vaccines, we can just immunize everyone assigned to receive half a vaccine or more, and end up with an easily calculated solution that requires at worst twice as many inoculations as in a perfect world. Maybe we can live with that.

Unless we're willing to spend eons striving for perfection every time we encounter a hitch, hard problems demand that instead of spinning our tires we imagine easier versions and tackle those first. When applied correctly, this is not just wishful thinking, not fantasy or idle daydreaming. It's one of our best ways of making progress.

9 | Randomness

When to Leave It to Chance

I must admit that after many years of work in this area, the efficacy of randomness for so many algorithmic problems is absolutely mysterious to me. It is efficient, it works; but why and how is absolutely mysterious.

—MICHAEL RABIN

Randomness seems like the opposite of reason—a form of giving up on a problem, a last resort. Far from it. The surprising and increasingly important role of randomness in computer science shows us that making use of chance can be a deliberate and effective part of approaching the hardest sets of problems. In fact, there are times when nothing else will do.

In contrast to the standard "deterministic" algorithms we typically imagine computers using, where one step follows from another in exactly the same way every time, a randomized algorithm uses randomly generated numbers to solve a problem. Recent work in computer science has shown that there are cases where randomized algorithms can produce good approximate answers to difficult questions faster than all known deterministic algorithms. And while they do not always guarantee the optimal solutions, randomized algorithms can get surprisingly close to them in a fraction of the time, just by strategically flipping a few coins while their deterministic cousins sweat it out.

There is a deep message in the fact that on certain problems, randomized approaches can outperform even the best deterministic ones. Sometimes the best solution to a problem is to turn to chance rather than trying to fully reason out an answer.

But merely knowing that randomness can be helpful isn't good enough. You need to know when to rely on chance, in what way, and to what extent. The recent history of computer science provides some answers—though the story begins a couple of centuries earlier.

Sampling

In 1777, George-Louis Leclerc, Comte de Buffon, published the results of an interesting probabilistic analysis. If we drop a needle onto a lined piece of paper, he asked, how likely is it to cross one of the lines? Buffon's work showed that if the needle is shorter than the gap between the lines, the answer is $\frac{2}{\pi}$ times the needle's length divided by the length of the gap. For Buffon, deriving this formula was enough. But in 1812, Pierre-Simon Laplace, one of the heroes of chapter 6, pointed out that this result has another implication: one could estimate the value of π simply by dropping needles onto paper.

Laplace's proposal pointed to a profound general truth: when we want to know something about a complex quantity, we can estimate its value by *sampling* from it. This is exactly the kind of calculation that his work on Bayes's Rule helps us to perform. In fact, following Laplace's suggestion, several people have carried out exactly the experiment he suggested, confirming that it is possible—although not particularly efficient—to estimate the value of π in this hands-on way.*

Throwing thousands of needles onto lined paper makes for an interesting pastime (for some), but it took the development of the computer to make sampling into a practical method. Before, when mathematicians and physicists tried using randomness to solve problems, their calculations had to be laboriously worked out by hand, so it was hard to generate enough

---

*Interestingly, some of these experiments appear to have produced a far better estimate of π than would be expected by chance—which suggests that they may have been deliberately cut short at a good stopping point, or faked altogether. For example, in 1901 the Italian mathematician Mario Lazzarini supposedly made 3,408 tosses and obtained an estimate of $\pi \approx \frac{355}{113} = 3.1415929$ (the actual value of π to seven decimal places is 3.1415927). But if the number of times the needle crossed the line had been off by just a single toss, the estimate would have been far less pretty—3.1398 or 3.1433—which makes Lazzarini's report seem suspicious. Laplace might have found it fitting that we can use Bayes's Rule to confirm that this result is unlikely to have arisen from a valid experiment.

samples to yield accurate results. Computers—in particular, the computer developed in Los Alamos during World War II—made all the difference.

Stanislaw "Stan" Ulam was one of the mathematicians who helped develop the atomic bomb. Having grown up in Poland, he moved to the United States in 1939, and joined the Manhattan Project in 1943. After a brief return to academia he was back at Los Alamos in 1946, working on the design of thermonuclear weapons. But he was also sick—he had contracted encephalitis, and had emergency brain surgery. And as he recovered from his illness he worried about whether he would regain his mathematical abilities.

While convalescing, Ulam played a lot of cards, particularly solitaire (a.k.a. Klondike). As any solitaire player knows, some shuffles of the deck produce games that just can't be won. So as Ulam played, he asked himself a natural question: what is the probability that a shuffled deck will yield a winnable game?

In a game like solitaire, reasoning your way through the space of possibilities gets almost instantly overwhelming. Flip over the first card, and there are fifty-two possible games to keep track of; flip over the second, and there are fifty-one possibilities for each first card. That means we're already up into thousands of possible games before we've even begun to play. F. Scott Fitzgerald once wrote that "the test of a first-rate intelligence is the ability to hold two opposing ideas in mind at the same time and still retain the ability to function." That may be true, but no first-rate intelligence, human or otherwise, can possibly hold the eighty unvigintillion possible shuffled-deck orders in mind and have any hope of functioning.

After trying some elaborate combinatorial calculations of this sort and giving up, Ulam landed on a different approach, beautiful in its simplicity: *just play the game.*

> I noticed that it may be much more practical to [try] . . . laying down the cards, or experimenting with the process and merely noticing what proportion comes out successfully, rather than to try to compute all the combinatorial possibilities which are an exponentially increasing number so great that, except in very elementary cases, there is no way to estimate it. This is intellectually surprising, and if not exactly humiliating, it gives one a feeling of modesty about the limits of rational or traditional thinking. In a sufficiently complicated problem, actual sampling is better than an examination of all the chains of possibilities.

When he says "better," note that he doesn't necessarily mean that sampling will offer you more *precise* answers than exhaustive analysis: there will always be some error associated with a sampling process, though you can reduce it by ensuring your samples are indeed random and by taking more and more of them. What he means is that sampling is better because it gives you an answer at all, in cases where nothing else will.

Ulam's insight—that sampling can succeed where analysis fails—was also crucial to solving some of the difficult nuclear physics problems that arose at Los Alamos. A nuclear reaction is a branching process, where possibilities multiply just as wildly as they do in cards: one particle splits in two, each of which may go on to strike others, causing them to split in turn, and so on. Exactly calculating the chances of some particular outcome of that process, with many, many particles interacting, is hard to the point of impossibility. But simulating it, with each interaction being like turning over a new card, provides an alternative.

Ulam developed the idea further with John von Neumann, and worked with Nicholas Metropolis, another of the physicists from the Manhattan Project, on implementing the method on the Los Alamos computer. Metropolis named this approach—replacing exhaustive probability calculations with sample simulations—the **Monte Carlo Method**, after the Monte Carlo casino in Monaco, a place equally dependent on the vagaries of chance. The Los Alamos team was able to use it to solve key problems in nuclear physics. Today the Monte Carlo Method is one of the cornerstones of scientific computing.

Many of these problems, like calculating the interactions of subatomic particles or the chances of winning at solitaire, are themselves intrinsically probabilistic, so solving them through a randomized approach like Monte Carlo makes a fair bit of sense. But perhaps the most surprising realization about the power of randomness is that it can be used in situations where chance seemingly plays no role at all. Even if you want the answer to a question that is strictly yes or no, true or false—no probabilities about it—rolling a few dice may still be part of the solution.

Randomized Algorithms

The first person to demonstrate the surprisingly broad applications of randomness in computer science was Michael Rabin. Born in 1931 in Breslau, Germany (which became Wrocław, Poland, at the end of World

War II), Rabin was the descendant of a long line of rabbis. His family left Germany for Palestine in 1935, and there he was diverted from the rabbinical path his father had laid down for him by the beauty of mathematics—discovering Alan Turing's work early in his undergraduate career at the Hebrew University and immigrating to the United States to begin a PhD at Princeton. Rabin would go on to win the Turing Award—the computer science equivalent of a Nobel—for extending theoretical computer science to accommodate "nondeterministic" cases, where a machine isn't forced to pursue a single option but has multiple paths it might follow. On sabbatical in 1975, Rabin came to MIT, searching for a new research direction to pursue.

He found it in one of the oldest problems of them all: how to identify prime numbers.

Algorithms for finding prime numbers date back at least as far as ancient Greece, where mathematicians used a straightforward approach known as the Sieve of Erastothenes. The Sieve of Erastothenes works as follows: To find all the primes less than n, begin by writing down all the numbers from 1 to n in sequence. Then cross out all the numbers that are multiples of 2, besides itself (4, 6, 8, ~~10~~, ~~12~~, and so on). Take the next smallest number that hasn't been crossed out (in this case, 3), and cross out all multiples of that number (6, 9, ~~12~~, ~~15~~). Keep going like this, and the numbers that remain at the end are the primes.

For millennia, the study of prime numbers was believed to be, as G. H. Hardy put it, "one of the most obviously useless branches" of mathematics. But it lurched into practicality in the twentieth century, becoming pivotal in cryptography and online security. As it happens, it is *much* easier to multiply primes together than to factor them back out. With big enough primes—say, a thousand digits—the multiplication can be done in a fraction of a second while the factoring could take literally millions of years; this makes for what is known as a "one-way function." In modern encryption, for instance, secret primes known only to the sender and recipient get multiplied together to create huge composite numbers that can be transmitted publicly without fear, since factoring the product would take any eavesdropper way too long to be worth attempting. Thus virtually all secure communication online—be it commerce, banking, or email—begins with a hunt for prime numbers.

This cryptographic application suddenly made algorithms for finding and checking primes incredibly important. And while the Sieve of Erastothenes is effective, it is not efficient. If you want to check whether a particular number is prime—known as testing its "primality"—then following the sieve strategy requires trying to divide it by all the primes up to its square root.* Checking whether a six-digit number is prime would require dividing by all of the 168 primes less than 1,000—not so bad. But checking a twelve-digit number involves dividing by the 78,498 primes less than 1 million, and all that division quickly starts to get out of control. The primes used in modern cryptography are hundreds of digits long; forget about it.

At MIT, Rabin ran into Gary Miller, a recent graduate from the computer science department at Berkeley. In his PhD thesis, Miller had developed an intriguingly promising, much faster algorithm for testing primality—but there was one small problem: it didn't always work.

Miller had found a set of equations (expressed in terms of two numbers, n and x) that are always true if n is prime, regardless of what values you plug in for x. If they come out false even for a single value of x, then there's no way n can be prime—in these cases, x is called a "witness" against primality. The problem, though, is false positives: even when n is not prime, the equations will still come out true some of the time. This seemed to leave Miller's approach hanging.

Rabin realized that this was a place where a step outside the usually deterministic world of computer science might be valuable. If the number n is actually nonprime, how many possible values of x would give a false positive and declare it a prime number? The answer, Rabin showed, is no more than one-quarter. So for a random value of x, if Miller's equations come out true, there's only a one-in-four chance that n isn't actually prime. And crucially, each time we sample a new random x and Miller's equations check out, the probability that n only seems prime, but isn't really, drops by another multiple of four. Repeat the procedure ten times, and the probability of a false positive is one in four to the tenth power—less than one in a million. Still not enough certainty? Check another five times and you're down to one in a billion.

---

*You don't need to check beyond the square root, because if a number has a factor greater than its square root then by definition it must also have a corresponding factor smaller than the square root—so you would have caught it already. If you're looking for factors of 100, for instance, every factor that's greater than 10 will be paired with a factor smaller than 10: 20 is matched up with 5, 25 with 4, and so on.

Vaughan Pratt, another computer scientist at MIT, implemented Rabin's algorithm and started getting results late one winter night, while Rabin was at home having friends over for a Hanukkah party. Rabin remembers getting a call around midnight:

> "Michael, this is Vaughan. I'm getting the output from these experiments. Take a pencil and paper and write this down." And so he had that $2^{400}-593$ is prime. Denote the product of all primes p smaller than 300 by k. The numbers $k \times 338 + 821$ and $k \times 338 + 823$ are twin primes.* These constituted the largest twin primes known at the time. My hair stood on end. It was incredible. It was just incredible.

The Miller-Rabin primality test, as it's now known, provides a way to quickly identify even gigantic prime numbers with an arbitrary degree of certainty.

Here we might ask a philosophical question—about what the meaning of "is" is. We're so used to mathematics being a realm of certainty that it's jarring to think that a number could be "probably prime" or "almost definitely prime." How certain is certain enough? In practice, modern cryptographic systems, the ones that encrypt Internet connections and digital transactions, are tuned for a false positive rate of less than one in a million billion billion. In other words, that's a decimal that begins with twenty-four zeros—less than one false prime for the number of grains of sand on Earth. This standard comes after a mere forty applications of the Miller-Rabin test. It's true that you are never fully certain—but you can get *awfully* close, awfully quick.

Though you may have never heard of the Miller-Rabin test, your laptop, tablet, and phone know it well. Several decades after its discovery, it is still the standard method used to find and check primes in many domains. It's working behind the scenes whenever you use your credit card online, and almost any time secure communications are sent through the air or over wires.

For decades after Miller and Rabin's work, it wasn't known whether there would ever be an efficient algorithm that allows testing primality in deterministic fashion, with absolute certainty. In 2002, one such method

---

*Twin primes are consecutive odd numbers that are both prime, like 5 and 7.

did get discovered by Manindra Agrawal, Neeraj Kayal, and Nitin Saxena at the Indian Institute of Technology—but randomized algorithms like Miller-Rabin are much faster and thus are still the ones used in practice today.

And for some other problems, randomness still provides the *only* known route to efficient solutions. One curious example from mathematics is known as "polynomial identity testing." If you have two polynomial expressions, such as $2x^3 + 13x^2 + 22x + 8$ and $(2x+1) \times (x+2) \times (x+4)$, working out whether those expressions are in fact the same function—by doing all the multiplication, then comparing the results—can be incredibly time-consuming, especially as the number of variables increases.

Here again randomness offers a way forward: just generate some random xs and plug them in. If the two expressions are *not* the same, it would be a big coincidence if they gave the same answer for some randomly generated input. And an even bigger coincidence if they also gave identical answers for a second random input. And a bigger coincidence still if they did it for three random inputs in a row. Since there is no known deterministic algorithm for efficiently testing polynomial identity, this randomized method—with multiple observations quickly giving rise to near-certainty—is the only practical one we have.

In Praise of Sampling

The polynomial identity test shows that sometimes our effort is better spent checking random values—sampling from the two expressions we want to know about—than trying to untangle their inner workings. To some extent this seems reasonably intuitive. Given a pair of nondescript gadgets and asked whether they are two different devices or two copies of the same one, most of us would start pushing random buttons rather than crack open the cases to examine the wiring. And we aren't particularly surprised when, say, a television drug lord knifes open a few bundles at random to be reasonably certain about the quality of the entire shipment.

There are cases, though, where we don't turn to randomness—and maybe we should.

Arguably the most important political philosopher of the twentieth century was Harvard's John Rawls, who set for himself the ambitious task of reconciling two seemingly opposite key ideas in his field: *liberty* and *equality.* Is a society more "just" when it's more free, or more equal? And do the two really have to be mutually exclusive? Rawls offered a way of approaching this set of questions that he called the "veil of ignorance." Imagine, he said, that you were about to be born, but didn't know as whom: male or female, rich or poor, urban or rural, sick or healthy. And before learning your status, you had to choose what kind of society you'd live in. What would you want? By evaluating various social arrangements from behind the veil of ignorance, argued Rawls, we'd more readily come to a consensus about what an ideal one would look like.

What Rawls's thought experiment does not take into account, however, is the computational cost of making sense of a society from behind such a veil. How could we, in this hypothetical scenario, possibly hope to hold all of the relevant information in our heads? Set aside grand questions of justice and fairness for a moment and try to apply Rawls's approach merely to, say, a proposed change in health insurance regulations. Take the probability of being born, perhaps, as someone who grows up to become a town clerk in the Midwest; multiply that by the distribution of the different health care plans available to government employees across various midwestern municipalities; multiply that by actuarial data that offer the probability of, for instance, a fractured tibia; multiply that by the average medical bill for the average procedure for a fractured tibia at a midwestern hospital given the distribution of possible insurance plans. . . . Okay, so would the proposed insurance revision be "good" or "bad" for the nation? We can barely hope to evaluate a single injured shin this way, let alone the lives of hundreds of millions.

Rawls's philosophical critics have argued at length about how exactly we are supposed to leverage the information obtained from the veil of ignorance. Should we be trying, for instance, to maximize mean happiness, median happiness, total happiness, or something else? Each of these approaches, famously, leaves itself open to pernicious dystopias—such as the civilization of Omelas imagined by writer Ursula K. Le Guin, in which prosperity and harmony abound but a single child is forced to live in abject misery. These are worthy critiques, and Rawls deliberately sidesteps them by leaving open the question of what to do with the information we get from behind the veil. Perhaps the bigger question, though, is how to *gather* that information in the first place.

The answer may well come from computer science. MIT's Scott Aaronson says he's surprised that computer scientists haven't yet had more influence on philosophy. Part of the reason, he suspects, is just their "*failure to communicate* what they can add to philosophy's conceptual arsenal." He elaborates:

> One might think that, once we know something is *computable*, whether it takes 10 seconds or 20 seconds to compute is obviously the concern of engineers rather than philosophers. But that conclusion would *not* be so obvious, if the question were one of 10 seconds versus $10^{10^{10}}$ seconds! And indeed, in complexity theory, the quantitative gaps we care about are usually so vast that one has to consider them qualitative gaps as well. Think, for example, of the difference between reading a 400-page book and reading *every possible* such book, or between writing down a thousand-digit number and counting to that number.

Computer science gives us a way to articulate the complexity of evaluating all possible social provisions for something like an injured shin. But fortunately, it also provides tools for dealing with that complexity. And the sampling-based Monte Carlo algorithms are some of the most useful approaches in that toolbox.

When we need to make sense of, say, national health care reform—a vast apparatus too complex to be readily understood—our political leaders typically offer us two things: cherry-picked personal anecdotes and aggregate summary statistics. The anecdotes, of course, are rich and vivid, but they're unrepresentative. Almost any piece of legislation, no matter how enlightened or misguided, will leave *someone* better off and *someone* worse off, so carefully selected stories don't offer any perspective on broader patterns. Aggregate statistics, on the other hand, are the reverse: comprehensive but thin. We might learn, for instance, whether average premiums fell nationwide, but not how that change works out on a more granular level: they might go down for most but, Omelas-style, leave some specific group—undergraduates, or Alaskans, or pregnant women—in dire straits. A statistic can only tell us part of the story, obscuring any underlying heterogeneity. And often we don't even know which statistic we need.

Since neither sweeping statistics nor politicians' favorite stories can truly guide us through thousands of pages of proposed legislation, a Monte Carlo–informed computer scientist would propose a different approach: sampling. A close examination of random samples can be one of the most effective means of making sense of something too complex to be comprehended directly. When it comes to handling a qualitatively unmanageable problem, something so thorny and complicated that it can't be digested whole—solitaire or atomic fission, primality testing or public policy—sampling offers one of the simplest, and also the best, ways of cutting through the difficulties.

We can see this approach at work with the charity GiveDirectly, which distributes unconditional cash transfers to people living in extreme poverty in Kenya and Uganda. It has attracted attention for rethinking conventional charity practices on a number of levels: not only in its unusual mission, but in the level of transparency and accountability it brings to its own process. And the latest element of the status quo that it's challenging is success stories.

"If you regularly check our website, blog, or Facebook page," writes program assistant Rebecca Lange, "you may have noticed something you *don't* often see: stories and photos of our recipients." The problem isn't that the glowing stories proffered by other charities aren't true. Rather, the very fact that they were deliberately chosen to showcase successes makes it unclear how much information can be gleaned from them. So GiveDirectly decided to put a twist on this conventional practice as well.

Every Wednesday, the GiveDirectly team selects a cash recipient at random, sends out a field officer to interview them, and publishes the field officer's notes verbatim, no matter what. For instance, here's their first such interview, with a woman named Mary, who used the money for a tin roof:*

> She was able to make a better house and that was a tinned house. She was also able to buy a sofa set for her own house. Her life has changed because she used to have a leaking roof soaking up everything in the house whenever it rained. But because of the transfer she was able to make a better tinned house.

"We hope that this gives you confidence in all types of information we share with you," Lange writes, "and maybe even inspires you to hold other organizations to a higher bar."

---

*Note that we deliberately took the very first story from the site—that is, we did not read through all of them to pick one to share, which would have defeated the purpose.

The Three-Part Tradeoff

At once it struck me what quality went to form a Man of Achievement, especially in Literature, and which Shakespeare possessed so enormously— I mean Negative Capability, that is, when a man is capable of being in uncertainties, mysteries, doubts, without any irritable reaching after fact and reason.

—JOHN KEATS

There is no such thing as absolute certainty, but there is assurance sufficient for the purposes of human life.

—JOHN STUART MILL

Computer science is often a matter of negotiating tradeoffs. In our discussion of sorting in chapter 3, for instance, we noted the tradeoff between time spent up front on sorting versus the time spent later on searching. And in the discussion of caching in chapter 4, we explored the tradeoff of taking up extra *space*—caches for caches for caches—to save time.

Time and space are at the root of the most familiar tradeoffs in computer science, but recent work on randomized algorithms shows that there's also another variable to consider: certainty. As Harvard's Michael Mitzenmacher puts it, "What we're going to do is come up with an answer which saves you in time and space and trades off this third dimension: error probability." Asked for his favorite example of this tradeoff into uncertainty, he doesn't hesitate. "A colleague just said that there should be a drinking game that every time this term appears on one of my slides, you have to take a drink. Have you ever heard of Bloom filters?"

To understand the idea behind a Bloom filter, Mitzenmacher says, consider a search engine like Google, trying to crawl the entire web and index every possible URL. The web is comprised of well over a trillion distinct URLs, and the average URL weighs in at about seventy-seven characters long. When the search engine looks at some URL, how can it check whether that page has already been processed? Just storing a list of all the URLs that have been visited would take a huge amount of space, and repeatedly searching that list (even if it were fully sorted) could prove a nightmare. In fact, it could well be that the cure is worse than the disease: in other words, checking every time to make sure that we're not reindexing a page might be more time-consuming than just indexing the occasional page twice.

But what if we only needed to be *mostly* sure this URL was new to us? That's where the Bloom filter comes in. Named for its inventor, Burton H. Bloom, a Bloom filter works much like the Rabin-Miller primality test: the URL is entered into a set of equations that esssentially check for "witnesses" to its novelty. (Rather than proclaim "*n* is not prime," these equations say "I have not seen *n* before.") If you're willing to tolerate an error rate of just 1% or 2%, storing your findings in a probabilistic data structure like a Bloom filter will save you significant amounts of both time and space. And the usefulness of such filters is not confined to search engines: Bloom filters have shipped with a number of recent web browsers to check URLs against a list of known malicious websites, and they are also an important part of cryptocurrencies like Bitcoin.

Says Mitzenmacher, "The idea of the error tradeoff space—I think the issue is that people don't associate that with computing. They think computers are supposed to give you the answer. So when you hear in your algorithms class, 'It's supposed to give you one answer; it might not be the right answer'—I like to think that when [students] hear that, it focuses them. I think people don't realize in their own lives how much they do that and accept that."

Hills, Valleys, and Traps

> *The river meanders because it can't think.*
>
> —RICHARD KENNEY

Randomness has also proven itself to be a powerful weapon for solving discrete optimization problems, like assembling the calendar for NCAA basketball or finding the shortest route for a traveling salesman. In the previous chapter we saw how relaxation can play a big role in cutting such problems down to size, but the tactical use of randomness has emerged as an arguably even more important technique.

Imagine you're putting together a globe-trotting ten-city vacation, your own version of the traveling salesman problem: you'll start and finish in San Francisco and visit Seattle, Los Angeles, New York, Buenos Aires, London, Amsterdam, Copenhagen, Istanbul, Delhi, and Kyoto. You might not be too worried about the total length of the route, but you probably do want to minimize the monetary cost of the trip. The first thing to note here is that even though ten cities hardly sounds like a lot, the number of possible itineraries is ten factorial: more than three and a half million. In other words, there's no practical way for you to simply check every permutation and pick the lowest price. You have to work smarter than that.

For your first attempt at an itinerary, you might look at taking the cheapest flight out of San Francisco (let's say it's Seattle), then taking the cheapest flight from there to any of the other remaining cities (call it Los Angeles), then the cheapest from there (say, New York), and so forth, until you're at your tenth city and you fly from there back to San Francisco. This is an example of a so-called greedy algorithm, which you can also think of as a "myopic algorithm": one that shortsightedly takes the best thing available every step of the way. In scheduling theory, as we saw in chapter 5, a greedy algorithm—for instance, always doing the shortest job available, without looking or planning beyond—can sometimes be all that a problem requires. In this case, for the traveling salesman problem, the solution given by the greedy algorithm probably isn't terrible, but it's likely to be far from the best you can do.

Once you've assembled a baseline itinerary, you might test some alternatives by making slight perturbations to the city sequence and seeing if that makes an improvement. For instance, if we are going first to Seattle, then to Los Angeles, we can try doing those cities in reverse order: L.A. first, then Seattle. For any given itinerary, we can make eleven such two-city flip-flops; let's say we try them all and then go with the one that gives us the best savings. From here we've got a new itinerary to work with, and we can start permuting *that* one, again looking for the best local improvement. This is an algorithm known as **Hill Climbing**—since the search through a space of solutions, some better and some worse, is commonly thought of in terms of a landscape with hills and valleys, where your goal is to reach the highest peak.

Eventually you will end up with a solution that is better than all of its permutations; no matter which adjacent stops you flip, nothing beats it. It's here that the hill climbing stops. Does this mean you've definitely found the single best possible itinerary, though? Sadly, no. You may have found only a so-called "local maximum," not the global maximum of all the possibilities. The hill-climbing landscape is a misty one. You can know that you're standing on a mountaintop because the ground falls away in

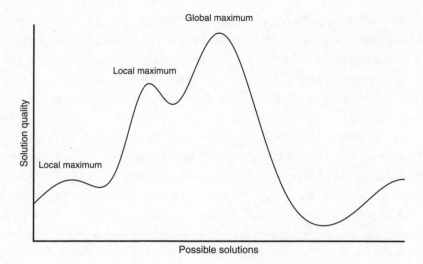

An "error landscape," which depicts how solution quality can vary across different possibilities.

all directions—but there might be a higher mountain just across the next valley, hidden behind clouds.

Consider the lobster stuck in the lobster trap: poor beast, he doesn't realize that exiting the cage means backtracking to the cage's center, that he needs to go *deeper* into the cage to make it out. A lobster trap is nothing other than a local maximum made of wire—a local maximum that kills.

In the case of vacation planning, local maxima are fortunately less fatal, but they have the same character. Even once we've found a solution that can't be improved by any small tweaks, it's possible that we are still missing the global maximum. The true best itinerary may require a radical over-haul of the trip: doing entire continents in a different order, for instance, or proceeding westward instead of eastward. We may need to temporarily worsen our solution if we want to continue searching for improvements. And randomness provides a strategy—actually, several strategies—for doing just that.

Out of the Local Maximum

One approach is to augment Hill Climbing with what's known as "jitter": if it looks like you're stuck, mix things up a little. Make a few random small changes (even if they are for the worse), then go back to Hill Climbing; see if you end up at a higher peak.

Another approach is to *completely* scramble our solution when we reach a local maximum, and start Hill Climbing anew from this random new starting point. This algorithm is known, appropriately enough, as "Random-Restart Hill Climbing"—or, more colorfully, as "Shotgun Hill Climbing." It's a strategy that proves very effective when there are lots of local maxima in a problem. For example, computer scientists use this approach when trying to decipher codes, since there are lots of ways to begin decrypting a message that look promising at first but end up being dead ends. In decryption, having a text that looks somewhat close to sensible English doesn't necessarily mean that you're even on the right track. So sometimes it's best not to get too attached to an initial direction that shows promise, and simply start over from scratch.

But there's also a third approach: instead of turning to full-bore randomness when you're stuck, use a little bit of randomness *every* time you make a decision. This technique, developed by the same Los Alamos team that came up with the Monte Carlo Method, is called the **Metropolis Algorithm**. The Metropolis Algorithm is like Hill Climbing, trying out different small-scale tweaks on a solution, but with one important difference: at any given point, it will potentially accept bad tweaks as well as good ones.

We can imagine applying this to our vacation planning problem. Again, we try to tweak our proposed solution by jiggling around the positions of different cities. If a randomly generated tweak to our travel route results in an improvement, then we always accept it, and continue tweaking from there. But if the alteration would make thing a little worse, there's still a chance that we go with it anyway (although the worse the alteration is, the smaller the chance). That way, we won't get stuck in any local maximum for very long: eventually we'll try another nearby solution, even though it's more expensive, and potentially be on our way to coming up with a new and better plan.

Whether it's jitter, random restarts, or being open to occasional worsening, randomness is incredibly useful for avoiding local maxima. Chance is not just a viable way of dealing with tough optimization problems; in many cases, it's essential. Some questions linger, however. How much randomness should you use? And when? And—given that strategies such as the Metropolis Algorithm can permute our itinerary pretty much ad infinitum—how do you ever know that you're done? For researchers working on optimization, a surprisingly definitive answer to these questions would come from another field entirely.

Simulated Annealing

In the late 1970s and early '80s, Scott Kirkpatrick considered himself a physicist, not a computer scientist. In particular, Kirkpatrick was interested in statistical physics, which uses randomness as a way to explain certain natural phenomena—for instance, the physics of annealing, the way that materials change state as they are heated and cooled. Perhaps the most interesting characteristic of annealing is that how quickly or slowly a material is cooled tends to have tremendous impact on its final structure. As Kirkpatrick explains:

> Growing a single crystal from a melt [is] done by careful annealing, first melting the substance, then lowering the temperature slowly, and spending a long time at temperatures in the vicinity of the freezing point. If this is not done, and the substance is allowed to get out of equilibrium, the resulting crystal will have many defects, or the substance may form a glass, with no crystalline order.

Kirkpatrick was then working at IBM, where one of the biggest, trickiest, and most hallowed problems was how to lay out the circuits on the chips that IBM was manufacturing. The problem was ungainly and intractable: there was an enormous range of possible solutions to consider, and some tricky constraints. It was better in general for the components to be close together, for instance—but not too close, or there would be no room for the wires. And any time you moved anything, you'd have to recompute how all the wires would run in the new hypothetical layout.

At the time, this process was led by something of a cryptic guru-type figure within IBM. As Kirkpatrick recalls, "The guy who was the best at IBM at squeezing more circuits on a chip . . . he had the most mysterious way of explaining what he was doing. He didn't like to really tell you."

Kirkpatrick's friend and IBM colleague Dan Gelatt was fascinated by the problem, and quickly hooked Kirkpatrick, who had a flash of insight. "The way to study [physical systems] was to warm them up then cool them down, and let the system organize itself. From that background, it seemed like a perfectly natural thing to treat all kinds of optimization problems as if the degrees of freedom that you were trying to organize were little atoms, or spins, or what have you."

In physics, what we call "temperature" is really velocity—random motion at the molecular scale. This was directly analogous, Kirkpatrick reasoned, to the random jitter that can be added to a hill-climbing algorithm to make it sometimes backtrack from better solutions to worse ones. In fact, the Metropolis Algorithm itself had initially been designed to model random behavior in physical systems (in that case, nuclear explosions). So what would happen, Kirkpatrick wondered, if you treated an optimization problem like an annealing problem—if you "heated it up" and then slowly "cooled it off"?

Taking the ten-city vacation problem from above, we could start at a "high temperature" by picking our starting itinerary entirely at random, plucking one out of the whole space of possible solutions regardless of price. Then we can start to slowly "cool down" our search by rolling a die whenever we are considering a tweak to the city sequence. Taking a superior variation always makes sense, but we would only take inferior ones when the die shows, say, a 2 or more. After a while, we'd cool it further by only taking a higher-price change if the die shows a 3 or greater—then 4, then 5. Eventually we'd be mostly hill climbing, making the inferior move just occasionally when the die shows a 6. Finally we'd start going *only* uphill, and stop when we reached the next local max.

This approach, called **Simulated Annealing**, seemed like an intriguing way to map physics onto problem solving. But would it work? The initial reaction among more traditional optimization researchers was that this whole approach just seemed a little too . . . metaphorical. "I couldn't convince math people that this messy stuff with temperatures, all this analogy-based stuff, was real," says Kirkpatrick, "because mathematicians are trained to really distrust intuition."

But any distrust regarding the analogy-based approach would soon vanish: at IBM, Kirkpatrick and Gelatt's simulated annealing algorithms started making better chip layouts than the guru. Rather than keep mum about their secret weapon and become cryptic guru figures themselves, they published their method in a paper in *Science*, opening it up to others. Over the next few decades, that paper would be cited a whopping thirty-two thousand times. To this day, simulated annealing remains one of the most promising approaches to optimization problems known to the field.

Randomness, Evolution, and Creativity

In 1943, Salvador Luria didn't know he was about to make a discovery that would lead to a Nobel Prize; he thought he was going to a dance. A recent immigrant to the United States from Mussolini's Italy, where his Sephardic Jewish family had lived, Luria was a researcher studying how bacteria developed immunity from viruses. But at this moment his research was far from his mind, as he attended a faculty gathering at a country club near Indiana University.

Luria was watching one of his colleagues play a slot machine:

> Not a gambler myself, I was teasing him about his inevitable losses, when he suddenly hit the jackpot, about three dollars in dimes, gave me a dirty look, and walked away. Right then I began giving some thought to the actual numerology of slot machines; in doing so it dawned on me that slot machines and bacterial mutations have something to teach each other.

In the 1940s, it wasn't known exactly why or how bacterial resistance to viruses (and, for that matter, to antibiotics) came about. Were they *reactions* within the bacteria to the virus, or were there simply ongoing mutations that occasionally produced resistance by accident? There seemed no way to devise an experiment that would offer a decisive answer one way or the other—that is, until Luria saw that slot machine and something clicked. Luria realized that if he bred several generations of different lineages of bacteria, then exposed the last generation to a virus, one of two radically different things would happen. If resistance was a response to the virus, he'd expect roughly the same amount of resistant bacteria to appear in every one of his bacterial cultures, regardless of their lineage. On the other hand, if resistance emerged from chance mutations, he'd expect to see something a lot more uneven—just like a slot machine's payouts. That is, bacteria from most lineages would show no resistance at all; some lineages would have a single "grandchild" culture that had mutated to become resistant; and on rare occasions, if the proper mutation had happened several generations up the "family tree," there would be a jackpot: *all* the "grandchildren" in the lineage would be resistant. Luria left the dance as soon as he could and set the experiment in motion.

After several days of tense, restless waiting, Luria returned to the lab to check on his colonies. *Jackpot.*

Luria's discovery was about the power of chance: about how random, haphazard mutations can produce viral resistance. But it was also, at least in part, *due to* the power of chance. He was in the right place at the right time, where seeing the slot machine triggered a new idea. Tales of discovery often feature a similar moment: Newton's (possibly apocryphal) apple, Archimedes' bathtub "Eureka!," the neglected petri dish that grew *Penicillium* mold. Indeed, it's a common enough phenomenon that a word was invented to capture it: in 1754, Horace Walpole coined the term "serendipity," based on the fairy tale adventures of *The Three Princes of Serendip* (Serendip being the archaic name of Sri Lanka), who "were always making discoveries, by accidents and sagacity, of things they were not in quest of."

This double role of randomness—a key part of biology, a key part of discovery—has repeatedly caught the eye of psychologists who want to explain human creativity. An early instance of this idea was offered by William James. In 1880, having recently been appointed assistant professor of psychology at Harvard, and ten years away from publishing his definitive *Principles of Psychology*, James wrote an article in the *Atlantic Monthly* called "Great Men, Great Thoughts, and the Environment." The article opens with his thesis:

> A remarkable parallel, which to my knowledge has never been noticed, obtains between the facts of social evolution and the mental growth of the race, on the one hand, and of zoölogical evolution, as expounded by Mr. Darwin, on the other.

At the time James was writing, the idea of "zoölogical evolution" was still fresh—*On the Origin of Species* having been published in 1859 and Mr. Darwin himself still alive. James discussed how evolutionary ideas might be applied to different aspects of human society, and toward the end of the article turned to the evolution of ideas:

> New conceptions, emotions, and active tendencies which evolve are originally *produced* in the shape of random images, fancies, accidental outbirths of spontaneous variation in the functional activity of the excessively unstable human brain, which the outer environment simply confirms or refutes, adopts or rejects, preserves or destroys—*selects*, in short, just as it selects morphological and social variations due to molecular accidents of an analogous sort.

James thus viewed randomness as the heart of creativity. And he believed it was magnified in the most creative people. In their presence, he wrote, "we seem suddenly introduced into a seething caldron of ideas, where everything is fizzling and bobbing about in a state of bewildering activity, where partnerships can be joined or loosened in an instant, treadmill routine is unknown, and the unexpected seems the only law." (Note here the same "annealing" intuition, rooted in metaphors of temperature, where wild permutation equals heat.)

The modern instantiation of James's theory appears in the work of Donald Campbell, a psychologist who lived a hundred years later. In 1960, Campbell published a paper called "Blind Variation and Selective Retention in Creative Thought as in Other Knowledge Processes." Like James, he opened with his central thesis: "A blind-variation-and-selective-retention process is fundamental to all inductive achievements, to all genuine increases in knowledge, to all increases in fit of system to environment." And like James he was inspired by evolution, thinking about creative innovation as the outcome of new ideas being generated randomly and astute human minds retaining the best of those ideas. Campbell supported his argument liberally with quotes from other scientists and mathematicians about the processes behind their own discoveries. The nineteenth-century physicists and philosophers Ernst Mach and Henri Poincaré both seemed to offer an account similar to Campbell's, with Mach going so far as to declare that "thus are to be explained the statements of Newton, Mozart, Richard Wagner, and others, when they say that thought, melodies, and harmonies had poured in upon them, and that they had simply retained the right ones."

When it comes to stimulating creativity, a common technique is introducing a random element, such as a word that people have to form associations with. For example, musician Brian Eno and artist Peter Schmidt created a deck of cards known as Oblique Strategies for solving creative problems. Pick a card, any card, and you will get a random new perspective on your project. (And if that sounds like too much work, you can now download an app that will pick a card for you.) Eno's account of why they developed the cards has clear parallels with the idea of escaping local maxima:

> When you're very in the middle of something, you forget the most obvious things. You come out of the studio and you think "why didn't we remember

to do this or that?" These [cards] really are just ways of throwing you out of the frame, of breaking the context a little bit, so that you're not a band in a studio focused on one song, but you're people who are alive and in the world and aware of a lot of other things as well.

Being randomly jittered, thrown out of the frame and focused on a larger scale, provides a way to leave what might be locally good and get back to the pursuit of what might be globally optimal.

And you don't need to be Brian Eno to add a little random stimulation to your life. Wikipedia, for instance, offers a "Random article" link, and Tom has been using it as his browser's default homepage for several years, seeing a randomly selected Wikipedia entry each time he opens a new window. While this hasn't yet resulted in any striking discoveries, he now knows a lot about some obscure topics (such as the kind of knife used by the Chilean armed forces) and he feels that some of these have enriched his life. (For example, he's learned that there is a word in Portuguese for a "vague and constant desire for something that does not and probably cannot exist," a problem we still can't solve with a search engine.) An interesting side effect is that he now also has a better sense not just of what sorts of topics are covered on Wikipedia, but also of what randomness really looks like. For example, pages that feel like they have some connection to him—articles about people or places he knows—show up with what seems like surprising frequency. (In a test, he got "Members of the Western Australian Legislative Council, 1962–1965" after just two reloads, and he grew up in Western Australia.) Knowing that these are actually randomly generated makes it possible to become better calibrated for evaluating other "coincidences" in the rest of his life.

In the physical world, you can randomize your vegetables by joining a Community-Supported Agriculture farm, which will deliver a box of produce to you every week. As we saw earlier, a CSA subscription does potentially pose a scheduling problem, but being sent fruits and vegetables you wouldn't normally buy is a great way to get knocked out of a local maximum in your recipe rotation. Likewise, book-, wine-, and chocolate-of-the-month clubs are a way to get exposed to intellectual, oenophilic, and gustatory possibilities that you might never have encountered otherwise.

You might worry that making every decision by flipping a coin could lead to trouble, not least with your boss, friends, and family. And it's true that mainlining randomness into your life is not necessarily a recipe for success. The cult classic 1971 novel *The Dice Man* by Luke Rhinehart (real name: George Cockcroft) provides a cautionary tale. Its narrator, a man who replaces decision-making with dice rolling, quickly ends up in situations that most of us would probably like to avoid.

But perhaps it's just a case of a little knowledge being a dangerous thing. If the Dice Man had only had a deeper grasp of computer science, he'd have had some guidance. First, from Hill Climbing: even if you're in the habit of sometimes acting on bad ideas, you should *always* act on good ones. Second, from the Metropolis Algorithm: your likelihood of following a bad idea should be inversely proportional to how bad an idea it is. Third, from Simulated Annealing: you should front-load randomness, rapidly cooling out of a totally random state, using ever less and less randomness as time goes on, lingering longest as you approach freezing. Temper yourself—literally.

This last point wasn't lost on the novel's author. Cockcroft himself apparently turned, not unlike his protagonist, to "dicing" for a time in his life, living nomadically with his family on a Mediterranean sailboat, in a kind of Brownian slow motion. At some point, however, his annealing schedule cooled off: he settled down comfortably into a local maximum, on a lake in upstate New York. Now in his eighties, he's still contentedly there. "Once you got somewhere you were happy," he told the *Guardian*, "you'd be stupid to shake it up any further."

10 | Networking

How We Connect

The term connection *has a wide variety of meanings. It can refer to a physical or logical path between two entities, it can refer to the flow over the path, it can inferentially refer to an action associated with the setting up of a path, or it can refer to an association between two or more entities, with or without regard to any path between them.*

—VINT CERF AND BOB KAHN

Only connect.

—E. M. FORSTER

The long-distance telegraph began with a portent—Samuel F. B. Morse, standing in the chambers of the US Supreme Court on May 24, 1844, wiring his assistant Alfred Vail in Baltimore a verse from the Old Testament: "WHAT HATH GOD WROUGHT." The first thing we ask of any new connection is how it began, and from that origin can't help trying to augur its future.

The first telephone call in history, made by Alexander Graham Bell to his assistant on March 10, 1876, began with a bit of a paradox. "Mr. Watson, come here; I want to see you"—a simultaneous testament to its ability *and* inability to overcome physical distance.

The cell phone began with a boast—Motorola's Martin Cooper walking down Sixth Avenue on April 3, 1973, as Manhattan pedestrians gawked, calling his rival Joel Engel at AT&T: "Joel, I'm calling you from a cellular phone. A real cellular phone: a handheld, portable, real cellular phone." ("I don't remember exactly what he said," Cooper recalls, "but it was really quiet for a while. My assumption was that he was grinding his teeth.")

And the text message began, on December 3, 1992, with cheer: Neil Papworth at Sema Group Telecoms wishing Vodafone's Richard Jarvis an early "Merry Christmas."

The beginnings of the Internet were, somehow fittingly, much humbler and more inauspicious than all of that. It was October 29, 1969, and Charley Kline at UCLA sent to Bill Duvall at the Stanford Research Institute the first message ever transmitted from one computer to another via the ARPANET. The message was "login"—or would have been, had the receiving machine not crashed after "lo."

Lo—verily, Kline managed to sound portentous and Old Testament despite himself.

The foundation of human connection is *protocol*—a shared convention of procedures and expectations, from handshakes and hellos to etiquette, politesse, and the full gamut of social norms. Machine connection is no different. Protocol is how we get on the same page; in fact, the word is rooted in the Greek *protokollon*, "first glue," which referred to the outer page attached to a book or manuscript.

In interpersonal affairs, these protocols prove a subtle but perennial source of anxiety. I sent so-and-so a message however many days ago; at what point do I begin to suspect they never received it? It's now 12:05 p.m. and our call was set for noon; are we both expecting each other to be the one calling? Your answer seems odd; did I mishear you or did you mishear me? Come again?

Most of our communication technology—from the telegraph to the text—has merely provided us with new conduits to experience these familiar person-to-person challenges. But with the Internet, computers became not only the conduit but also the endpoints: the ones doing the talking. As such, they've needed to be responsible for solving their own communication issues. These machine-to-machine problems—and their solutions—at once mimic and illuminate our own.

Packet Switching

What we now think of as "the Internet" is actually a collection of many protocols, but the chief among them (so much so that it's often referred to more or less synonymously with the Internet) is what's known as Transmission Control Protocol, or TCP. It was born from a 1973 talk and a 1974 paper by Vinton "Vint" Cerf and Robert "Bob" Kahn, who laid out a proposal for the language of—as they imagined calling it—an "internetwork."

TCP initially used telephone lines, but it's more appropriately regarded as the evolution of the *mail* rather than the phone. Phone calls use what's called "circuit switching": the system opens a channel between the sender and the receiver, which supplies constant bandwidth between the parties in both directions as long as the call lasts. Circuit switching makes plenty of sense for human interaction, but as early as the 1960s it was clear that this paradigm wasn't going to work for machine communications.

As UCLA's Leonard Kleinrock recalls,

> I knew that computers, when they talk, they don't talk the way I am now—continuously. They go *blast!* and they're quiet for a while. A little while later, they suddenly come up and blast again. And you can't afford to dedicate a communications connection to something which is almost never talking, but when it wants to talk it wants immediate access. So we had to not use the telephone network, which was designed for continuous talking—the circuit switching network—but something else.

The telephone companies, for their part, did not seem especially amenable to talk of a fundamental shift in their protocols. Moving away from circuit switching was considered lunatic—"utter heresy," in the words of networking researcher Van Jacobson. Kleinrock reminisces about his own encounters with the telecommunications industry:

> I went to AT&T, the biggest network of the time, and I explained to them, you guys ought to give us good data communications. And their answer was, what are you talking about? The United States is a copper mine, it's full of telephone wires, use that. I said no, no, you don't understand. It takes 35 seconds to set up a call, you charge me a minimum of 3 minutes, and I want to send 100 milliseconds of data! And their answer was, "Little boy, go away." So little boy went away and, with others, developed this technology which ate their lunch.

The technology that ate circuit switching's lunch would become known as *packet* switching. In a packet-switched network, rather than using a dedicated channel for each connection, senders and receivers atomize their messages into tiny shards known as "packets," and merge them into the communal flow of data—a bit like postcards moving at the speed of light.

In such a network, "what you might call a connection is a consensual illusion between the two endpoints," explains Apple networking expert Stuart Cheshire. "There are no connections in the Internet. Talking about a connection in the Internet is like talking about a connection in the US Mail system. You write letters to people and each letter goes independently— and you may have a correspondence that goes back and forth and has some continuity to it, but the US Mail doesn't need to know about that. . . . They just deliver the letters."

Efficient use of bandwidth wasn't the only consideration driving research into packet switching in the 1960s; the other was nuclear war. Paul Baran at the RAND Corporation was trying to solve the problem of network robustness, so that military communications could survive a nuclear attack that took out a sizable fraction of the network. Inspired by algorithms developed in the 1950s for navigating mazes, Baran imagined a design in which every piece of information could independently make its own way to its destination, even as the network was changing dynamically— or being torn to tatters.

This was the second demerit against circuit switching and its dedicated, stable connections: that very stability meant that a dropped call stayed dropped. Circuit switching just wasn't flexible or adaptable enough to be robust. And here, too, packet switching could offer just what the times were calling for. In circuit-switched networks, a call fails if any one of its links gets disrupted—which means that reliability goes down exponentially as a network grows larger. In packet switching, on the other hand, the proliferation of paths in a growing network becomes a virtue: there are now that many more ways for data to flow, so the reliability of the network *increases* exponentially with its size.

Still, as Van Jacobson tells it, even after packet switching was devised, the phone companies were unimpressed. "All the telco people said, with very loud voices, that's not a network! That's just a crummy way to use *our* network! You're taking *our* wires, you're sending on the paths that *we* create! And you're putting a lot of extra gunk on it so that you use it really inefficiently." But from a packet-switching point of view, the phone wires are just a means to an end; the sender and receiver don't actually care how the packets get delivered. The ability to operate agnostically over any number of diverse media would be packet switching's great virtue. After early networks in the late '60s and early '70s, such as the ARPANET, proved the viability of the concept, networks of all types began sprouting across the country, doing packet switching not only over copper phone wires, but over satellites and over radio. In 2001, a group of computer scientists in the Norwegian city of Bergen briefly even implemented a packet-switching network over "Avian Carriers"—that is, packets written down on paper and tied to pigeons' feet.

Of course, packet switching would not be without its own problems. For starters, one of the first questions for any protocol, human or machine, is, quite simply: how do you know your messages are getting through?

Acknowledgment

No transmission can be 100 percent reliable.

—VINT CERF AND BOB KAHN

"WHAT HATH GOD WROUGHT" wasn't just the first long-distance telegraph message sent in the United States. It was also the second: Alfred Vail sent the quotation back to Morse in the Supreme Court chambers as a way of confirming receipt.

Now, Vail's reply could make Morse, and the US legislators gathered around him, confident that Morse's message had been received—presuming, of course, that Vail hadn't known the choice of message in advance. But what would make Vail confident that his *confirmation* had been received?

Computer scientists know this concept as the "Byzantine generals problem." Imagine two generals, on opposite sides of a valley that contains their common enemy, attempting to coordinate an attack. Only by perfect synchronization will they succeed; for either to attack alone is suicide. What's worse, any messages from one general to the other must be delivered by hand across the very terrain that contains the enemy, meaning there's a chance that any given message will never arrive.

The first general, say, suggests a time for the attack, but won't dare go for it unless he knows for sure that his comrade is moving, too. The second general receives the orders and sends back a confirmation—but won't dare attack unless he knows that the first general received that confirmation (since otherwise the first general won't be going). The first general receives the confirmation—but won't attack until he's certain that the second general *knows* he did. Following this chain of logic requires an infinite series of messages, and obviously that won't do. Communication is one of those delightful things that work only in practice; in theory it's impossible.

In most scenarios the consequences of communication lapses are rarely so dire, and the need for certainty rarely so absolute. In TCP, a failure generally leads to retransmission rather than death, so it's considered enough for a session to begin with what's called a "triple handshake." The visitor says hello, the server acknowledges the hello and says hello back, the visitor acknowledges that, and if the server receives this third message, then no further confirmation is needed and they're off to the races. Even after this initial connection is made, however, there's still a risk that some later packets may be damaged or lost in transit, or arrive out of order. In the postal mail, package delivery can be confirmed via return receipts; online, packet delivery is confirmed by what are called acknowledgment packets, or ACKs. These are critical to the functioning of the network.

The way that ACKs work is both simple and clever. Behind the scenes of the triple handshake, each machine provides the other with a kind of serial number—and it's understood that every packet sent after that will increment those serial numbers by one each time, like checks in a checkbook. For instance, if your computer initiates contact with a web server, it might send that server, say, the number 100. The ACK sent by the server will in turn specify the serial number at which the server's own packets will begin (call it 5,000), and also will say "Ready for 101." Your machine's ACK will carry the number 101 and will convey in turn "Ready for 5,001." (Note that these two numbering schemes are totally independent, and the number that begins each sequence is typically chosen at random.)

This mechanism offers a ready way to pinpoint when packets have gone astray. If the server is expecting 101 but instead gets 102, it will send an ACK to packet 102 that still says "Ready for 101." If it next gets packet 103, it will say, again, "Ready for 101." Three such redundant ACKs in a row would signal to your machine that 101 isn't just delayed but hopelessly gone, so it will resend that packet. At that point, the server (which has kept packets 102 and 103) will send an ACK saying "Ready for 104" to signal that the sequence has been restored.

All those acknowledgments can actually add up to a considerable amount of traffic. We think of, say, a large file transfer as a one-way operation, but in fact the recipient is sending hundreds of "control messages" back to the sender. A report from the second half of 2014 showed that almost 10% of upstream Internet traffic during peak hours was due to Netflix—which we tend to think of as sending data almost exclusively *downstream*, to users. But all that video generates an awful lot of ACKs.

In the human sphere, the anxiety that the message is indeed being received similarly pervades conversation. A speaker might subconsciously append "You know?" to the end of every sentence, and a listener, for their part, can't help but make a steady stream of nods, yeahs, aye-ayes, roger-thats, ten-fours, uh-huhs. We do this even face-to-face, but on a phone call sometimes it's the only way to know the call is even still in progress. No wonder that the most successful twenty-first-century marketing campaign for a wireless carrier featured a network engineer's quality-control catchphrase, repeated again and again: "Can you hear me now?"

When something goes wrong in that back-and-forth, we're often left with a question mark. As software blogger Tyler Treat says,

> In a distributed system, we try to guarantee the delivery of a message by waiting for an acknowledgement that it was received, but all sorts of things can go wrong. Did the message get dropped? Did the ack get dropped? Did the receiver crash? Are they just slow? Is the network slow? Am *I* slow?

The issues faced by the Byzantine generals, as he reminds us, "are not design complexities, they are *impossibility results*."

Earlier networking research, Vint Cerf notes, had been founded "on the assumption that you could build a reliable underlying net." On the other hand, "the Internet was based on the assumption that *no* network was necessarily reliable, and you had to do end-to-end retransmissions to recover."

Ironically, one of the few exceptions to this is in transmitting the human voice. Real-time voice communications, such as Skype, typically do *not* use TCP, which underlies most of the rest of the Internet. As researchers discovered in the early days of networking, using reliable, robust protocols—with all their ACKs and retransmission of lost packets—to transmit the human voice is overkill. The humans provide the robustness themselves. As Cerf explains, "In the case of voice, if you lose a packet, you just say, 'Say that again, I missed something.'"

For this reason, phone services that automatically reduce background noise to silence are doing their users a major disservice. Background static is a continual reassurance that the call is still connected and that any silence is a deliberate choice by the other party. Without it, one must constantly confront the possibility that the call has dropped, and constantly offer reassurances that it has not. This, too, is the anxiety of all packet-switching protocols, indeed of any medium rooted in asynchronous turn-taking—be it letter writing, texting, or the tentative back-and-forths of online dating. Every message could be the last, and there is often no telling the difference between someone taking their time to respond and someone who has long since ended the conversation.

So how exactly should we handle a person—or a computer—that's *unreliable*?

The first question is how long a period of nonresponsiveness we should take to constitute a breakdown. Partly this depends on the nature of the network: we start to worry in a matter of seconds over the phone, days over email, and weeks over postal mail. The longer the round-trip time between sender and receiver, the longer it takes a silence to be significant—and the more information can be potentially "in flight" before the sender realizes there's a problem. In networking, having the parties properly tune their expectations for the timeliness of acknowledgments is crucial to the system functioning correctly.

The second question, of course, once we do recognize a breakdown, is what exactly we should do about it.

Exponential Backoff: The Algorithm of Forgiveness

> *The world's most difficult word to translate has been identified as "ilunga," from the Tshiluba language spoken in south-eastern DR Congo. . . . Ilunga means "a person who is ready to forgive any abuse for the first time, to tolerate it a second time, but never a third time."*
>
> —BBC NEWS

> *If at first you don't succeed, / Try, try again.*
>
> —T. H. PALMER

Today we expect our devices to communicate wirelessly even when wires would be easy—our keyboard and mouse, for instance, talking wirelessly with a computer sitting inches away. But wireless networking began as a matter of necessity, in a place where no wires could do the job: Hawaii. In the late '60s and early '70s, Norman Abramson at the University of Hawaii in Honolulu was trying to link together the university's seven campuses and many research institutes, spread across four islands and hundreds of miles. He hit upon the idea of implementing packet switching via radio rather than the phone system, connecting the islands with a loose chain of transmitters and receivers. This system would come to be known as the ALOHAnet.

The biggest hurdle that the ALOHAnet had to overcome was interference. Sometimes two stations would transmit at the same moment, inadvertently jamming one another's signals. (This is, of course, a familiar feature in human conversation as well.) If both stations simply retransmitted right away to try to get their message across, they'd run the risk of getting stuck in perpetual interference forever. Clearly the ALOHAnet protocol was going to need to tell competing signals how to give each other space, how to yield and make way for one another.

The first thing that the senders need to do here is what's called "breaking symmetry." As any sidewalk pedestrian knows, dodging right as an oncoming walker dodges left, and then having both of you simultaneously dodge back the other way, doesn't solve anything. It's the same story when two speakers both pause, make gestures of deference to the other, and then start to speak again at the same time; or when two cars at an intersection, each having stopped to yield to the other, try to accelerate in sync. This is an area where the use of randomness becomes essential—indeed, networking wouldn't be possible without it.

One straightforward solution is to have each station flip a coin. Heads, it retransmits; tails, it waits a turn, then retransmits. Surely one of them will get through uncontested before long. This works well enough when there are only two senders. But what if there are three simultaneous signals? Or four? It would take a one-in-four chance for the network to get even a single packet through at that point (after which you'd still have three conflicting stations left, and perhaps even more competing signals arriving meanwhile). As the number of conflicts increases further, the network's throughput could simply fall off a cliff. A 1970 report on the ALOHAnet said that above a mere 18.6% average utilization of the airwaves, "the channel becomes unstable . . . and the average number of retransmissions becomes unbounded." Not good.

So, what to do? Is there a way to make a system that could avoid this fate?

The breakthrough turned out to be increasing the average delay after every successive failure—specifically, *doubling* the potential delay before trying to transmit again. So after an initial failure, a sender would randomly retransmit either one or two turns later; after a second failure, it would try again anywhere from one to four turns later; a third failure in a row would mean waiting somewhere between one and eight turns, and so on. This elegant approach allows the network to accommodate potentially *any* number of competing signals. Since the maximum delay length (2, 4, 8, 16 . . .) forms an exponential progression, it's become known as **Exponential Backoff**.

Exponential Backoff was a huge part of the successful functioning of the ALOHAnet beginning in 1971, and in the 1980s it was baked into TCP, becoming a critical part of the Internet. All these decades later, it still is. As one influential paper puts it, "For a transport endpoint embedded in a network of unknown topology and with an unknown, unknowable and constantly changing population of competing conversations, only one scheme has any hope of working—exponential backoff."

But it is the algorithm's other uses that suggest something both more prescriptive and more profound. Beyond just collision avoidance, Exponential Backoff has become the default way of handling almost all cases of networking failure or unreliability. For instance, when your computer is trying to reach a website that appears to be down, it uses Exponential Backoff—trying again one second later, again a few seconds after that, and so forth. This is good for everyone: it prevents a host server that's down from getting slammed with requests as soon as it comes back online, and it prevents your own machine from wasting too much effort trying to get blood from a stone. But interestingly, it also does not force (or allow) your machine to ever completely give up.

Exponential Backoff is also a critical part of networking security, when successive password failures in logging into an account are punished by an exponentially increasing lockout period. This prevents a hacker from using a "dictionary attack" against an account, cycling through potential password after password until eventually they get lucky. At the same time it also solves another problem: the account's real owner, no matter how forgetful, is never permanently locked out after some arbitrary cutoff.

In human society, we tend to adopt a policy of giving people some finite number of chances in a row, then giving up entirely. Three strikes, you're out. This pattern prevails by default in almost any situation that requires forgiveness, lenience, or perseverance. Simply put, maybe we're doing it wrong.

A friend of ours recently mused about a childhood companion who had a disconcerting habit of flaking on social plans. What to do? Deciding once and for all that she'd finally had enough and giving up entirely on the relationship seemed arbitrary and severe, but continuing to persist in perpetual rescheduling seemed naïve, liable to lead to an endless amount of disappointment and wasted time. Solution: Exponential Backoff on the invitation rate. Try to reschedule in a week, then two, then four, then eight. The rate of "retransmission" goes toward zero—yet you never have to completely give up.

Another friend of ours agonized about whether to offer shelter and financial assistance to a family member with a history of drug addiction. She couldn't bear to give up hope that he would turn things around, and couldn't bear the thought of turning her back on him for good. But she also couldn't bring herself to do all that it required to have him in her house—buying him clothes and cooking for him, reopening bank accounts for him, and driving him to work each morning—when at some mysterious and abrupt moment he would take all the money and disappear, only to call again several weeks later and ask to be forgiven and taken back in. It seemed like a paradox, a cruel and impossible choice.

Exponential Backoff isn't a magic panacea in cases like this, but it does offer a possible way forward. Requiring an exponentially increasing period of sobriety, for instance, would offer a disincentive to violate the house rules again. It would make the family member prove ever more assiduously that he was serious about returning, and would protect the host from the otherwise continuous stress of the cycle. Perhaps most importantly, the host would never have to tell her relative that she'd given up on him for good or that he was beyond redemption. It offers a way to have finite patience and infinite mercy. Maybe we don't have to choose.

In fact, the past decade has seen the beginnings of a quiet revolution in the way the justice system itself handles community supervision for drug offenders. That revolution is being spearheaded by a pilot program called

HOPE, which uses the Exponential Backoff principles of the ALOHAnet—and which, in a striking coincidence, began at the birthplace of the ALOHAnet itself: Honolulu.

Shortly after being sworn in to Hawaii's First Circuit Court, Judge Steven Alm noticed a remarkable pattern. Probationers would repeatedly violate their probation terms, and circuit judges would routinely use their discretion to let them off with a warning. But at some point, perhaps after a dozen or more violations, the judge would decide to be strict, and assign the violator a prison sentence measured in years. Says Alm, "I thought, what a crazy way to try to change anybody's behavior." So Alm proposed almost exactly the opposite. In place of violation hearings scheduled a long time into the future, requiring uncertain judgment calls, and occasionally producing huge penalties, HOPE is based on immediate, predefined punishments that begin with just one day in jail and increase after each incident. A five-year study by the Department of Justice reported that HOPE probationers were half as likely as regular probationers to be arrested for a new crime or have their probation revoked. And they were 72% less likely to use drugs. Seventeen states have followed Hawaii's lead and launched their own versions of HOPE.

Flow Control and Congestion Avoidance

The first efforts at computer networking focused on establishing reliable transmissions over unreliable links. These efforts proved to be so successful that a second concern immediately arose: making sure that an overloaded network could avoid catastrophic meltdown. No sooner had TCP solved the problem of getting data from point A to point B than it was confronted with the problem of gridlock.

The most significant early warning came in 1986, on a line connecting the Lawrence Berkeley Laboratory and the UC Berkeley campus, which are separated by about the length of a football field. (At Berkeley, the space happens to be filled with an actual football field.) One day, the bandwidth of that line dropped abruptly from its typical 32,000 bits per second to just 40 bits per second. The victims, Van Jacobson at LBL and Michael Karels at UCB, "were fascinated by this sudden factor-of-thousand drop in bandwidth and embarked on an investigation of why things had gotten so bad."

Meanwhile, they heard murmurings from other networking groups across the country who were running into the same thing. Jacobson began looking into the underlying code. "Is there some mistake in the protocol?" he wondered. "This thing was working on smaller-scale tests, and then it suddenly fell apart."

One of the biggest differences between circuit switching and packet switching emerges in how they deal with congestion. In circuit switching, the system either approves a channel request, or denies it outright if the request cannot be accommodated. That's why, if you've ever tried using a phone system during some peak time, you may have encountered the "special information tone" and message proclaiming that "all circuits are busy."

Packet switching is radically different. The phone system gets *full*; the mail system gets *slow*. There's nothing in the network to explicitly tell a sender how many other senders there are, or how congested the network is at any given moment, and the amount of congestion is constantly changing. Therefore, the sender and receiver must not only communicate but metacommunicate: they need to figure out how fast the data should be sent. Somehow, assorted packet flows—without explicit management or coordination—must both get out of each other's way and quickly take advantage of any newly available space.

The result of Jacobson and Karels's detective work was a revised set of flow control and congestion-avoidance algorithms—one of the biggest modifications to TCP in forty years.

At the heart of TCP congestion control is an algorithm called **Additive Increase, Multiplicative Decrease**, or AIMD. Before AIMD kicks in, a new connection will ramp up its transmission rate aggressively: if the first packet is received successfully it sends out two more, if both of those get through it sends out a batch of four, and so on. But as soon as any packet's ACK does not come back to the sender, the AIMD algorithm takes over. Under AIMD, any fully received batch of packets causes the number of packets in flight not to double but merely to increase by 1, and dropped packets cause the transmission rate to cut back by half (hence the name Additive Increase, Multiplicative Decrease). Essentially, AIMD takes the form of someone saying, "A little more, a little more, a little more, whoa, too much, cut way back, okay a little more, a little more . . ." Thus it leads to a characteristic bandwidth shape known as the "TCP sawtooth"—steady upward climbs punctuated by steep drops.

Why such a sharp, asymmetrical decrease? As Jacobson and Karels explain, the first time AIMD kicks in is when a connection has experienced the first dropped packet in its initial aggressive ramping-up phase. Because that initial phase involved doubling the rate of transmission with every successful volley, cutting the speed back by half as soon as there's been a problem is entirely appropriate. And once a transmission is in progress, if it starts to falter again that's likely to be because some new connection is competing for the network. The most conservative assessment of that situation—namely, assuming you were the only person using the network and now there's a second person taking half the resources—also leads to cutting back by half. Conservatism here is essential: a network can stabilize only if its users pull back at least as fast as the rate at which it is being overloaded. For the same reason, a merely additive increase helps stabilize things for everyone, preventing rapid overload-and-recovery cycles.

Though such a strict distinction between addition and multiplication is the kind of thing unlikely to be found in nature, the TCP sawtooth does find resonance in various domains where the idea is to take as much as one can safely get away with.

In a serendipitous 2012 collaboration, for instance, Stanford ecologist Deborah Gordon and computer scientist Balaji Prabhakar discovered that ants appear to have developed flow control algorithms millions of years before humans did. Like a computer network, an ant colony faces an allocation problem in trying to manage its "flow"—in this case, the flow of ants heading out to forage for food—under variable conditions that may sharply affect the rate at which the ants make successful round-trips. And like computers on the Internet, ants must solve this shared problem without the benefit of a central decision maker, instead developing what Gordon calls "control without hierarchy." It turns out the ants' solution is similar, too: a feedback cycle where successful foragers prompt more to leave the nest, while unsuccessful returnees result in a diminishment of foraging activity.

Other animal behavior also evokes TCP flow control, with its characteristic sawtooth. Squirrels and pigeons going after human food scraps will creep forward a step at a time, occasionally leap back, then steadily creep forward again. And it may be that human communications themselves mirror the very protocols that transmit them: every text message or email reply encourages yet another, while every unreturned message stanches the flow.

More broadly, AIMD suggests an approach to the many places in life where we struggle to allocate limited resources in uncertain and fluctuating conditions.

The satirical "Peter Principle," articulated in the 1960s by education professor Laurence J. Peter, states that "every employee tends to rise to his level of incompetence." The idea is that in a hierarchical organization, anyone doing a job proficiently will be rewarded with a promotion into a new job that may involve more complex and/or different challenges. When the employee finally reaches a role in which they *don't* perform well, their march up the ranks will stall, and they will remain in that role for the rest of their career. Thus it stands to reason, goes the ominous logic of the Peter Principle, that eventually every spot in an organization will come to be filled by someone doing that job badly. Some fifty years before Peter's formulation, Spanish philosopher José Ortega y Gasset in 1910 voiced the same sentiment. "Every public servant should be demoted to the immediately lower rank," he wrote, "because they were advanced until they became incompetent."

Some organizations have attempted to remediate the Peter Principle by simply firing employees who don't advance. The so-called Cravath System, devised by leading law firm Cravath, Swaine & Moore, involves hiring almost exclusively recent graduates, placing them into the bottom ranks, and then routinely either promoting or firing them over the following years. In 1980, the US Armed Forces adopted a similar "up or out" policy with the Defense Officer Personnel Management Act. The United Kingdom has likewise pursued what they call "manning control," to great controversy.

Is there any alternative, any middle path between the institutional stagnation of the Peter Principle and the draconian severity of the "up or out" system? The AIMD algorithm can offer just such an approach, since it is explicitly designed to handle the demands of a volatile environment. A computer network must manage its own maximum transmission capacity, plus the transmission rates of its clients, all of which may be fluctuating unpredictably. Likewise, in a business setting, a company has a limited pool of funds to pay for its operations, and each worker or vendor has a limited capacity for the amount of work they can do and the amount of responsibility they can handle. Everyone's needs, capacities, and partnerships are always in flux.

The lesson of the TCP sawtooth is that in an unpredictable and changing environment, pushing things to the point of failure is indeed sometimes the best (or the only) way to use all the resources to their fullest. What matters is making sure that the response to failure is both sharp and resilient. Under AIMD, every connection that isn't dropping the ball is accelerated until it is—and then it's cut in half, and immediately begins accelerating again. And though it would violate almost every norm of current corporate culture, one can imagine a corporation in which, annually, every employee is always either promoted a single step up the org chart or sent part of the way back down.

As Laurence J. Peter saw it, the insidious Peter Principle arises in corporations because of "the first commandment of hierarchical life: the hierarchy must be preserved." TCP, in contrast, teaches the virtues of flexibility. Companies speak of "flat" hierarchies and "tall" hierarchies, but they might consider speaking of *dynamic* ones. Under an AIMD system, no one is long anxious about being overtaxed, nor long resentful about a missed promotion; both are temporary and frequent correctives, and the system hovers near its equilibrium despite everything changing all the time. Perhaps one day we'll speak not of the arc of one's career, but rather of its sawtooth.

Backchannels: Flow Control in Linguistics

Looking into networking's flow control makes it clear that upstream ACK packets not only acknowledge and confirm transmissions, but shape the contours of the entire interaction, its pace and cadence. This offers us both a reminder and an insight into how important feedback is to communication. In TCP, as we've seen, there's no such thing as a one-way transmission: without consistent feedback, the sender will slow down almost immediately.

Curiously, the rising awareness of the critical role of feedback in the field of networking mirrored an almost identical set of developments going on around the same time in the linguistics community. In the middle of the twentieth century, linguistics was dominated by the theories of Noam Chomsky, which considered language in its most perfect and ideal state— perfectly fluent, grammatical, uninterrupted sentences, as if all communication were written text. But starting in the 1960s and '70s, a surge of interest in the practical aspects of spoken language revealed just how elaborate and subtle the processes are that govern turn-taking, interruption, and composing a sentence or story on the fly while being attuned to a listener's reactions every step of the way. What emerged was a vision of even ostensibly one-way communication as a collaborative act. As linguist Victor Yngve would write in 1970, "In fact, both the person who has the turn and his partner are simultaneously engaged in both speaking and listening. This is because of the existence of what I call the back channel, over which the person who has the turn receives short messages such as 'yes' and 'uh-huh' without relinquishing the turn."

An examination of human "backchannels" opened a whole new horizon for the field of linguistics, prompting a complete re-evaluation of the dynamics of communication—specifically, the role of the listener. In one illustrative study, a team led by Janet Bavelas at the University of Victoria investigated what would happen when someone listening to a personal story got distracted: not what would happen to the listener's comprehension, but what would happen to the *story*. With poor feedback, they discovered, the story falls apart.

> Narrators who told close-call stories to distracted listeners . . . told them less well overall and particularly poorly at what should have been the dramatic conclusion. Their story endings were abrupt or choppy, or they circled around and retold the ending more than once, and they often justified their story by explaining the obvious close call.

We've all had the experience of talking to someone whose eyes drifted away—to their phone, perhaps—making us wonder whether our lackluster storytelling was to blame. In fact, it's now clear that the cause and effect are often the reverse: a poor listener destroys the tale.

Understanding the exact function and meaning of human backchannels continues to be an active area of research. In 2014, for instance, UC Santa Cruz's Jackson Tolins and Jean Fox Tree demonstrated that those inconspicuous "uh-huhs" and "yeahs" and "hmms" and "ohs" that pepper our speech perform distinct, precise roles in regulating the flow of information from speaker to listener—both its rate and level of detail. Indeed, they are every bit as critical as ACKs are in TCP. Says Tolins, "Really, while some people may be worse than others, 'bad storytellers' can at least partly blame their audience." This realization has had the unexpected side effect of taking off some of the pressure when he gives lectures—including, of course, lectures about that very result. "Whenever I give these backchannel talks, I always tell the audience that the way they are backchanneling to my talk right now is changing what I say," he jokes, "so they're responsible for how well I do."

Bufferbloat: It's the Latency, Stupid

Developing effective active queue management has been hampered by misconceptions about the cause and meaning of queues.
—KATHLEEN NICHOLS AND VAN JACOBSON

It was the summer of 2010, and like many parents, Jim Gettys was fielding frequent complaints from his children that the family wi-fi network was running slowly. Unlike most parents, though, Gettys has worked at HP, Alcatel-Lucent, the World Wide Web Consortium, and the Internet Engineering Task Force. He was literally the editor, in 1999, of the HTTP specification still in use today. So where most geek dads would look into the problem, Gettys *looked into the problem*.

As Gettys would explain to a roomful of Google engineers, with networking jargon giving way to an urgent and unmistakable conviction:

> I happened to be copying, or rsyncing, the old X Consortium archives from my house to MIT over this ten-millisecond-long path. . . . SmokePing [was] reporting latencies averaging well over one second, along with bad packet loss, just while copying a file. . . . I took Wireshark, and there were these *bursts* of really strange behavior. . . . This looked like no TCP [sawtooth] I expected at all. It should never occur that way.

In plain English, he saw something . . . very weird. As the saying goes, "the most exciting phrase to hear in science, the one that heralds new discoveries, is not 'Eureka!' but 'That's funny.'"

At first Gettys thought that something was wrong with his cable modem. What his family had been calling a problem in the Internet seemed like a traffic jam at their own wall socket. Packets meant for Boston weren't getting stuck midway there; they were getting stuck in the house.

But the deeper Gettys looked into it, the more concerned he grew. The problem didn't affect just his home router and modem, but *every* home router and modem. And the problem wasn't just in networking devices—it was in computers themselves, in desktops, laptops, tablets, and smartphones, woven into Linux, Windows, and OS X. And it wasn't just in end-user hardware, either: it touched the very infrastructure of the Internet itself. Gettys sat down to lunches with key players at Comcast, Verizon, Cisco, and Google, including Van Jacobson and Vint Cerf, and slowly started to piece the puzzle together.

The problem was everywhere. And the problem was bufferbloat.

A buffer is essentially a queue whose function is to smooth out bursts. If you walked into a doughnut shop at roughly the same time as another customer, it wouldn't do for the very momentarily overwhelmed cashier to make one of you *leave the store* and come back another time. Customers wouldn't have it, of course, but neither would management: such a policy is virtually guaranteed to underutilize the cashier. Putting the customers in a queue instead ensures that the *average* throughput of the store approaches its *maximum* throughput. That's a good thing.

This superior resource utilization comes with a very real cost, however: delay. When Tom took his daughter to a Cinco de Mayo festival in Berkeley, she set her heart on a chocolate banana crêpe, so they got in line and waited. Eventually—after twenty minutes—Tom got to the front of the line and placed his order. But after paying, they then had to wait *forty* more minutes to actually get the crêpe. (Like Jim Gettys, Tom quickly found himself fielding a substantial volume of familial complaints.) Taking orders turned out to take less time than making crêpes, so the queue to order was just the first part of the problem. At least it was visible, though; customers knew what they were in for. The second, longer queue was invisible. So in this case it would have been a much happier outcome for all if the crêpe stand had just cut off the line at some point and put up a sign that they weren't taking orders for a bit. Turning customers away would have made everyone better off—whether they ended up in a shorter crêpe line or went elsewhere. And wouldn't have cost the crêpe stand a dime of lost sales, because either way they can only sell as many crêpes as they can make in a day, regardless of how long their customers are waiting.

This is precisely the phenomenon that Jim Gettys was observing in his home cable modem. Because he was uploading a file, his computer was sending the modem as many upstream packets as it could handle. And the modem was pretending to handle a lot more than it actually could, turning none away while building up a massive queue. So when Gettys tried to download something at the same time—to visit a webpage or check email—his ACK packets would get stuck behind the upload, having to wait in line at the modem to leave the house. Because his ACKs then took forever to return to the web and email servers, the servers would in turn throttle their own downstream connection speeds to a corresponding crawl.

It was like trying to have a conversation where every time you say "uh-huh" it is delayed by ten or twenty seconds. The speaker is going to slow way down, assuming you aren't comprehending them, and there's nothing you can do about it.

When a networking buffer fills up, what typically happens is called **Tail Drop**: an unceremonious way of saying that every packet arriving after that point is simply rejected, and effectively deleted. (Turning new customers away from the crêpe stand once the line gets too long would be a version of Tail Drop in a human context.) Given the postal metaphor for packet switching, it might seem a bit odd to imagine a mail carrier who simply vaporizes every parcel that doesn't fit onto the truck that morning. Yet it's precisely such "packet drops" that lead a computer to notice that one of its packets hasn't been acknowledged, prompting AIMD to start halving the bandwidth. Dropped packets are the Internet's primary feedback mechanism. A buffer that's too large—a restaurant taking every order no matter how short-staffed the kitchen, a modem taking every packet that comes in regardless of how long it'll take to send them on—prevents this moderation from happening as it should.

Fundamentally, buffers use delay—known in networking as "latency"—in order to maximize throughput. That is, they cause packets (or customers) to wait, to take advantage of later periods when things are slow. But a buffer that's operating permanently full gives you the worst of both worlds: all the latency and none of the give. Smoothing out bursts is great if you are, *on average*, clearing things at least as quickly as they're arriving—but if your average workload exceeds your average work rate, no buffer can work miracles. And the bigger the buffer is, the further behind you'll get before you start signaling for help. One of the fundamental principles of buffers, be they for packets or patrons, is that they only work correctly when they are routinely zeroed out.

For decades, computer memory was sufficiently expensive that there was simply no reason to build modems with oodles of unnecessary memory capacity. Thus, there had simply been no way for a modem to build up a queue bigger than it could handle. But at some point, as economies of scale in the computer industry radically lowered the cost of memory, modem manufacturers started giving their machines gigabytes of RAM because that was effectively the smallest amount of RAM they could get. As a result, the ubiquitous device buffers—in modems, routers, laptops, smartphones, and in the backbone of the Internet itself—became *thousands* of times too big, before people like Jim Gettys sounded the alarm to do something about it.

Better Never than Late

> Take your most basic problem as a single person . . . someone likes you, you don't like them back. At one point, that used to be kind of an awkward situation. You had to have a conversation, it was weird. Now what do you do? Someone likes you, you don't like them back? You just pretend to be busy . . . forever.
>
> —AZIZ ANSARI

> Now is better than never.
> Although never is often better than right now.
>
> —THE ZEN OF PYTHON

Singer Katy Perry has 107% more Twitter followers than her home state of California has people. The most-followed person on Twitter, as of early 2016 she counts some 81.2 million accounts among her fans. This means that even if 99% of her fans never message her at all—and even if that most devoted 1% who message her do so only once per year—then she still gets 2,225 messages a day. Every single day.

Imagine if Perry were committed to answering each fan message in the order received. If she could reply to 100 a day, then the fans' expected wait time for a response would soon be measured in *decades*. It's fair to imagine that most fans would prefer a slim chance of getting a reply right away to a guaranteed reply ten or twenty years hence.

Note that Perry doesn't have this problem when she leaves a venue and must run a gauntlet of fans expecting an autograph or a few words. Perry does what she can, moves on, and the lost opportunities dissipate. The body is its own flow control. We can't be in more than one place at one time. At a crowded party we inevitably participate in less than 5% of the conversation, and cannot read up or catch up on the remainder. Photons that miss the retina aren't queued for later viewing. In real life, packet loss is almost total.

We use the idiom of "dropped balls" almost exclusively in a derogatory sense, implying that the person in question was lazy, complacent, or forgetful. But the tactical dropping of balls is a critical part of getting things done under overload.

The most prevalent critique of modern communications is that we are "always connected." But the problem isn't that we're always connected; we're not. The problem is that we're always *buffered*. The difference is enormous.

The feeling that one needs to look at everything on the Internet, or read all possible books, or see all possible shows, is bufferbloat. You miss an episode of your favorite series and watch it an hour, a day, a decade later. You go on vacation and come home to a mountain of correspondence. It used to be that people knocked on your door, got no response, and went away. Now they're effectively waiting in line when you come home.

Heck, email was deliberately designed to overcome Tail Drop. As its inventor, Ray Tomlinson, puts it:

> At the time there was no really good way to leave messages for people. The telephone worked up to a point, but someone had to be there to receive the call. And if it wasn't the person you wanted to get, it was an administrative assistant or an answering service or something of that sort. That was the mechanism you had to go through to leave a message, so everyone latched onto the idea that you could leave messages on the computer.

In other words, we asked for a system that would never turn a sender away, and for better or worse we got one. Indeed, over the past fifteen years, the move from circuit switching to packet switching has played itself out across society. We used to request dedicated circuits with others; now we send them packets and wait expectantly for ACKs. We used to *reject*; now we *defer*.

The much-lamented "lack of idleness" one reads about is, perversely, the primary *feature* of buffers: to bring average throughput up to peak throughput. Preventing idleness is what they do. You check email from the road, from vacation, on the toilet, in the middle of the night. You are never, ever bored. This is the mixed blessing of buffers, operating as advertised.

Vacation email autoresponders explicitly tell senders to expect latency; a better one might instead tell senders to expect Tail Drop. Rather than warning senders of above-average queue times, it might warn them that it was simply rejecting all incoming messages. And this doesn't need to be limited to vacations: one can imagine an email program set to auto-reject all incoming messages once the inbox reached, say, a hundred items. This is ill-advised for bills and the like, but not an unreasonable approach to, say, social invitations.

The idea of encountering a "full" inbox or "full" voicemail is an anachronism now, a glaring throwback to the late twentieth century and the early 2000s. But if the networks that connect our newfangled phones and computers, with their effectively infinite storage, are still deliberately dropping packets when things get fast and furious, then maybe there's reason to think of Tail Drop not as the lamentable consequence of limited memory space but as a purposeful strategy in its own right.

As for network bufferbloat, the ongoing story is a complicated but happy one, involving large-scale efforts by hardware and operating system manufacturers to make fundamental changes to network queues. There's also a proposal for a new backchannel for TCP, the first such modification in many years: Explicit Congestion Notification, or ECN. Fully extricating the Internet from bufferbloat will draw on all of these changes and require the patience of many years. "This is a long-term swamp," says Gettys.

But there's a lot to look forward to about a post-bufferbloat future. With their inherent latency, buffers are bad for most interactive processes. When we speak via Skype, for example, we generally prefer an occasionally staticky signal now to a clear recording of what our caller said three seconds ago. For gamers, even a 50-millisecond lag could be the difference between fragging and being fragged; in fact, gaming is so sensitive to latency that all important gaming honors are still contested in person, with players boarding airplanes to gather and compete over a network serving just a single room. And much the same is true for anything else where being in sync matters. "If you want to play music with your friends, even in [your] metropolitan area, you care about tens of milliseconds," Gettys notes, imagining a whole host of new applications and businesses that might spring forth to take advantage of the interactive potential of low latencies. "A generalization I take away from this whole experience is that engineers should think about time as a first-class citizen."

Apple's Stuart Cheshire concurs that it's high time for latency to become a top priority for network engineers. He's appalled that companies who advertise "fast" Internet connections refer only to high bandwidth, not to low delay. By analogy, he notes that a Boeing 737 and a Boeing 747 both fly at about five hundred miles per hour; the former can hold 120 passengers, while the latter carries three times as many. So "would you say that a Boeing 747 is three times 'faster' than a Boeing 737? Of course not," Cheshire exclaims. Capacity does matter sometimes: for transferring large files, bandwidth is key. (If you've got a huge amount of cargo to move, a container ship may well trump thousands of trips by a 747.) For interhuman applications, however, a quick turnaround time is often far more important, and what we really need are more Concordes. And indeed, bringing latencies down is one of the current frontiers of networking research, and it will be interesting to see what that brings.

Meanwhile, there are other battles to be waged. Gettys snaps his attention away for a second, looking out of the frame. "It's not working for you? I'm talking to someone at the moment, and I'll deal with it when I'm finished. We're wrapping up here—uh, no, the 5 GHz is working at the moment, the 2.4 GHz channel has hung. It's the infamous bug. I'll reboot the router." Which seems an opportune moment to say our good-byes and release our bandwidth to the commons, to the myriad flows making their additive increase.

Game Theory

The Minds of Others

I'm an optimist in the sense that I believe humans are noble and honorable,
and some of them are really smart. . . . I have a somewhat more pessimistic
view of people in groups.

—STEVE JOBS

An investor sells a stock to another, one convinced it's headed down and
the other convinced it's going up; I think I know what you think but have
no idea what you think I think; an economic bubble bursts; a prospective
lover offers a gift that says neither "I want to be more than friends" nor "I
don't want to be more than friends"; a table of diners squabbles over who
should treat whom and why; someone trying to be helpful unintentionally
offends; someone trying hard to be cool draws snickers; someone trying
to break from the herd finds, dismayingly, the herd following his lead. "I
love you," says one lover to another; "I love you, too," the other replies; and
both wonder what exactly the other means by that.

What does computer science have to say about all this?

Schoolchildren are taught to conceive of literary plots as belonging to
one of several categories: man vs. nature, man vs. self, man vs. man, man
vs. society. Thus far in this book we have considered primarily cases in
the first two categories—that is to say, computer science has thus far been
our guide to problems created by the fundamental structure of the world,
and by our limited capacities for processing information. Optimal stopping problems spring from the irreversibility and irrevocability of time; the

explore/exploit dilemma, from time's limited supply. Relaxation and randomization emerge as vital and necessary strategies for dealing with the ineluctable complexity of challenges like trip planning and vaccinations.

In this chapter we shift the focus and consider the remaining two genres—that is, man vs. man and man vs. society: in effect, the problems that we pose and cause each other. Our best guide to this terrain comes from a branch of mathematics known as game theory, a field that in its classical incarnation had an enormous impact on the twentieth century. In the past couple of decades, cross-pollination between game theory and computer science has produced the field of *algorithmic* game theory— which has already begun to have an impact on the twenty-first.

Recursion

> Now, a clever man would put the poison into his own goblet because he would know that only a great fool would reach for what he was given. I am not a great fool, so I can clearly not choose the wine in front of you. But you must have known I was not a great fool—you would have counted on it— so I can clearly not choose the wine in front of me.
>
> —THE PRINCESS BRIDE

Arguably the most influential economist of the twentieth century, John Maynard Keynes, once said that "successful investing is anticipating the anticipations of others." For a share of stock to be sold at, say, $60, the buyer must believe he can sell it later for $70—to someone who believes he can sell it for $80 to someone who believes he can sell it for $90 to someone who believes he can sell it for $100 to someone else. In this way, the value of a stock isn't what people think it's worth but what people *think* people think it's worth. In fact, even that's not going far enough. As Keynes put it, making a crucial distinction between beauty and popularity:

> Professional investment may be likened to those newspaper competitions in which the competitors have to pick out the six prettiest faces from a hundred photographs, the prize being awarded to the competitor whose choice most nearly corresponds to the average preferences of the competitors as a whole; so that each competitor has to pick, not those faces which he himself finds prettiest, but those which he thinks likeliest to catch the fancy of the other competitors, all of whom are looking at the problem from the

same point of view. It is not a case of choosing those which, to the best of one's judgment, are really the prettiest, nor even those which average opinion genuinely thinks the prettiest. We have reached the third degree where we devote our intelligences to anticipating what average opinion expects the average opinion to be. And there are some, I believe who practice the fourth, fifth, and higher degrees.

Computer science illustrates the fundamental limitations of this kind of reasoning with what's called the "halting problem." As Alan Turing proved in 1936, a computer program can never tell you for sure whether another program might end up calculating forever without end—except by simulating the operation of that program and thus potentially going off the deep end itself. (Accordingly, programmers will never have automated tools that can tell them whether their software will freeze.) This is one of the foundational results in all of computer science, on which many other proofs hang.* Simply put, any time a system—be it a machine or a mind—simulates the workings of something as complex as itself, it finds its resources totally maxed out, more or less by definition. Computer scientists have a term for this potentially endless journey into the hall of mirrors, minds simulating minds simulating minds: "recursion."

"In poker, you never play your hand," James Bond says in *Casino Royale*; "you play the man across from you." In fact, what you really play is a theoretically infinite recursion. There's your own hand and the hand you believe your opponent to have; then the hand you believe your opponent believes you have, and the hand you believe your opponent believes you to believe *he* has . . . and on it goes. "I don't know if this is an actual game-theory term," says the world's top-rated poker player, Dan Smith, "but poker players call it 'leveling.' Level one is 'I know.' Two is 'you know that I know.' Three, 'I know that you know that I know.' There are situations where it just comes up where you are like, 'Wow, this is a really silly spot to bluff but if he knows that it is a silly spot to bluff then he won't call me and that's where it's the clever spot to bluff.' Those things happen."

One of the most memorable bluffs in high-level poker occurred when Tom Dwan wagered $479,500 on Texas Hold 'Em's absolute worst possible hand, the 2–7—while literally telling his opponent, Sammy George, that he

---

*Indeed, it's the origin of all modern computers—it was the halting problem that inspired Turing to formally define computation, via what we now call the Turing machine.

was holding it. "You don't have deuce-seven," George replied. "You don't have deuce-seven." George folded, and Dwan—with, yes, deuce-seven—took the pot.

In poker, recursion is a dangerous game. You don't want to get caught one step behind your opponent, of course—but there's also an imperative not to get too far ahead of them either. "There's a rule that you really only want to play one level above your opponent," explains poker professional Vanessa Rousso. "If you play too far above your opponent, you're going to think they have information that they don't actually have—[and] they won't be able to glean the information that you want them to glean from your actions." Sometimes poker pros will deliberately bait their opponent into a convoluted recursion, meanwhile playing completely by-the-book, unpsychological poker themselves. This is known as luring them into "a leveling war against themselves."

(Luring an opponent into fruitless recursion can be an effective strategy in other games, too. One of the most colorful, bizarre, and fascinating episodes in the history of man-vs.-machine chess came in a 2008 blitz showdown between American grandmaster Hikaru Nakamura and leading computer chess program Rybka. In a game where each side got just three minutes on the clock to play all of their moves or automatically lose, the advantage surely seemed to be on the side of the computer—capable of evaluating millions of positions every second, and of making its move without twitching a muscle. But Nakamura immediately gridlocked the board, and proceeded to make repetitive, meaningless moves as fast as he could click. Meanwhile, the computer wasted precious moments fruitlessly searching for winning variations that didn't exist and doggedly trying to anticipate all the possible future moves by Nakamura, who himself was simply doing the chess equivalent of twiddling his thumbs. When the computer had nearly exhausted its time and began flailing so as not to lose by the clock, Nakamura finally opened the position and crashed through.)

Given recursion's dangers, how do poker professionals break out of it? They use game theory. "Sometimes you can come up with reasons to make exploitive [leveling] plays, but a lot of the time you are just making inferior plays for reasons that are really just noise," Dan Smith explains. "I try really hard to have a base level of theory understanding in most situations. . . . I always start by knowing or trying to know what Nash is."

So what *is* Nash?

Reaching Equilibrium

You know the rules, and so do I. . . .
We know the game and we're gonna play it.

—RICK ASTLEY

Game theory covers an incredibly broad spectrum of scenarios of cooperation and competition, but the field began with those resembling heads-up poker: two-person contests where one player's gain is another player's loss. Mathematicians analyzing these games seek to identify a so-called *equilibrium*: that is, a set of strategies that both players can follow such that neither player would want to change their own play, given the play of their opponent. It's called an equilibrium because it's stable—no amount of further reflection by either player will bring them to different choices. I'm content with my strategy, given yours, and you're content with your strategy, given mine.

In rock-paper-scissors, for example, the equilibrium tells us, perhaps unexcitingly, to choose one of the eponymous hand gestures completely at random, each roughly a third of the time. What makes this equilibrium stable is that, once both players adopt this $\frac{1}{3}$-$\frac{1}{3}$-$\frac{1}{3}$ strategy, there is nothing better for either to do than stick with it. (If we tried playing, say, more rock, our opponent would quickly notice and start playing more paper, which would make us play more scissors, and so forth until we both settled into the $\frac{1}{3}$-$\frac{1}{3}$-$\frac{1}{3}$ equilibrium again.)

In one of the seminal results in game theory, the mathematician John Nash proved in 1951 that *every* two-player game has at least one equilibrium. This major discovery would earn Nash the Nobel Prize in Economics in 1994 (and lead to the book and film *A Beautiful Mind*, about Nash's life). Such an equilibrium is now often spoken of as the "Nash equilibrium"—the "Nash" that Dan Smith always tries to keep track of.

On the face of it, the fact that a Nash equilibrium always exists in two-player games would seem to bring us some relief from the hall-of-mirrors recursions that characterize poker and many other familiar contests. When we feel ourselves falling down the recursive rabbit hole, we always have an option to step out of our opponent's head and look for the equilibrium, going directly to the best strategy, assuming rational play. In rock-paper-scissors, scrutinizing your opponent's face for signs of what they might throw next may not be worthwhile, if you know that simply throwing at random is an unbeatable strategy in the long run.

More generally, the Nash equilibrium offers a prediction of the stable long-term outcome of any set of rules or incentives. As such, it provides an invaluable tool for both predicting and shaping economic policy, as well as social policy in general. As Nobel laureate economist Roger Myerson puts it, the Nash equilibrium "has had a fundamental and pervasive impact in economics and the social sciences which is comparable to that of the discovery of the DNA double helix in the biological sciences."

Computer science, however, has complicated this story. Put broadly, the object of study in mathematics is *truth*; the object of study in computer science is *complexity*. As we've seen, it's not enough for a problem to have a solution if that problem is intractable.

In a game-theory context, knowing that an equilibrium exists doesn't actually tell us what it is—or how to get there. As UC Berkeley computer scientist Christos Papadimitriou writes, game theory "predicts the agents' equilibrium behavior typically with no regard to the ways in which such a state will be reached—a consideration that would be a computer scientist's foremost concern." Stanford's Tim Roughgarden echoes the sentiment of being unsatisfied with Nash's proof that equilibria always exist. "Okay," he says, "but we're computer scientists, right? Give us something we can use. Don't just tell me that it's there; tell me how to find it." And so, the original field of game theory begat algorithmic game theory—that is, the study of theoretically ideal strategies for games became the study of how machines (and people) *come up* with strategies for games.

As it turns out, asking too many questions about Nash equilibria gets you into computational trouble in a hurry. By the end of the twentieth century, determining whether a game has more than one equilibrium, or an equilibrium that gives a player a certain payoff, or an equilibrium that involves taking a particular action, had all been proved to be intractable problems. Then, from 2005 to 2008, Papadimitriou and his colleagues proved that simply *finding* Nash equilibria is intractable as well.

Simple games like rock-paper-scissors may have equilibria visible at a glance, but in games of real-world complexity it's now clear we cannot take for granted that the participants will be able to discover or reach the game's equilibrium. This, in turn, means that the game's designers can't necessarily use the equilibrium to predict how the players will behave. The ramifications of this sobering result are profound: Nash equilibria have held a hallowed place within economic theory as a way to model and predict market behavior, but that place might not be deserved. As Papadimitriou explains, "If an equilibrium concept is not efficiently computable, much of its credibility as a prediction of the behavior of rational agents is lost." MIT's Scott Aaronson agrees. "In my opinion," he says, "if the theorem that Nash equilibria exist is considered relevant to debates about (say) free markets versus government intervention, then the theorem that finding those equilibria is [intractable] should be considered relevant also." The predictive abilities of Nash equilibria only matter if those equilibria can actually be found by the players. To quote eBay's former director of research, Kamal Jain, "If your laptop cannot find it, neither can the market."

Dominant Strategies, for Better or Worse

Even when we can reach an equilibrium, just because it's *stable* doesn't make it *good*. It may seem paradoxical, but the equilibrium strategy— where neither player is willing to change tack—is by no means necessarily the strategy that leads to the best outcomes for the players. Nowhere is that better illustrated than in game theory's most famous, provocative, and controversial two-player game: "the prisoner's dilemma."

The prisoner's dilemma works as follows. Imagine that you and a co-conspirator have been arrested after robbing a bank, and are being held in separate jail cells. Now you must decide whether to "cooperate" with each other—by remaining silent and admitting nothing—or to "defect" from your partnership by ratting out the other to the police. You know that if you both cooperate with each other and keep silent, the state doesn't have enough evidence to convict either one of you, so you'll both walk free, splitting the loot—half a million dollars each, let's say. If one of you defects and informs on the other, and the other says nothing, the informer goes free and gets the entire million dollars, while the silent one is convicted as the sole perpetrator of the crime and receives a ten-year sentence. If you both inform on each other, then you'll share the blame and split the sentence: five years each.

Here's the problem. No matter what your accomplice does, it's always better for you to defect.

If your accomplice has ratted you out, ratting them out in turn will give you five years of your life back—you'll get the shared sentence (five years) rather than serving the whole thing yourself (ten years). And if your accomplice has stayed quiet, turning them in will net you the full million dollars—you won't have to split it. No matter what, you're always better off defecting than cooperating, regardless of what your accomplice decides. To do otherwise will always make you worse off, no matter what.

In fact, this makes defection not merely the equilibrium strategy but what's known as a *dominant* strategy. A dominant strategy avoids recursion altogether, by being the best response to all of your opponent's possible strategies—so you don't even need to trouble yourself getting inside their head at all. A dominant strategy is a powerful thing.

But now we've arrived at the paradox. If everyone does the rational thing and follows the dominant strategy, the story ends with both of you serving five years of hard time—which, compared to freedom and a cool half million apiece, is dramatically worse for *everyone* involved. How could that have happened?

This has emerged as one of the major insights of traditional game theory: the equilibrium for a set of players, all acting rationally in their own interest, may not be the outcome that is actually best for those players.

Algorithmic game theory, in keeping with the principles of computer science, has taken this insight and quantified it, creating a measure called "the price of anarchy." The price of anarchy measures the gap between cooperation (a centrally designed or coordinated solution) and competition (where each participant is independently trying to maximize the outcome for themselves). In a game like the prisoner's dilemma, this price is effectively infinite: increasing the amount of cash at stake and lengthening the jail sentences can make the gap between possible outcomes arbitrarily wide, even as the dominant strategy stays the same. There's no limit to how painful things can get for the players if they don't coordinate. But in other games, as algorithmic game theorists would discover, the price of anarchy is not nearly so bad.

For instance, consider traffic. Whether it's individual commuters trying to make their way through the daily bumper-to-bumper, or routers shuffling TCP packets across the Internet, everyone in the system merely wants what's easiest for them personally. Drivers just want to take the fastest route, whatever it is, and routers just want to shuffle along their packets with minimal effort—but in both cases this can result in overcrowding along critical pathways, creating congestion that harms everyone. How much harm, though? Surprisingly, Tim Roughgarden and Cornell's Éva Tardos proved in 2002 that the "selfish routing" approach has a price of anarchy that's a mere 4/3. That is, a free-for-all is only 33% worse than perfect top-down coordination.

Roughgarden and Tardos's work has deep implications both for urban planning of physical traffic and for network infrastructure. Selfish routing's low price of anarchy may explain, for instance, why the Internet works as well as it does without any central authority managing the routing of individual packets. Even if such coordination were possible, it wouldn't add very much.

When it comes to traffic of the human kind, the low price of anarchy cuts both ways. The good news is that the lack of centralized coordination is making your commute at most only 33% worse. On the other hand, if you're hoping that networked, self-driving autonomous cars will bring us a future of traffic utopia, it may be disheartening to learn that today's selfish, uncoordinated drivers are already pretty close to optimal. It's true that self-driving cars should reduce the number of road accidents and may be able to drive more closely together, both of which would speed up traffic. But from a congestion standpoint, the fact that anarchy is only 4/3 as congested as perfect coordination means that perfectly coordinated commutes will only be 3/4 as congested as they are now. It's a bit like the famous line by James Branch Cabell: "The optimist proclaims that we live in the best of all possible worlds; and the pessimist fears this is true." Congestion will always be a problem solvable more by planners and by overall demand than by the decisions of individual drivers, human or computer, selfish or cooperative.

Quantifying the price of anarchy has given the field a concrete and rigorous way to assess the pros and cons of decentralized systems, which has broad implications across any number of domains where people find themselves involved in game-playing (whether they know it or not). A low price of anarchy means the system is, for better or worse, about as good on its own as it would be if it were carefully managed. A high price of anarchy, on the other hand, means that things have the potential to turn out fine if they're carefully coordinated—but that without some form of intervention, we are courting disaster. The prisoner's dilemma is clearly of this latter type. Unfortunately, so are many of the most critical games the world must play.

The Tragedy of the Commons

In 1968, the ecologist Garrett Hardin took the two-player prisoner's dilemma and imagined scaling it up to involve all the members of a farming village. Hardin invited his readers to picture a "commons" of public lawn—available to be grazed by everyone's livestock, but with finite capacity. In theory, all the villagers should graze only as many animals as would leave some grass for everyone. In practice, though, the benefits of grazing a little bit more than that accrue directly to you, while the harms seem too small to be of consequence. Yet if everyone follows this logic of using just slightly more of the commons than they should, a dreadful equilibrium results: a completely devastated lawn, and no grass for *anyone's* livestock thereafter.

Hardin called this the "tragedy of the commons," and it has become one of the primary lenses through which economists, political scientists, and the environmental movement view large-scale ecological crises like pollution and climate change. "When I was a kid, there was this thing called leaded gasoline," says Avrim Blum, Carnegie Mellon computer scientist and game theorist. "Leaded was ten cents cheaper or something, but it pollutes the environment. . . . Given what everyone else is doing, how much worse really are you personally [health-wise] if you put leaded gasoline in your own car? Not that much worse. It's the prisoner's dilemma." The same is true at the corporate and national levels. A recent newspaper headline put the trouble succinctly: "Stable climate demands most fossil fuels stay in the ground, but whose?" Every corporation (and, to some degree, every nation) is better off being a bit more reckless than their peers for the sake of competitive advantage. Yet if they *all* act more recklessly, it leads to a ravaged Earth, and all for nothing: there's no economic advantage for anyone relative to where they started.

The logic of this type of game is so pervasive that we don't even have to look to misdeeds to see it running amok. We can just as easily end up in a terrible equilibrium with a clean conscience. How? Look no further than your company vacation policy. In America, people work some of the longest hours in the world; as the *Economist* put it, "nowhere is the value of work higher and the value of leisure lower." There are few laws mandating that employers provide time off, and even when American employees do get vacation time they don't use it. A recent study showed that the average worker takes only half of the vacation days granted them, and a stunning 15% take no vacation at all.

At the present moment, the Bay Area (where the two of us live) is attempting to remedy this sorry state of affairs by going through a radical paradigm shift when it comes to vacation policy—a shift that is very well meaning and completely, apocalyptically doomed. The premise sounds innocent enough: instead of metering out some fixed arbitrary number of days for each employee, then wasting HR man-hours making sure no one goes over their limit, why not just let your employees free? Why not simply allow them unlimited vacation? Anecdotal reports thus far are mixed—but from a game-theoretic perspective, this approach is a nightmare. All employees want, in theory, to take as much vacation as possible. But they also all want to take just slightly less vacation than each other, to be perceived as more loyal, more committed, and more dedicated (hence more promotion-worthy). Everyone looks to the others for a baseline, and will take just slightly less than that. The Nash equilibrium of this game is *zero*. As the CEO of software company Travis CI, Mathias Meyer, writes, "People will hesitate to take a vacation as they don't want to seem like that person who's taking the most vacation days. It's a race to the bottom."

This is the tragedy of the commons in full effect. And it's just as bad between firms as within them. Imagine two shopkeepers in a small town. Each of them can choose either to stay open seven days a week or to be open only six days a week, taking Sunday off to relax with their friends and family. If both of them take a day off, they'll retain their existing market share and experience less stress. However, if one shopkeeper decides to open his shop seven days a week, he'll draw extra customers—taking them away from his competitor and threatening his livelihood. The Nash equilibrium, again, is for everyone to work all the time.

This exact issue became a flash point in the United States during the 2014 holiday season, as retailer after retailer, unwilling to cede market share to competitors who were getting ahead of the usual post-Thanksgiving shopping rush, caved in toward the lousy equilibrium. "Stores are opening earlier than ever before," the *International Business Times* reported. Macy's decided to open two hours earlier than the year before, as did Target. Kmart, for its part, opened at 6:00 a.m. on Thanksgiving morning, and was continuously open for forty-two hours.

So what can we, as players, do when we find ourselves in such a situation—either the two-party prisoner's dilemma, or the multi-party tragedy of the commons? In a sense, nothing. The very stability that these bad equilibria have, the thing that *makes* them equilibria, becomes damnable. By and large we cannot shift the dominant strategies from within. But this doesn't mean that bad equilibria can't be fixed. It just means that the solution is going to have to come from somewhere else.

Mechanism Design: Change the Game

> Don't hate the player, hate the game.
>
> —ICE-T

> Don't ever take sides with anyone against the family again—ever.
>
> —THE GODFATHER

The prisoner's dilemma has been the focal point for generations of debate and controversy about the nature of human cooperation, but University College London game theorist Ken Binmore sees at least some of that controversy as misguided. As he argues, it's "just plain wrong that the Prisoner's Dilemma captures what matters about human cooperation. On the contrary, it represents a situation in which the dice are as loaded against the emergence of cooperation as they could possibly be."*

Well, if the rules of the game force a bad strategy, maybe we shouldn't try to change strategies. Maybe we should try to change the game.

This brings us to a branch of game theory known as "mechanism design." While game theory asks what behavior will emerge given a set of rules, mechanism design (sometimes called "reverse game theory") works in the other direction, asking: what rules will give us the behavior we want to see? And if game theory's revelations—like the fact that an equilibrium strategy might be rational for each player yet bad for everyone—have proven counterintuitive, the revelations of mechanism design are even more so.

Let's return you and your bank-robbing co-conspirator to the jail cell for another go at the prisoner's dilemma, with one crucial addition: the Godfather. Now you and your fellow thief are members of a crime syndicate, and the don has made it, shall we say, all too clear that any informants will sleep with the fishes. This alteration of the game's payoffs has the

---

*Binmore adds another insight: games like the prisoner's dilemma seemingly obliterate Immanuel Kant's argument that rationality consists of what he called the "categorical imperative," acting the way you wish everyone else would act. The categorical imperative would give us a better outcome in the prisoner's dilemma than the equilibrium strategy, but there's no getting around the fact that this outcome isn't a stable one.

effect of limiting the actions you can take, yet ironically makes it far more likely that things will end well, both for you and your partner. Since defection is now less attractive (to put it mildly), both prisoners are induced to cooperate, and both will confidently walk away half a million dollars richer. Minus, of course, a nominal tithe to the don.

The counterintuitive and powerful thing here is we can worsen *every* outcome—death on the one hand, taxes on the other—yet make everyone's lives better by shifting the equilibrium.

For the small-town shopkeepers, a verbal truce to take Sundays off would be unstable: as soon as either shopkeeper needed some extra cash he'd be liable to violate it, prompting the other to start working Sundays as well so as not to lose market share. This would land them right back in the bad equilibrium where they get the worst of both worlds—they're exhausted and don't get any competitive advantage for it. But they might be able to act as their own don by signing a legally binding contract to the effect that, say, any proceeds earned by either shop on a Sunday go to the other shop. By worsening the unsatisfactory equilibrium, they'd make a new and better one.

On the other hand, a change to the game's payoffs that *doesn't* change the equilibrium will typically have a much smaller effect than desired. The CEO of the software firm Evernote, Phil Libin, made headlines with a policy of offering Evernote employees a thousand dollars cash for taking a vacation. This sounds like a reasonable approach to getting more employees to take vacation, but from a game-theoretic perspective it's actually misguided. Increasing the cash on the table in the prisoner's dilemma, for instance, misses the point: the change doesn't do anything to alter the bad equilibrium. If a million-dollar heist ends up with both thieves in jail, so does a ten-million-dollar heist. The problem isn't that vacations aren't attractive; the problem is that everyone wants to take slightly less vacation than their peers, producing a game whose only equilibrium is no vacation at all. A thousand bucks sweetens the deal but doesn't change the principle of the game—which is to take as much vacation as possible while still being perceived as slightly more loyal than the next guy or gal, therefore getting a raise or promotion over them that's worth many thousands of dollars.

Does this mean that Libin needs to offer tens of thousands of dollars per employee per vacation? No. Mechanism design tells us that Libin can get the happy employees he wants with the stick, rather than the carrot; he can get a better equilibrium without spending a dime. For instance, he could simply make a certain minimal amount of vacation *compulsory*. If he can't change the race, he can still change the bottom. Mechanism design makes a powerful argument for the need for a designer—be it a CEO, a contract binding all parties, or a don who enforces *omertà* by garroted carotid.

A league commissioner is this kind of a designer as well. Imagine how pathetic a sight the NBA would be if there were no games as such, and teams could simply score on each other at literally any time between the start and end of the season: 3:00 a.m. on a Sunday, noon on Christmas, you name it. What you'd see would be haggard, cadaverous players, in extreme sleep debt, forcing vigilance with chemical stimulants, almost losing their minds. War is like this. On the other hand, even Wall Street, ruthless cut-throat capitalists trading by the microsecond in the "city that never sleeps," comes to a cease-fire every day at 4:00 p.m. sharp, so that brokers *can* sleep at predictable hours every night without getting too badly ambushed by competitors pushing toward a sleepless equilibrium. In this sense, the stock market is more a sport than a war.

Scaling up this logic results in a potent argument for the role of government. In fact, many governments do have laws on the books mandating minimum vacations and limiting shop hours. And while the United States is one of the only developed nations without federal requirements for paid vacation, Massachusetts, Maine, and Rhode Island do have state-level prohibitions on Thanksgiving commerce.

Laws like these often stem from the colonial era and were initially religious in nature. Indeed, religion itself provides a very direct way of modifying the structure of games of this kind. In particular, a religious law such as "Remember the Sabbath day" neatly solves the problem faced by the shopkeepers, whether enforced by an all-powerful God or by the more proximate members of a religious community. And adding divine force to injunctions against other kinds of antisocial behavior, such as murder, adultery, and theft, is likewise a way to solve some of the game-theoretic problems of living in a social group. God happens to be even better than government in this respect, since omniscience and omnipotence provide a particularly strong guarantee that taking bad actions will have dire consequences. It turns out there's no Godfather quite like God the Father.

Religion seems like the kind of thing a computer scientist rarely talks about; in fact, it's literally the subject of a book called *Things a Computer Scientist Rarely Talks About*. But by reducing the number of options that people have, behavioral constraints of the kind imposed by religion don't just make certain kinds of decisions less computationally challenging—they can also yield better outcomes.

Mechanism Design by Evolution

How selfish soever man may be supposed, there are evidently some principles in his nature, which interest him in the fortune of others, and render their happiness necessary to him, though he derives nothing from it, except the pleasure of seeing it.

—ADAM SMITH, *THE THEORY OF MORAL SENTIMENTS*

The heart has its reasons which reason knows nothing of.

—BLAISE PASCAL

The redwoods of California are some of the oldest and most majestic living things on the planet. From a game-theoretic standpoint, though, they're something of a tragedy. The only reason they're so tall is that they're trying to be taller than *each other*—up to the point where the harms of over-extension are finally even worse than the harms of getting shaded out. As Richard Dawkins puts it,

> The canopy can be thought of as an aerial meadow, just like a rolling grassland prairie, but raised on stilts. The canopy is gathering solar energy at much the same rate as a grassland prairie would. But a substantial portion of the energy is "wasted" by being fed straight into the stilts, which do nothing more useful than loft the "meadow" high in the air, where it picks up exactly the same harvest of photons as it would—at far lower cost—if it were laid flat on the ground.

If the forest could only somehow agree to a kind of truce, the ecosystem could enjoy the photosynthetic bounty without the wood-making arms race wasting it all. But as we've seen, good outcomes in these scenarios tend only to arise in the context of an authority outside the game—someone changing the payoffs from the top down. It would seem as though in nature, then, there is simply no way of establishing good equilibria between individuals.

On the other hand, if cooperation really does lead to better outcomes in certain games, then we'd expect that cooperatively minded species would prevail evolutionarily. But then where would the cooperation come from if it's only rational at the group level, not the individual level? Maybe it would have to come from something that individuals can't entirely control. Something, for instance, like *emotions*.

Consider two seemingly unrelated scenarios: (1) A man buys a vacuum cleaner, it breaks within a few weeks, and he spends ten minutes online leaving a vindictive review. (2) A woman shopping at a convenience store notices someone steal an elderly man's wallet and bolt for the door; she tackles the thief and wrestles the wallet free.

Though the latter protagonist seems clearly heroic, and the former merely angry, what these vignettes have in common—albeit in very different ways—is involuntary selflessness. The unhappy consumer isn't trying to get the vacuum cleaner replaced or his money back; he's after a highly indirect kind of retribution, from which—in a rational, game-theoretic sense—he stands to gain little other than the spiteful satisfaction of writing the review itself. In the convenience store, the heroic woman metes out vigilante justice at enormous personal cost; she risks injury or even death to return, say, $40 to a man who is a total stranger to her. Even if she wanted to help, she could have simply taken two twenties out of her own pocket and given them to him without risking a trip to the ER! In this sense, both protagonists are acting irrationally. On the other hand, their actions are good for their society: we all want to live in a world in which pickpocketing doesn't pay and in which businesses that sell poor-quality products get a bad reputation.

Perhaps each of us, individually, would be better off being the kind of person who can always make a detached, calculated decision in their own best interest, not willing to lose time fuming over a sunk cost, let alone lose a tooth over $40. But *all* of us are better off living in a society in which such defiant stands are common.

So what has acted up in these people, in the absence of an external authority, to make them buck the selfish equilibrium? Anger, for one thing. Whether prompted by a shoddy business or a petty thief, outrage can override rationality. And in these instances, it may be that the hand of evolution has done what it would otherwise have taken an authority outside the game to accomplish.

Nature is full of examples of individuals being essentially hijacked to serve the goals of another species. The lancet liver fluke (*Dicrocoelium dendriticum*), for instance, is a parasite that makes ants deliberately climb to the tops of grass blades so that they'll be eaten by sheep—the lancet fluke's preferred host. Likewise, the parasite *Toxoplasma gondii* makes mice permanently lose their fear of cats, with similar results.

Emotion, for the bitter, retaliatory consumer and for the convenience-store hero alike, is our *own* species taking over the controls for a minute. "Morality is herd instinct in the individual," wrote Nietzsche. Paraphrasing slightly, we might hazard that emotion is mechanism design in the species. Precisely because feelings are involuntary, they enable contracts that need no outside enforcement. Revenge almost never works out in favor of the one who seeks it, and yet someone who will respond with "irrational" vehemence to being taken advantage of is for that very reason more likely to get a fair deal. As Cornell economist Robert Frank puts it, "If people *expect* us to respond irrationally to the theft of our property, we will seldom *need* to, because it will not be in their interests to steal it. Being predisposed to respond irrationally serves much better here than being guided only by material self-interest."

(Lest you think that civilized modern humans have legal contracts and rule of law *instead* of retribution, recall that it's often more work and suffering to sue or prosecute someone than the victim could ever hope to recover in material terms. Lawsuits are the *means* for self-destructive retaliation in a developed society, not the substitute.)

As for anger, so for compassion and guilt—and love.

As odd as it might sound, the prisoner's dilemma also has a lot to tell us about marriage. In our discussion of optimal stopping problems, such as the secretary problem, back in chapter 1, we looked at both dating and apartment hunting as cases where we must make a commitment with possible future options yet unseen. In both love and housing, though, we continue to encounter more options even after our optimal-stopping decision is made—so why not be ready to jump ship? Of course, knowing that the other party (be it spouse or landlord) is in turn prepared to jump ship would prevent many of the long-term investments (having children together, or laboriously moving in one's belongings) that make those agreements worthwhile.

In both cases this so-called commitment problem can be at least partially addressed by a contract. But game theory suggests that in the case of dating, the voluntary bonds of the law are less relevant to an enduring partnership than the involuntary bonds of love itself. As Robert Frank puts it, "The worry that people will leave relationships because it may later become rational for them to do so is largely erased if it is not rational assessment that binds them in the first place." He explains:

> Yes, people search for objective characteristics they care about. Everybody wants somebody who's kind and intelligent and interesting and healthy and maybe physically attractive, good earning power, the whole laundry list of features, but that's the first pass. . . . After you've spent enough time together, it's not those things that make you want to stay together. It's just the fact that it's *that* particular person—that is what's valuable to you, so you don't really need the contract so much as you need a feeling that makes you not want to separate, even though objectively there might be a better option available to you.

Said differently: Love is like organized crime. It changes the structure of the marriage game so that the equilibrium becomes the outcome that works best for everybody.

Playwright George Bernard Shaw once wrote of marriage that "If the prisoner is happy, why lock him in? If he is not, why pretend that he is?" Game theory offers a subtle answer to this particular riddle. Happiness *is* the lock.

A game-theoretic argument for love would highlight one further point: marriage is a prisoner's dilemma in which you get to *choose* the person with whom you're in cahoots. This might seem like a small change, but it potentially has a big effect on the structure of the game you're playing. If you knew that, for some reason, your partner in crime would be miserable if you weren't around—the kind of misery even a million dollars couldn't cure—then you'd worry much less about them defecting and leaving you to rot in jail.

So the rational argument for love is twofold: the emotions of attachment not only spare you from recursively overthinking your partner's intentions, but by changing the payoffs actually enable a better outcome altogether. What's more, being able to fall involuntarily in love makes you, in turn, a more attractive partner to have. Your capacity for heartbreak, for sleeping with the emotional fishes, is the very quality that makes you such a trusty accomplice.

Information Cascades: The Tragic Rationality of Bubbles

Whenever you find yourself on the side of the majority, it is time to pause and reflect.

—MARK TWAIN

Part of the reason why it's a good idea to pay attention to the behavior of others is that in doing so, you get to add their information about the world to your own. A popular restaurant is probably good; a half-empty concert hall is probably a bad sign; and if someone you're talking to abruptly yanks their gaze toward something you can't see, it's probably not a bad idea to turn your head, too.

On the other hand, learning from others doesn't always seem particularly rational. Fads and fashions are the result of following others' behavior without being anchored to any underlying objective truth about the world. What's worse, the assumption that other people's actions are a useful guide can lead to the sort of herd-following that precipitates economic disaster. If everybody else is investing in real estate, it seems like a good idea to buy a house; after all, the price is only going to go up. Isn't it?

An interesting aspect of the 2007–2009 mortgage crisis is that everybody involved seemed to feel like they were unfairly punished for simply doing what they were supposed to. A generation of Americans who grew up believing that houses were fail-safe investments, and who saw everyone around them buying houses despite (or because of) rapidly rising prices, were badly burned when those prices finally started to tumble. Bankers, meanwhile, felt they were unfairly blamed for doing what they had always done—offering opportunities, which their clients could accept or decline. In the wake of an abrupt market collapse, the temptation is always to assign blame. Here game theory offers a sobering perspective: catastrophes like this can happen even when no one's at fault.

Properly appreciating the mechanics of financial bubbles begins with understanding auctions. While auctions may seem like niche corners of the economy—evoking either million-dollar oil paintings at Sotheby's and

Christie's, or Beanie Babies and other collectibles on eBay—they actually power a substantial portion of the economy. Google, for instance, makes more than 90% of its revenue from selling ads, and those ads are all sold via auctions. Meanwhile, governments use auctions to sell rights to bands of the telecommunications spectrum (such as cell phone transmission frequencies), raising tens of billions of dollars in revenue. In fact, many global markets, in everything from homes to books to tulips, operate via auctions of various styles.

One of the simplest auction formats has each participant write down their bid in secret, and the one whose bid is highest wins the item for whatever price they wrote down. This is known as a "sealed-bid first-price auction," and from an algorithmic game theory perspective there's a big problem with it—actually, several. For one thing, there's a sense in which the winner always overpays: if you value an item at $25 and I value it at $10, and we both bid our true valuations ($25 and $10), then you end up buying it for $25 when you could have had it for just a hair over $10. This problem, in turn, leads to another one, which is that in order to bid properly—that is, in order not to overpay—you need to predict the true valuation of the other players in the auction and "shade" your bid accordingly. That's bad enough—but the other players aren't going to bid their true valuations either, because they're shading their bids based on their prediction of yours! We are back in the land of recursion.

Another classic auction format, the "Dutch auction" or "descending auction," gradually lowers an item's price until someone is willing to buy it. The name references the Aalsmeer Flower Auction, the largest flower auction in the world, which takes place daily in the Netherlands—but Dutch auctions are more prevalent than they might initially seem. A store marking down its unsold items, and landlords listing apartments at the highest price they think the market will bear, both share its basic quality: the seller is likely to begin optimistically and nudge the price down until a buyer is found. The descending auction resembles the first-price auction in that you're more likely to win by paying near the top of your range (i.e., you'll be poised to bid as the price falls to $25), and therefore will want to shade your offer by some complexly strategic amount. Do you buy at $25, or stay your hand and try to wait for a lower price? Every dollar you save risks losing out altogether.

The inverse of a Dutch or descending auction is what's known as an "English auction" or "ascending auction"—the most familiar auction format. In an English auction, bidders alternate raising the price until all but one of them drop out. This seems to offer something closer to what we want: here, if you value an item at $25 and I value it at $10, you'll win it for just over $10 without either having to go all the way to $25 or disappearing down the strategic rabbit hole.

Both the Dutch auction and English auction introduce an extra level of complexity when compared to a sealed-bid auction, however. They involve not only the private information that each bidder has but also the public flow of bidding behavior. (In a Dutch auction, it is the absence of a bid that reveals information, by making it clear that none of the other bidders value the item at the current price level.) And under the right circumstances, this mixing of private and public data can prove toxic.

Imagine the bidders are doubtful about their own estimations of the value of an auction lot—say, the right to drill for oil in some part of the ocean. As University College London game theorist Ken Binmore notes, "the amount of oil in a tract is the same for everybody, but the buyers' estimates of how much oil is likely to be in a tract will depend on their differing geological surveys. Such surveys aren't only expensive, but notoriously unreliable." In such a situation, it seems natural to look closely at your opponents' bids, to augment your own meager private information with the public information.

But this public information might not be nearly as informative as it seems. You don't actually get to know the other bidders' *beliefs*—only their *actions*. And it is entirely possible that their behavior is based on your own, just as your behavior is being influenced by theirs. It's easy to imagine a bunch of people all going over a cliff together because "everyone else" was acting as though it'd all be fine—when in reality each person had qualms, but suppressed them because of the apparent confidence of everyone else in the group.

Just as with the tragedy of the commons, this failure is not necessarily the players' fault. An enormously influential paper by the economists Sushil Bikhchandani, David Hirshleifer, and Ivo Welch has demonstrated that under the right circumstances, a group of agents who are all behaving perfectly rationally and perfectly appropriately can nonetheless fall prey to what is effectively infinite misinformation. This has come to be known as an "information cascade."

To continue the oil drilling rights scenario, imagine there are ten companies that might bid on the rights for a given tract. One of them has a geological survey suggesting the tract is rich with oil; another's survey is inconclusive; the reconnaissance of the other eight suggests it's barren. But being competitors, of course, the companies do not share their survey results with each other, and instead can only watch each other's actions. When the auction begins, the first company, with the promising report, makes a high initial bid. The second company, encouraged by this bid to take an optimistic view of their own ambiguous survey, bids even higher. The third company has a weak survey but now doesn't trust it in light of what they take to be two independent surveys that suggest it's a gold mine, so they make a new high bid. The fourth company, which also has a lackluster survey, is now even more strongly inclined to disregard it, as it seems like *three* of their competitors all think it's a winner. So they bid too. The "consensus" unglues from reality. A cascade has formed.

No single bidder has acted irrationally, yet the net result is catastrophe. As Hirshleifer puts it, "Something very important happens once somebody decides to follow blindly his predecessors independently of his own information signal, and that is that his action becomes uninformative to all later decision makers. Now the public pool of information is no longer growing. That welfare benefit of having public information . . . has ceased."

To see what happens in the real world when an information cascade takes over, and the bidders have almost nothing but one another's behavior to estimate an item's value, look no further than Peter A. Lawrence's developmental biology text *The Making of a Fly*, which in April 2011 was selling for $23,698,655.93 (plus $3.99 shipping) on Amazon's third-party marketplace. How and why had this—admittedly respected—book reached a sale price of more than $23 million? It turns out that two of the sellers were setting their prices algorithmically as constant fractions of each other: one was always setting it to 0.99830 times the competitor's price, while the competitor was automatically setting their own price to 1.27059 times the other's. Neither seller apparently thought to set any limit on the resulting numbers, and eventually the process spiraled totally out of control.

It's possible that a similar mechanism was in play during the enigmatic and controversial stock market "flash crash" of May 6, 2010, when, in a matter of minutes, the price of several seemingly random companies in the S&P 500 rose to more than $100,000 a share, while others dropped precipitously—sometimes to $0.01 a share. Almost $1 trillion of value instantaneously went up in smoke. As CNBC's Jim Cramer reported live, dumbfounded, "That . . . it can't be there. That is not a real price. Oh well, just go buy Procter! Just go buy Procter & Gamble, they reported a decent quarter, just go buy it. . . . I mean, this is ridi—this is a good opportunity." Cramer's incredulity is his private information holding up against the public information. He's seemingly the only person in the world willing to pay, in this case, $49 for a stock that the market is apparently valuing at under $40, but he doesn't care; he's seen the quarterly reports, he's certain in what he knows.

Investors are said to fall into two broad camps: "fundamental" investors, who trade on what they perceive as the underlying value of a company, and "technical" investors, who trade on the fluctuations of the market. The rise of high-speed algorithmic trading has upset the balance between these two strategies, and it's frequently complained that computers, unanchored to the real-world value of goods—unbothered at pricing a texbook at tens of millions of dollars and blue-chip stocks at a penny—worsen the irrationality of the market. But while this critique is typically leveled at computers, people do the same kind of thing too, as any number of investment bubbles can testify. Again, the fault is often not with the players but with the game itself.

Information cascades offer a rational theory not only of bubbles, but also of fads and herd behavior more generally. They offer an account of how it's easily possible for any market to spike and collapse, even in the absence of irrationality, malevolence, or malfeasance. The takeaways are several. For one, be wary of cases where public information seems to exceed private information, where you know more about what people are doing than why they're doing it, where you're more concerned with your judgments fitting the consensus than fitting the facts. When you're mostly looking to others to set a course, they may well be looking right back at you to do the same. Second, remember that actions are not beliefs; cascades get caused in part when we misinterpret what others *think* based on what they *do*. We should be especially hesitant to overrule our own doubts—and if we do, we might want to find some way to broadcast those doubts even as we move forward, lest others fail to distinguish the reluctance in our minds from the implied enthusiasm in our actions. Last, we should remember from the prisoner's dilemma that sometimes a game can have irredeemably lousy rules. There may be nothing we can do once we're in it, but the theory of information cascades may help us to avoid such a game in the first place.

And if you're the kind of person who always does what you think is right, no matter how crazy others think it is, take heart. The bad news is that you *will* be wrong more often than the herd followers. The good news is that sticking to your convictions creates a positive externality, letting people make accurate inferences from your behavior. There may come a time when you will save the entire herd from disaster.

To Thine Own Self Compute

The application of computer science to game theory has revealed that being obligated to strategize is itself a part—often a big part—of the price we pay in competing with one another. And as the difficulties of recursion demonstrate, nowhere is that price as high as when we're required to get inside each other's heads. Here, algorithmic game theory gives us a way to rethink mechanism design: to take into account not only the outcome of the games, but also the computational effort required of the players.

We've seen how seemingly innocuous auction mechanisms, for instance, can run into all sorts of problems: overthinking, overpaying, runaway cascades. But the situation is not completely hopeless. In fact, there's one auction design in particular that cuts through the burden of mental recursion like a hot knife through butter. It's called the Vickrey auction.

Named for Nobel Prize–winning economist William Vickrey, the Vickrey auction, just like the first-price auction, is a "sealed bid" auction process. That is, every participant simply writes down a single number in secret, and the highest bidder wins. However, in a Vickrey auction, the winner ends up paying not the amount of their own bid, but that of the *second*-place bidder. That is to say, if you bid $25 and I bid $10, you win the item at *my* price: you only have to pay $10.

To a game theorist, a Vickrey auction has a number of attractive properties. And to an algorithmic game theorist in particular, one property especially stands out: the participants are incentivized to be honest. In fact, there is no better strategy than just bidding your "true value" for the item— exactly what you think the item is worth. Bidding any more than your true value is obviously silly, as you might end up stuck buying something for more than you think it's worth. And bidding any less than your true value (i.e., shading your bid) risks losing the auction for no good reason, since it doesn't save you any money—because if you win, you'll only be paying the value of the second-highest bid, regardless of how high your own was. This makes the Vickrey auction what mechanism designers call "strategy-proof," or just "truthful." In the Vickrey auction, honesty is literally the best policy.

Even better, honesty remains the best policy regardless of whether the other bidders are honest themselves. In the prisoner's dilemma, we saw how defection turned out to be the "dominant" strategy—the best move no matter whether your partner defected or cooperated. In a Vickrey auction, on the other hand, honesty is the dominant strategy. This is the mechanism designer's holy grail. You do not need to strategize or recurse.

Now, it seems like the Vickrey auction would cost the seller some money compared to the first-price auction, but this isn't necessarily true. In a first-price auction, every bidder is shading their bid down to avoid overpaying; in the second-price Vickrey auction, there's no need to—in a sense, the auction itself is optimally shading their bid *for them*. In fact, a game-theoretic principle called "revenue equivalence" establishes that over time, the average expected sale price in a first-price auction will converge to precisely the same as in a Vickrey auction. Thus the Vickrey equilibrium involves the same bidder winning the item for the same price—without any strategizing by any of the bidders whatsoever. As Tim Roughgarden tells his Stanford students, the Vickrey auction is "awesome."

For Hebrew University algorithmic game theorist Noam Nisan, this awesomeness has an air to it that's nearly utopian. "You would like to get some kind of rules of society where it's not worthwhile to lie, and then people won't lie so much, right? That's the basic idea. From my point of view, the amazing thing about Vickrey is that you wouldn't expect that in general it's possible to do that, right? Especially in things like an auction, where of course I want to pay less, how can you *ever* get— And then yet Vickrey shows, here is the way to do that. I think that's really fantastic."

In fact, the lesson here goes far beyond auctions. In a landmark finding called the "revelation principle," Nobel laureate Roger Myerson proved that *any* game that requires strategically masking the truth can be transformed into a game that requires nothing but simple honesty. Paul Milgrom, Myerson's colleague at the time, reflects: "It's one of those results that as you look at it from different sides, on the one side, it's just absolutely shocking and amazing, and on the other side, it's trivial. And that's totally wonderful, it's so awesome: that's how you know you're looking at one of the best things you can see."

The revelation principle may seem hard to accept on its face, but its proof is actually quite intuitive. Imagine that you have an agent or a lawyer who will be playing the game for you. If you trust them to represent your interests, you're going to simply tell them exactly what you want, and let them handle all of the strategic bid-shading and the recursive strategizing on your behalf. In the Vickrey auction, the game itself performs this function. And the revelation principle just expands this idea: *any* game that can be played for you by agents to whom you'll tell the truth, it says, will become an honesty-is-best game if the behavior you want from your agent is incorporated into the rules of the game itself. As Nisan puts it, "The basic thing is if you don't want your clients to optimize against you, you'd better optimize for them. That's the whole proof.... If I design an algorithm that already optimizes for you, there is nothing you can do."

Algorithmic game theory has made huge contributions to a number of practical applications over the past twenty years: helping us understand packet routing on the Internet, improving FCC spectrum auctions that allocate precious (if invisible) public goods, and enhancing the matching algorithms that pair medical students with hospitals, among others. And this is likely just the beginning of a much larger transformation. "We are just scratching the surface," says Nisan. "Even in the *theory* we are just starting to understand it. And there is another generation probably until what I completely understand today theoretically will successfully be applied to humans. It's a generation; I think not more than that. It will take a generation."

French existentialist philosopher Jean-Paul Sartre famously wrote that "Hell is other people." He didn't mean that others are inherently malicious or unpleasant, but rather that they complicate our own thoughts and beliefs:

> When we think about ourselves, when we try to know ourselves ... we use the knowledge of us which other people already have. We judge ourselves with the means other people have and have given us for judging ourselves. Into whatever I say about myself someone else's judgment always enters. Into whatever I feel within myself someone else's judgment enters.... But that does not at all mean that one cannot have relations with other people. It simply brings out the capital importance of all other people for each one of us.

Perhaps, given what we've seen in this chapter, we might endeavor to revise Sartre's statement. Interacting with others doesn't have to be a nightmare—although in the wrong game it surely can be. As Keynes observed, popularity is complicated, intractable, a recursive hall of mirrors; but beauty, in the eye of the beholder, is not. Adopting a strategy that doesn't require anticipating, predicting, reading into, or changing course because of the tactics of others is one way to cut the Gordian knot of recursion. And sometimes that strategy is not just easy—it's optimal.

If changing strategies doesn't help, you can try to change the game. And if that's not possible, you can at least exercise some control about which games you choose to play. The road to hell is paved with intractable recursions, bad equilibria, and information cascades. Seek out games where honesty is the dominant strategy. Then just be yourself.

Conclusion

Computational Kindness

I firmly believe that the important things about humans are social in character and that relief by machines from many of our present demanding intellectual functions will finally give the human race time and incentive to learn how to live well together.

—MERRILL FLOOD

Any dynamic system subject to the constraints of space and time is up against a core set of fundamental and unavoidable problems. These problems are computational in nature, which makes computers not only our tools but also our comrades. From this come three simple pieces of wisdom.

First, there are cases where computer scientists and mathematicians have identified good algorithmic approaches that can simply be transferred over to human problems. The 37% Rule, the Least Recently Used criterion for handling overflowing caches, and the Upper Confidence Bound as a guide to exploration are all examples of this.

Second, knowing that you are using an optimal algorithm should be a relief even if you don't get the results you were looking for. The 37% Rule fails 63% of the time. Maintaining your cache with LRU doesn't guarantee that you will always find what you're looking for; in fact, neither would clairvoyance. Using the Upper Confidence Bound approach to the explore/exploit tradeoff doesn't mean that you will have *no* regrets, just that those regrets will accumulate ever more slowly as you go through life. Even the best strategy sometimes yields bad results—which is why computer scientists take care to distinguish between "process" and "outcome." If you followed the best possible process, then you've done all you can, and you shouldn't blame yourself if things didn't go your way.

Outcomes make news headlines—indeed, they make the world we live in—so it's easy to become fixated on them. But processes are what we have control over. As Bertrand Russell put it, "it would seem we must take account of probability in judging of objective rightness. . . . The objectively right act is the one which will *probably* be most fortunate. I shall define this as the *wisest* act." We can hope to be fortunate—but we should strive to be wise. Call it a kind of computational Stoicism.

Finally, we can draw a clear line between problems that admit straightforward solutions and problems that don't. If you wind up stuck in an intractable scenario, remember that heuristics, approximations, and strategic use of randomness can help you find workable solutions. A theme that came up again and again in our interviews with computer scientists was: sometimes "good enough" really is good enough. What's more, being aware of complexity can help us pick our problems: if we have control over which situations we confront, we should choose the ones that are tractable.

But we don't only pick the problems that we pose to ourselves. We also pick the problems we pose each other, whether it's the way we design a city or the way we ask a question. This creates a surprising bridge from computer science to ethics—in the form of a principle that we call *computational kindness.*

◆

There's a certain paradox the two of us observed when it came to scheduling the interviews that went into this book. Our interviewees were on average more likely to be available when we requested a meeting, say, "next Tuesday between 1:00 and 2:00 p.m. PST" than "at a convenient time this coming week." At first this seems absurd, like the celebrated studies where people on average donate more money to save the life of one penguin than eight thousand penguins, or report being more worried about dying in an act of terrorism than about dying from any cause, terrorism included. In the case of interviews, it seems that people preferred receiving a constrained problem, even if the constraints were plucked out of thin air, than a wide-open one. It was seemingly less difficult for them to accommodate *our* preferences and constraints than to compute a better option based on their own. Computer scientists would nod knowingly here, citing the complexity gap between "verification" and "search"—which is about as wide as the gap between knowing a good song when you hear it and writing one on the spot.

One of the implicit principles of computer science, as odd as it may sound, is that computation is *bad*: the underlying directive of any good algorithm is to minimize the labor of thought. When we interact with other people, we present them with computational problems—not just explicit requests and demands, but implicit challenges such as interpreting our intentions, our beliefs, and our preferences. It stands to reason, therefore, that a computational understanding of such problems casts light on the nature of human interaction. We can be "computationally kind" to others by framing issues in terms that make the underlying computational problem easier. This matters because many problems—especially social ones, as we've seen—are intrinsically and inextricably hard.

Consider this all-too-common scenario. A group of friends are standing around, trying to figure out where to go for dinner. Each of them clearly has some preferences, albeit potentially weak ones. But none of them wants to state those preferences explicitly, so they politely navigate the social hazards with guesses and half-hints instead.

They may well come to a resolution that is satisfying to all. But this procedure can easily go awry. The summer after college, for instance, Brian and two friends took a trip to Spain. They negotiated the trip itinerary on the fly, and at one point it became clear that they wouldn't have time to go to the bullfight they'd researched and planned. Only then, as each of the three attempted to console the others, did they suddenly discover that in fact none of them had wanted to see the bullfight in the first place. Each had just gamely adopted what they'd perceived to be the others' level of enthusiasm, thereby *producing* the level of enthusiasm that the others gamely adopted in turn.

Likewise, seemingly innocuous language like "Oh, I'm flexible" or "What do you want to do tonight?" has a dark computational underbelly that should make you think twice. It has the veneer of kindness about it, but it does two deeply alarming things. First, it passes the cognitive buck: "Here's a problem, you handle it." Second, by not stating your preferences, it invites the others to simulate or imagine them. And as we have seen, the simulation of the minds of others is one of the biggest computational challenges a mind (or machine) can ever face.

In such situations, computational kindness and conventional etiquette diverge. Politely withholding your preferences puts the computational problem of inferring them on the rest of the group. In contrast, politely *asserting* your preferences ("Personally, I'm inclined toward x. What do you think?") helps shoulder the cognitive load of moving the group toward resolution.

Alternatively, you can try to reduce, rather than maximize, the number of options that you give other people—say, offering a choice between two or three restaurants rather than ten. If each person in the group eliminates their least preferred option, that makes the task easier for everyone. And if you're inviting somebody out to lunch, or scheduling a meeting, offering one or two concrete proposals that they can accept or decline is a good starting point.

None of these actions is necessarily "polite," but all of them can significantly lower the computational cost of interaction.

◆

Computational kindness isn't just a principle of behavior; it's also a design principle.

In 2003, University of Waterloo computer scientist Jeffrey Shallit investigated the question of what coin, if put into circulation in the United States, would most help to minimize the number of coins needed on average to make change. Delightfully, the answer turned out to be an 18-cent piece—but Shallit was somewhat stayed from making a policy recommendation by computational concerns.

At present, change-making is dead simple: for any given amount, just use as many quarters as you can without going over, then as many dimes as possible, and so on down the denominations. For instance, fifty-four cents is two quarters, then four pennies. With an 18-cent piece, that simple algorithm is no longer optimal: fifty-four cents is then best made with three 18-cent pieces—and no quarters at all. In fact, Shallit observed that ungainly denominations turn change-making into something "at least as hard . . . as the traveling salesman problem." That's a lot to ask of a cashier. If ease of computation is taken into account, Shallit found, then what the US money supply could best make use of is either a 2-cent or a 3-cent piece. Not quite as exciting as an 18-cent coin—but almost as good, and computationally kinder by a long shot.

The deeper point is that subtle changes in design can radically shift the kind of cognitive problem posed to human users. Architects and urban planners, for instance, have choices about how they construct our environment—which means they have choices about how they will structure the computational problems we have to solve.

Consider a large parking lot, with an array of different lanes, of the kind often found at stadiums and shopping centers. You may drive in one lane toward the destination, see a spot, decide to let it go in favor of (hopefully) a better one farther ahead—but then, finding no such luck, reach the destination and head away down a neighboring lane. After a certain amount of driving, you must decide whether another space is good enough to take, or so far away that you'll try searching in a third lane instead.

An algorithmic perspective here is useful not just for the driver but also for the architect. Contrast the hairy, messy decision problem posed by one of those lots to a single linear path going *away* from one's destination. In that case, one simply takes the first available space—no game theory, no analysis, no look-then-leap rule needed. Some parking garages are structured this way, with a single helix winding upward from the ground level. Their computational load is zero: one simply drives forward until the first space appears, then takes it. Whatever the other possible factors for and against this kind of construction, we can definitely say that it's cognitively humane to its drivers—computationally kind.

One of the chief goals of design ought to be protecting people from unnecessary tension, friction, and mental labor. (This is not just an abstract concern; when mall parking becomes a source of stress, for instance, shoppers may spend less money and return less frequently.) Urban planners and architects routinely weigh how different lot designs will use resources such as limited space, materials, and money. But they rarely account for the way their designs tax the computational resources of the people who use them. Recognizing the algorithmic underpinnings of our daily lives— in this case, optimal stopping—would not only allow drivers to make the best decisions when they're in a particular scenario, but also encourage planners to be more thoughtful about the problems they're forcing drivers into in the first place.

There are a number of other cases where computationally kinder designs suggest themselves. For example, consider restaurant seating policies. Some restaurants have an "open seating" policy, where waiting customers simply hover until a table opens up, and the first to sit down gets the table. Others will take your name, let you have a drink at the bar, and notify you when a table is ready. These approaches to the management of scarce shared resources mirror the distinction in computer science between "spinning" and "blocking." When a processing thread requests a resource and can't get it, the computer can either allow that thread to "spin"—to continue checking for the resource in a perpetual "Is it ready yet?" loop— or it can "block": halt that thread, work on something else, and then come back around whenever the resource becomes free. To a computer scientist, this is a practical tradeoff: weighing the time lost to spinning against the time lost in context switching. But at a restaurant, not all of the resources being traded off are their own. A policy of "spinning" fills empty tables faster, but the CPUs being worn out in the meantime are the minds of their customers, trapped in a tedious but consuming vigilance.

As a parallel example, consider the computational problem posed by a bus stop. If there is a live display saying that the next bus is "arriving in 10 minutes," then you get to decide *once* whether to wait, rather than taking the bus's continued not-coming as a stream of inferential evidence, moment by moment, and having to redecide and redecide. Moreover, you can take your attention away from squinting down the road—spinning—for those ten minutes straight. (For cities that aren't up to the implementation necessary to predict the next arrival, we saw how Bayesian inference can even make knowing when the last bus *left* a useful proxy.) Such subtle acts of computational kindness could do as much for ridership, if not more, as subsidizing the fares: think of it as a *cognitive* subsidy.

◆

If we can be kinder to others, we can also be kinder to ourselves. Not just computationally kinder—all the algorithms and ideas we have discussed will help with that. But also more forgiving.

The intuitive standard for rational decision-making is carefully considering all available options and taking the best one. At first glance, computers look like the paragons of this approach, grinding their way through complex computations for as long as it takes to get perfect answers. But as we've seen, that is an outdated picture of what computers do: it's a *luxury* afforded by an easy problem. In the hard cases, the best algorithms are all about doing what makes the most sense in the least amount of time, which by no means involves giving careful consideration to every factor and pursuing every computation to the end. Life is just too complicated for that.

In almost every domain we've considered, we have seen how the more real-world factors we include—whether it's having incomplete information when interviewing job applicants, dealing with a changing world when trying to resolve the explore/exploit dilemma, or having certain tasks depend on others when we're trying to get things done—the more likely we are to end up in a situation where finding the perfect solution takes unreasonably long. And indeed, people are almost always confronting what computer science regards as the hard cases. Up against such hard cases, effective algorithms make assumptions, show a bias toward simpler solutions, trade off the costs of error against the costs of delay, and take chances.

These aren't the concessions we make when we can't be rational. They're what being rational means.

Notes

INTRODUCTION

3   *al-Jabr wa'l-Muqābala:* *Al-Jabr wa'l-Muqābala* brought with it a truly disruptive technology—the Indian decimal system—and the fact that we refer to this system somewhat erroneously as *Arabic* numerals is testament to the book's influence. The introduction of Arabic numerals, and the algorithms they support, kicked off a medieval showdown between the advocates of this newfangled math (the "algorists") and more traditional accountants who favored Roman numerals backed up by an abacus (the "abacists"). It got pretty intense: the city of Florence passed a law in 1399 that banned the use of Arabic numerals by banks. Ironically, Roman numerals were themselves a controversial innovation when they were offered as an alternative to just writing out numbers with words, being declared "unfitted for showing a sum, since names have been invented for that purpose." See Murray, *Chapters in the History of Bookkeeping.*

3   **four-thousand-year-old Sumerian clay tablet:** A detailed analysis appears in Knuth, "Ancient Babylonian Algorithms." Further information on the history of algorithms, with an emphasis on mathematical algorithms, appears in Chabert, Barbin, and Weeks, *A History of Algorithms.*

4   **strikes with the end of an antler:** This technique is known as "soft hammer percussion."

4   **"Science is a way of thinking":** Sagan, *Broca's Brain.*

4   **the way we think about human rationality:** The limitations of a classical conception of rationality—which assumes infinite computational capacity and infinite time to solve a problem—were famously pointed out by the psychologist, economist, and artificial intelligence pioneer Herbert Simon in the 1950s (Simon, *Models of Man*), ultimately leading to a Nobel Prize. Simon argued that "bounded rationality" could provide a better account of human behavior. Simon's insight has been echoed in mathematics and computer science. Alan Turing's colleague I. J. Good (famous for the concept of "the singularity" and for advising Stanley Kubrick about HAL 9000 for *2001: A Space Odyssey*) called this sort of thinking "Type II Rationality." Whereas classic, old-fashioned Type I Rationality just worries about getting the right answer, Type II Rationality takes into account the cost of

getting that answer, recognizing that time is just as important a currency as accuracy. See Good, *Good Thinking*.

Artificial intelligence experts of the twenty-first century have also argued that "bounded optimality"—choosing the algorithm that best trades off time and error—is the key to developing functional intelligent agents. This is a point made by, for instance, UC Berkeley computer scientist Stuart Russell—who literally cowrote the book on artificial intelligence (the bestselling textbook *Artificial Intelligence: A Modern Approach*)—and by Eric Horvitz, managing director at Microsoft Research. See, for example, Russell and Wefald, *Do the Right Thing*, and Horvitz and Zilberstein, "Computational Tradeoffs Under Bounded Resources." Tom and his colleagues have used this approach to develop models of human cognition; see Griffiths, Lieder, and Goodman, "Rational Use of Cognitive Resources."

5 **analogy to a human mathematician:** In section 9 of Turing, "On Computable Numbers," Turing justifies the choices made in defining what we now call a Turing machine by comparing them to operations that a person might carry out: a two-dimensional piece of paper becomes a one-dimensional tape, the person's state of mind becomes the state of the machine, and symbols are written and read as the person or machine moves around on the paper. Computation is what a computer does, and at the time the only "computers" were people.

5 **we are irrational and error-prone:** For example, see Gilovich, *How We Know What Isn't So*; Ariely and Jones, *Predictably Irrational*; and Marcus, *Kluge*.

1. OPTIMAL STOPPING

9 **"Though all Christians start":** From Kepler's letter to "an unknown nobleman" on October 23, 1613; see, e.g., Baumgardt, *Johannes Kepler*.

9 **such a common phenomenon:** The turkey drop is mentioned, among many other places, in http://www.npr.org/templates/story/story.php?storyId=120913056 and http://jezebel.com/5862181/technology-cant-stop-the-turkey-drop.

10 **In any optimal stopping problem:** For more about the mathematics of optimal stopping, Ferguson, *Optimal Stopping and Applications*, is a wonderful reference.

10 **optimal stopping's most famous puzzle:** A detailed treatment of the nature and origins of the secretary problem appears in Ferguson, "Who Solved the Secretary Problem?"

10 **its first appearance in print:** What Gardner writes about is a parlor game called the "Game of Googol," apparently devised in 1958 by John Fox of the Minneapolis-Honeywell Regulator Company and Gerald Marnie of MIT. Here's how it was described by Fox in his original letter to Gardner on May 11, 1959 (all letters to Gardner we quote are from Martin Gardner's papers at Stanford University, series 1, box 5, folder 19):

> The first player writes down as many unique positive numbers on different slips of paper as he wishes. Then he shuffles them and turns them over one at a time. If the second player tells him to stop at a certain slip and the number on that slip is the largest number in the collection then the second player wins. If not, the first player wins.

Fox further noted that the name of the game comes from the fact that the number "one googol" is often written on one of the slips (presumably to trick the opponent into thinking it's the largest number, with "two googol" appearing somewhere else). He then claimed that the optimal strategy for the second player was to wait until half the slips had been turned over and then choose the first number larger than the largest in the first half, converging on a 34.7% chance of winning.

Gardner wrote to Leo Moser, a mathematician at the University of Alberta, to get more information about the problem. Moser had written a journal article in 1956 that addressed a closely related problem (Moser, "On a Problem of Cayley"), originally proposed in 1875 by the influential British mathematician Arthur Cayley (Cayley, "Mathematical Questions"; Cayley, *Collected Mathematical Papers*). Here's the version proposed by Cayley:

> A lottery is arranged as follows: There are n tickets representing a, b, c pounds respectively. A person draws once; looks at his ticket; and if he pleases, draws again (out of the remaining $n-1$ tickets); looks at his ticket, and if he pleases draws again (out of the remaining $n-2$ tickets); and so on, drawing in all not more than k times; and he receives the value of the last drawn ticket. Supposing that he regulates his drawings in the manner most advantageous to him according to the theory of probabilities, what is the value of his expectation?

Moser added one more piece of information—that the tickets were equally likely to take on any value between 0 and 1.

In Cayley's problem and Moser's slight reframing thereof (sometimes collectively referred to as the Cayley-Moser problem), the payoff is the value of the chosen ticket and the challenge is to find the strategy that gives the highest average payoff. It's here that the problem explored by Cayley and Moser differs from the secretary problem (and the Game of Googol) by focusing on maximizing the *average value* of the number chosen, rather than the probability of finding the *single largest* number (when nothing but the best will do). Moser's 1956 paper is notable not just for the neat solution it provides to this problem, but also because it's the first place we see mention of the real-world consequences of optimal stopping. Moser talks about two possible scenarios:

1. The tourist's problem: A tourist traveling by car wants to stop for the night at one of n motels indicated on his road guide. He seeks the most comfortable accommodation but naturally does not want to retrace any part of his journey. What criterion should he use for stopping?

2. The bachelor's dilemma: A bachelor meets a girl who is willing to marry him and whose "worth" he can estimate. If he rejects her she will have none of him later but he is likely to meet other girls in the future and he estimates that he will have n chances in all. Under what circumstances should he marry?

The idea of entertaining a series of suitors—with the sexes of the protagonists reversed—duly made an appearance in Gardner's 1960 column on the Game of Googol.

Moser provided the correct solution—the 37% Rule—to Gardner, but his letter of August 26, 1959, suggested that the problem might have an earlier origin: "I also found it in some notes that R. E. Gaskell (of Boeing Aircraft in Seattle) distributed in January, 1959. He credits the problem to Dr. G. Marsaglia."

Gardner's charitable interpretation was that Fox and Marnie were claiming the creation of the specific Game of Googol, not of the broader problem of which that game was an instance, a point that was carefully made in his column. But he received a variety of letters citing earlier instances of similar problems, and it's clear that the problem was passed around among mathematicians.

10 **origins of the problem are surprisingly mysterious:** Even Gilbert and Mosteller, "Recognizing the Maximum of a Sequence," one of the most authoritative scientific papers on the secretary problem, admits that "efforts to discover the originator of this problem have been unsuccessful." Ferguson, "Who Solved the Secretary Problem?," provides an amusing

and mathematically detailed history of the secretary problem, including some of its variants. Ferguson argued that in fact the problem described by Gardner hadn't been solved. It should already be clear that lots of people solved the secretary problem of maximizing the probability of selecting the best from a sequence of applicants distinguished only by their relative ranks, but Ferguson pointed out that this is not actually the problem posed in the Game of Googol. First of all, the Googol player knows the values observed on each slip of paper. Second, it's a competitive game—with one player trying to select numbers and a sequence that will deceive the other. Ferguson has his own solution to this more challenging problem, but it's complex enough that you will have to read the paper yourself!

11 **Mosteller recalled hearing about the problem:** Gilbert and Mosteller, "Recognizing the Maximum of a Sequence."

11 **Roger Pinkham of Rutgers wrote:** Letter from Roger Pinkham to Martin Gardner, January 29, 1960.

11 **Flood's influence on computer science:** See Cook, *In Pursuit of the Traveling Salesman*; Poundstone, *Prisoner's Dilemma*; and Flood, "Soft News."

11 **considering the problem since 1949:** Flood made this claim in a letter he wrote to Gardner on May 5, 1960. He enclosed a letter from May 5, 1958, in which he provided the correct solution, although he also indicated that Andrew Gleason, David Blackwell, and Herbert Robbins were rumored to have solved the problem in recent years.

   In a letter to Tom Ferguson dated May 12, 1988, Flood went into more detail about the origin of the problem. (The letter is on file in the Merrill Flood archive at the University of Michigan.) His daughter, recently graduated from high school, had entered a serious relationship with an older man, and Flood and his wife disapproved. His daughter was taking the minutes at a conference at George Washington University in January 1950, and Flood presented what he called the "fiancé problem" there. In his words, "I made no attempt to solve the problem at that time, but introduced it simply because I hoped that [she] would think in those terms a bit and it sounded like it might be a nice little easy mathematical problem." Flood indicates that Herbert Robbins provided an approximate solution a few years later, before Flood himself figured out the exact solution.

12 **appears to be in a 1964 paper:** The paper is Chow et al., "Optimal Selection Based on Relative Rank."

12 **the best you've seen so far:** In the literature, what we call "best yet" applicants are referred to (we think somewhat confusingly) as "candidates."

13 **settles to 37% of the pool:** The 37% Rule is derived by doing the same analysis for n applicants—working out the probability that setting a standard based on the first k applicants results in choosing the best applicant overall. This probability can be expressed in terms of the ratio of k to n, which we can call p. As n gets larger, the probability of choosing the best applicant converges to the mathematical function $-p \log p$. This is maximized when $p = 1/e$. The value of e is 2.71828 . . . , so $1/e$ is 0.367879441 . . . , or just under 37%. And the mathematical coincidence—that the probability of success is the same as p—arises because $\log e$ is equal to 1. So if $p = 1/e$, $-p \log p$ is just $1/e$. A well-explained version of the full derivation appears in Ferguson, "Who Solved the Secretary Problem?"

14 **one of the problem's curious mathematical symmetries:** Mathematicians John Gilbert and Frederick Mosteller call this symmetry "amusing" and discuss it at slightly greater length in Gilbert and Mosteller, "Recognizing the Maximum of a Sequence."

15 **"The passion between the sexes":** Malthus, *An Essay on the Principle of Population*.

15 **"married the first man I ever kissed":** Attributed by many sources, e.g., Thomas, *Front Row at the White House.*

15 **a graduate student, looking for love:** Michael Trick's blog post on meeting his wife is "Finding Love Optimally," *Michael Trick's Operations Research Blog*, February 27, 2011, http://mat.tepper.cmu.edu/blog/?p=1392.

15 **the number of applicants or the *time*:** The 37% Rule applies directly to the time period of one's search only when the applicants are uniformly distributed across time. Otherwise, you'll want to aim more precisely for 37% of the *distribution* over time. See Bruss, "A Unified Approach to a Class of Best Choice Problems."

15 **the 37% Rule gave age 26.1 years:** The analysis of waiting until at least age 26 to propose (37% of the way from 18 to 40) first appears in Lindley, "Dynamic Programming and Decision Theory," which is presumably where Trick encountered this idea.

16 **courting a total of eleven women:** Kepler's story is covered in detail in Koestler, *The Watershed*, and in Baumgardt, *Johannes Kepler*, as well as in Connor, *Kepler's Witch*. Most of what we know about Kepler's search for a second wife comes from one letter in particular, which Kepler wrote to "an unknown nobleman" from Linz, Austria, on October 23, 1613.

16 **propose early and often:** Smith, "A Secretary Problem with Uncertain Employment," showed that if the probability of a proposal being rejected is q, then the strategy that maximizes the probability of finding the best applicant is to look at a proportion of applicants equal to $q^{1/(1-q)}$ and then make offers to each applicant better than those seen so far. This proportion is always less than $1/e$, so you're making your chances better by making more offers. Unfortunately, those chances are still worse than if you weren't getting rejected—the probability of ending up with the best applicant is also $q^{1/(1-q)}$, and hence less than that given by the 37% Rule.

17 **until you've seen 61% of applicants:** If delayed proposals are allowed, the optimal strategy depends on the probability of an immediate proposal being accepted, q, and the probability of a delayed proposal being accepted, p. The proportion of candidates to initially pass over is given by the fairly daunting formula $\left(\frac{q^2}{q-p(1-q)}\right)^{1/(1-q)}$. This integrated formula for rejection and recall comes from Petruccelli, "Best-Choice Problems Involving Uncertainty," although recalling past candidates was considered earlier by Yang, "Recognizing the Maximum of a Random Sequence."

This formula simplifies when we make particular choices for q and p. If $p=0$, so delayed proposals are always rejected, we get back the rule for the secretary problem with rejection. As we approach $q=1$, with immediate proposals always being accepted, the proportion at which to begin making offers tends toward e^{p-1}, which is always greater than $1/e$ (which can be rewritten as e^{-1}). This means that having the potential to make offers to applicants who have been passed over should result in spending more time passing over applicants—something that is quite intuitive. In the main text we assume that immediate proposals are always accepted ($q=1$) but delayed proposals are rejected half the time ($p=0.5$). Then you should pass over 61% of applicants and make an offer to the best yet who follows, going back at the end and making an offer to the best overall if necessary.

Another possibility considered by Petruccelli is that the probability of rejection increases with time, as the ardor of applicants decreases. If the probability of an offer being accepted by an applicant is qp^s, where s is the number of "steps" into the past required to reach that applicant, then the optimal strategy depends on q, p, and the number of applicants, n. If $q/(1-p)$ is more than $n-1$ then it's best to play a waiting game, observing all applicants and then making an offer to the best. Otherwise, observe a proportion equal to $q^{1/(1-q)}$ and make an offer to the next applicant better than those seen so far. Interestingly, this is exactly the same strategy (with the same probability of success) as that when $p=0$, meaning that if the probability of rejection increases with time, there is no benefit to being able to go back to a previous candidate.

18 **"No buildup of experience is needed":** Gilbert and Mosteller, "Recognizing the Maximum of a Sequence."

18 **use the Threshold Rule:** The general strategy for solving optimal stopping problems like the full information game is to start at the end and reason backward—a principle that is called "backward induction." For instance, imagine a game where you roll a die, and have the option either to stick with that number or roll again a maximum of k times (we took this example from Hill, "Knowing When to Stop"). What's the optimal strategy? We can figure it out by working backward. If $k=0$, you don't have an option—you have to stick with your roll, and you will average 3.5 points (the average value of a die roll, $(1+2+3+4+5+6)/6$). If $k=1$, then you should only keep a roll that beats that average—a 4 or higher. If you get a 1, 2, or 3, you're better off chancing that final roll. Following this strategy, there's a 50% chance you stop with a 4, 5, or 6 (for an average of 5) and a 50% chance you go on to the final roll (for an average of 3.5). So your average score at $k=1$ is 4.25, and you should only keep a roll at $k=2$ if it beats that score—a 5 or higher. And so on.

Backward induction thus answers an age-old question. "A bird in the hand is worth two in the bush," we say, but is 2.0 the right coefficient here? The math suggests that the right number of birds in the bush actually depends on the quality of the bird in the hand. Replacing birds with dice for convenience, a roll of 1, 2, or 3 isn't even worth as much as a single die "in the bush." But a roll of 4 is worth one die in the bush, while a roll of 5 is worth two, three, or even four dice in the bush. And a roll of 6 is worth even more than the entire contents of an *infinitely large* dice bush—whatever that is.

Gilbert and Mosteller used the same approach to derive the series of thresholds that should be used in the full-information secretary problem. The thresholds themselves are not described by a simple mathematical formula, but some approximations appear in their paper. The simplest approximation gives a threshold of $t_k = 1/(1+0.804/k+0.183/k^2)$ for applicant $n-k$. If the probability of a random applicant being better than applicant $n-k$ is less than t_k, then you should take that applicant. Because the denominator increases—at an increasing rate—as k increases, you should be rapidly lowering your threshold as time goes on.

20 **many more variants of the secretary problem:** Freeman, "The Secretary Problem and Its Extensions" summarizes a large number of these variants. Here's a quick tour of some of the most useful results.

If the number of applicants is equally likely to be any number from 1 to n, then the optimal rule is to view the first n/e^2 (which is approximately 13.5% of n) and take the next candidate better than the best seen so far, with a chance of success of $2/e^2$ (Presman and Sonin, "The Best Choice Problem for a Random Number of Objects").

If the number of applicants is potentially infinite, but the search stops after each applicant with probability p, the optimal rule is to view the first $0.18/p$ applicants, with a 23.6% chance of success (ibid.).

Imagine you want to find the best secretary, but the value of doing so decreases the longer you search. If the payoff for finding the best secretary after viewing k applicants is d^k, then the strategy that maximizes the expected payoff sets a threshold based on a number of applicants that is guaranteed to be less than $1/(1-d)$ as the total number of applicants becomes large (Rasmussen and Pliska, "Choosing the Maximum"). If d is close to 1, then an approximation to the optimal strategy is to view the first

−0.4348/log d applicants and then take the next candidate better than any seen so far. Following this strategy can result in viewing only a handful of applicants, regardless of the size of the pool.

One way in which real life differs from idealized recruitment scenarios is that the goal might not be to maximize the probability of getting the best secretary. A variety of alternatives have been explored. Chow et al., "Optimal Selection Based on Relative Rank," showed that if the goal is to maximize the average rank of the selected candidate, a different kind of strategy applies. Rather than a single threshold on the relative rank of the applicant, there is a sequence of thresholds. These thresholds increase as more candidates are observed, with the interviewer becoming less stringent over time. For example, with four applicants, the minimum relative rank a candidate needs to have to stop the search is 0 for the first applicant (never stop on the first), 1 for the second (stop only if they are better than the first), 2 for the third (stop if best or second best), and 4 for the fourth (just stop already!). Following this strategy yields an average expected rank of $1\frac{7}{8}$, better than the $(1 + 2 + 3 + 4)/4 = 2\frac{1}{2}$ that would result from picking an applicant at random. The formula for the optimal thresholds is found by backward induction, and is complicated—we refer interested readers to the original paper.

You can think about the difference between the classical secretary problem and the average-rank case in terms of how they assign payoffs to different ranks. In the classical problem, you get a payoff of 1 for picking the best and 0 for everybody else. In the average-rank case, you get a payoff equal to the number of applicants minus the rank of the selected applicant. There are obvious ways to generalize this, and multi-threshold strategies similar to the one that maximizes the average rank work for any payoff function that decreases as the rank of the applicant increases (Mucci, "On a Class of Secretary Problems"). Another interesting generalization—with important implications for discerning lovers—is that if the payoff is 1 for choosing the best but −1 for choosing anybody else (with 0 for making no choice at all), you should go through a proportion of applicants given by $1/\sqrt{e} \approx 60.7\%$, then take the first person better than all seen so far (or nobody if they all fail this criterion) (Sakaguchi, "Bilateral Sequential Games"). So think hard about your payoff function before getting ready to commit!

But what if you don't just care about finding the best person, but about how much time you have together? Ferguson, Hardwick, and Tamaki, in "Maximizing the Duration of Owning a Relatively Best Object," examined several variants on this problem. If you just care about maximizing the time you spend with the very best person in your set of n, then you should look at the first $0.204n + 1.33$ people and leap for the next person better than all of them. But if you care about maximizing the amount of time you spend with somebody who is the best of all the people seen so far, you should just look at a proportion corresponding to $1/e^2 \approx 13.5\%$. These shorter looking periods are particularly relevant in contexts—such as dating—where the search for a partner might take up a significant proportion of your life.

It turns out that it's harder to find the second-best person than it is to find the best. The optimal strategy is to pass over the first half of the applicants, then choose the next applicant who is second best relative to those seen so far (Rose, "A Problem of Optimal Choice and Assignment"). The probability of success is just 1/4 (as opposed to 1/e for the best). So you're better off not trying to settle.

Finally, there are also variants that recognize the fact that while you are looking for a secretary, your applicants are themselves looking for a job. The added symmetry—which is particularly relevant when the scenario concerns dating—makes the problem even more complicated. Peter Todd, a cognitive scientist at Indiana University, has explored this complexity (and how to simplify it) in detail. See Todd and Miller, "From Pride and Prejudice to Persuasion Satisficing in Mate Search," and Todd, "Coevolved Cognitive Mechanisms in Mate Search."

20 **Selling a house is similar:** The house-selling problem is analyzed in Sakaguchi, "Dynamic Programming of Some Sequential Sampling Design"; Chow and Robbins, "A Martingale System Theorem and Applications"; and Chow and Robbins, "On Optimal Stopping Rules." We focus on the case where there are potentially infinitely many offers, but these authors also provide optimal strategies when the number of potential offers is known and finite (which are less conservative—you should have a lower threshold if you only have finitely many opportunities). In the infinite case, you should set a threshold based on the expected value of waiting for another offer, and take the first offer that exceeds that threshold.

21 **stopping price as a function of the cost of waiting:** Expressing both the offer price p and cost of waiting for another offer c as fractions of our price range (with 0 as the bottom of the range and 1 as the top), the chance that our next offer is better than p is simply $1 - p$. If (or when) a better offer arrives, the average amount we'd expect to gain relative to p is just $\frac{1-p}{2}$. Multiplying these together gives us the expected outcome of entertaining another offer, and this should be greater than or equal to the cost c to be worth doing. This equation $(1 - p)\left(\frac{1-p}{2}\right) \geq c$ can be simplified to $\frac{1}{2}(1 - p)^2 \geq c$, and solving it for p gives us the answer $p \geq 1 - \sqrt{2c}$, as charted on page 22.

22 **"The first offer we got was great":** Laura Albert McLay, personal interview, September 16, 2014.

22 **to model how people look for jobs:** The formulation of job search as an optimal stopping problem is dealt with in Stigler, "The Economics of Information," and Stigler, "Information in the Labor Market." McCall, "Economics of Information and Job Search," proposed using a model equivalent to the solution to the house-selling problem, and Lippman and McCall, "The Economics of Job Search," discusses several extensions to this model. Just as the secretary problem has inspired a vast array of variants, economists have refined this simple model in a variety of ways to make it more realistic: allowing multiple offers to arrive on the same day, tweaking the costs for the seller, and incorporating fluctuation in the economy during the search. A good review of optimal stopping in a job-seeking context can be found in Rogerson, Shimer, and Wright, *Search-Theoretic Models of the Labor Market*.

23 **won't be above your threshold now:** As a survey of the job-search problem puts it: "Assume previously rejected offers cannot be recalled, although this is actually not restrictive because the problem is stationary, so an offer that is not acceptable today will not be acceptable tomorrow" (ibid.).

23 **"parking for the faculty":** Clark Kerr, as quoted in "Education: View from the Bridge," *Time*, November 17, 1958.

23 **"plan on expected traffic":** Donald Shoup, personal correspondence, June 2013.

24 **implemented in downtown San Francisco:** More information on the SFpark system developed by the SFMTA, and its Shoup-inspired dynamic pricing, can be found at http://sfpark.org/how-it-works/pricing/. (Shoup himself is involved in an advisory role.) This program began taking effect in 2011, and is the first project of its kind in the world. For a recent analysis of the effects of the program, see Millard-Ball, Weinberger, and Hampshire, "Is the Curb 80% Full or 20% Empty?"

24 **when occupancy goes from 90% to 95%:** Donald Shoup, personal interview, June 7,

2013. To be precise, the increase from 90% to 95% occupancy reflects an increase of 5.555 . . . percent.

24 **Assume you're on an infinitely long road:** The basic parking problem, as formulated here, was presented as a problem in DeGroot, *Optimal Statistical Decisions*. The solution is to take the first empty spot less than $-\log 2 / \log(1-p)$ spots from the destination, where p is the probability of any given space being available.

25 **you don't need to start seriously looking:** Chapter 17 of Shoup's *The High Cost of Free Parking* discusses the optimal on-street parking strategy when pricing creates an average of one free space per block, which, as Shoup notes, "depends on the conflict between greed and sloth" (personal correspondence). The question of whether to "cruise" for cheap on-street spots or to pay for private parking spaces is taken up in Shoup's chapter 13.

25 **a variety of tweaks to this basic scenario:** Tamaki, "Adaptive Approach to Some Stopping Problems," allowed the probability of a spot being available to vary based on location and considered how these probabilities could be estimated on the fly. Tamaki, "Optimal Stopping in the Parking Problem with U-Turn," added the possibility of U-turns. Tamaki, "An Optimal Parking Problem," considered an extension to DeGroot's model where parking opportunities are not assumed to be a discrete set of spots. Sakaguchi and Tamaki, "On the Optimal Parking Problem in Which Spaces Appear Randomly," used this continuous formulation and allowed the destination to be unknown. MacQueen and Miller, "Optimal Persistence Policies," independently considered a continuous version of the problem that allows circling the block.

26 **"I ride my bike":** Donald Shoup, personal interview, June 7, 2013.

26 *Forbes* **magazine identified Boris Berezovsky:** *Forbes*, "World's Billionaires," July 28, 1997, p. 174.

26 **one of a new class of oligarchs:** Paul Klebnikov, "The Rise of an Oligarch," *Forbes*, September 9, 2000.

26 **"to hit just once, but on the head":** Vladimir Putin, interview with the French newspaper *Le Figaro*, October 26, 2000.

26 **book entirely devoted to the secretary problem:** Berezovsky and Gnedin, *Problems of Best Choice*.

26 **analyzed under several different guises:** There are various ways to approach the problem of quitting when you're ahead. The first is maximizing the length of a sequence of wins. Assume you're tossing a coin that has a probability p of coming up heads. You pay c dollars for each chance to flip the coin, and you get $1.00 when it comes up heads but lose all your accumulated gains when it comes up tails. When should you stop tossing the coin? The answer, as shown by Norman Starr in 1972, is to stop after r heads, where r is the smallest number such that $p^{r+1} \leq c$. So if it's a regular coin with $p = 1/2$, and it costs $0.10 to flip the coin, you should stop as soon as you get four heads in a row. The analysis of runs of heads appears in Starr, "How to Win a War if You Must," where it is presented as a model for winning a war of attrition. A more comprehensive analysis is presented in Ferguson, "Stopping a Sum During a Success Run."

Maximizing the length of a run of heads is a pretty good analogy for some kinds of business situations—for a sequence of deals that cost c to set up, have a probability p of working out, and pay d on success but wipe out your gains on failure, you should quit after making r dollars such that $p^{r/d+1} \leq c/d$. Ambitious drug dealers, take note.

In the burglar problem discussed in the text, assume the average amount gained from each robbery is m and the probability of getting away with the robbery is q. But if the burglar is caught, which happens with probability $1 - q$, he loses everything. The solution: quit when the accumulated gains are greater than or equal to $mq/(1-q)$. The burglar problem appears in Haggstrom, "Optimal Sequential Procedures When More Than One Stop Is Required," as part of a more complex problem in which the burglar is also trying to decide which city to move to.

27 **found by a bodyguard:** See, e.g., "Boris Berezovsky 'Found with Ligature Around His Neck,'" *BBC News*, March 28, 2013, http://www.bbc.com/news/uk-21963080.

27 **official conclusion of a postmortem examination:** See, e.g., Reuters, "Berezovsky Death Consistent with Hanging: Police," March 25, 2013, http://www.reuters.com/article/2013/03/25/us-britain-russia-berezovsky-postmortem-idUSBRE92O12320130325.

27 **"Berezovsky would not give up":** Hoffman, *The Oligarchs*, p. 128.

27 **there *is* no optimal stopping rule:** One condition for an optimal stopping rule to exist is that the average reward for stopping at the best possible point be finite (see Ferguson, *Optimal Stopping and Applications*). The "triple or nothing" game violates this condition—if heads come up k times followed by one tail, the best possible player gets $3^k - 1$ as a payoff, stopping right before that tail. The probability of this is $1/2^{k+1}$. The average over k is thus infinite.

If you're thinking that this could be resolved by assuming that people value money less the more they have—that tripling the monetary reward may not be tripling the utility people assign to that money—then there's a simple work-around: you still get a game with no optimal stopping rule just by offering rewards that triple in their utility. For example, if the utility you assign to money increases as a logarithmic function of the amount of money, then the game becomes "cube or nothing"—the amount of money you could receive on the next gamble is raised to the power of three each time you win.

Intriguingly, while there is no optimal stopping rule for "triple or nothing," where your entire fortune is always on the line, there are nonetheless good strategies for playing games like this when you can choose how much of your bankroll to bet. The so-called Kelly betting scheme, named after J. L. Kelly Jr. and first described in Kelly, "A New Interpretation of Information Rate," is one example. In this scheme, a player can maximize his rate of return by betting a proportion of $\frac{p(b+1)-1}{b}$ of his bankroll on each of a sequence of bets that pay off $b+1$ times the original stake with probability p. For our triple or nothing game, $b=2$ and $p=0.5$, so we should bet a quarter of our bankroll each time—not the whole thing, which inevitably leads to bankruptcy. An accessible history of Kelly betting appears in Poundstone, *Fortune's Formula*.

28 **"pass through this world but once":** The provenance of this quotation is not fully certain, although it has been cited as a Quaker saying since the second half of the nineteenth century, and appears to have been attributed to Grellet since at least 1893. For more, see W. Gurney Benham, *Benham's Book of Quotations, Proverbs, and Household Words*, 1907.

28 **"Spend the afternoon":** Dillard, *Pilgrim at Tinker Creek*.

28 **most closely follows the classical secretary problem:** Seale and Rapoport, "Sequential Decision Making with Relative Ranks."

29 **leapt sooner than they should have:** Ibid. The typical place where people switched from looking to leaping was 13 applicants out of 40, and 21 applicants out of 80, or 32% and 26%, respectively.

29 **"by nature I am very impatient":** Amnon Rapoport, personal interview, June 11, 2013.

29 **Seale and Rapoport showed:** Seale and Rapoport, "Sequential Decision Making with Relative Ranks."

29 **"It's not irrational to get bored":** Neil Bearden, personal correspondence, June 26, 2013. See also Bearden, "A New Secretary Problem."

29 **turns *all* decision-making into optimal stopping:** This kind of argument was first made by Herbert Simon, and it was one of the contributions for which he received the Nobel Prize. Simon began his remarkable career as a political scientist, writing a dissertation on the perhaps unpromising topic of administrative behavior. As he dug into the problem of understanding how organizations composed of real people make decisions, he experienced a growing dissatisfaction with the abstract models of decision-making offered by mathematical economics—models that line up with the intuition that rational action requires exhaustive consideration of our options.

Simon's investigation of how decisions actually get made in organizations made it clear to him that these assumptions were incorrect. An alternative was needed. As he put it in "A Behavioral Model of Rational Choice," "the task is to replace the global rationality of economic man with a kind of rational behavior that is compatible with the access to information and the computational capacities that are actually possessed by organisms, including man, in the kinds of environments in which such organisms exist."

The kind of solution that Simon proposed as a more realistic account of human choice—what he dubbed "satisficing"—uses experience to set some threshold for a satisfactory, "good enough" outcome, then takes the first option to exceed that threshold. This algorithm has the same character as the solutions to the optimal stopping problems we have considered here, where the threshold is either determined by spending some time getting a sense for the range of options (as in the secretary problem) or based on knowing the probability of different outcomes. Indeed, one of the examples Simon used in his argument was that of selling a house, with a similar kind of solution to that presented here.

29 **the definitive textbook on optimal stopping:** That's Ferguson, *Optimal Stopping and Applications.*

2. EXPLORE/EXPLOIT

32 **"Make new friends":** Joseph Parry, "New Friends and Old Friends," in *The Best Loved Poems of the American People,* ed. Hazel Felleman (Garden City, NY: Doubleday, 1936), 58.

32 **"life so rich and rare":** Helen Steiner Rice, "The Garden of Friendship," in *The Poems and Prayers of Helen Steiner Rice,* ed. Virginia J. Ruehlmann (Grand Rapids, MI: Fleming H. Revell), 47.

32 **"You try to find spaces":** Scott Plagenhoef, personal interview, September 5, 2013.

33 **The odd name comes from:** In a letter to Merrill Flood dated April 14, 1955 (available in the Merrill Flood archive at the University of Michigan), Frederick Mosteller tells the story of the origin of the name. Mosteller and his collaborator Robert Bush were working on mathematical models of learning—one of the earliest instances of what came to be known as mathematical psychology, informing the research that Tom does today. They were particularly interested in a series of experiments that had been done with a T-shaped maze, where animals are put into the maze at the bottom of the T and then have to decide whether to go left or right. Food—the payoff—may or may not appear on either side of the maze. To explore this behavior with humans they commissioned a machine with two levers that people could pull, which Mosteller dubbed the two-armed bandit. He then introduced the mathematical form of the problem to his colleagues, and it ultimately became generalized to the multi-armed bandit.

A comprehensive introduction to multi-armed bandits appears in Berry and Fristed,

Bandit Problems. Our focus in this chapter is on bandits where each arm either produces a payoff or doesn't, with different probabilities but the same payoff amount on all arms. This is known as a Bernoulli bandit in the literature, as the probability distribution that describes a coin flip is called the Bernoulli distribution (after the seventeenth-century Swiss mathematician Jacob Bernoulli). Other kinds of multi-armed bandits are also possible, with unknown distributions of different kinds characterizing the payoffs from each arm.

33 **how good the second machine might actually be:** The "myopic" strategy of pulling the arm with higher expected value is actually optimal in some cases. Bradt, Johnson, and Karlin, "On Sequential Designs for Maximizing the Sum of N Observations," showed that if the probabilities of a payoff for a two-armed bandit (with p_1 for one arm, p_2 for the other) satisfy $p_1 + p_2 = 1$, then this strategy is optimal. They conjectured that this also holds for pairs of probabilities where (p_1, p_2) either take on the values (a, b) or (b, a) (i.e., if p_1 is a, then p_2 is b, and vice versa). This was proved to be true by Feldman, "Contributions to the 'Two-Armed Bandit' Problem." Berry and Fristed, *Bandit Problems,* has further details on myopic strategies, including a result showing that choosing the highest expected value is optimal when p_1 and p_2 are restricted to take on just two possible values (e.g., either or both of p_1 or p_2 could be 0.4 or 0.7, but we don't know which of these possibilities is true).

34 **"embodies in essential form":** Whittle, *Optimization over Time.*

34 **"Eat, drink, and be merry":** "Eat, drink, and be merry, for tomorrow we die," an idiom in common parlance and in pop culture (e.g., forming the chorus of "Tripping Billies" by the Dave Matthews Band, among many other references), appears to be a conflation of two biblical verses: Ecclesiastes 8:15 ("A man hath no better thing under the sun, than to eat, and to drink, and to be merry") and Isaiah 22:13 ("Let us eat and drink, for tomorrow we die").

34 **"why take the risk?":** Chris Stucchio, personal interview, August 15, 2013.

35 **"a sixth helping of X-Men":** Nick Allen, "Hollywood makes 2013 the year of the sequel" http://www.telegraph.co.uk/culture/film/film-news/9770154/Hollywood-makes-2013 -the-year-of-the-sequel.html. See also http://www.shortoftheweek.com/2012/01/05/has -hollywood-lost-its-way/ and http://boxofficemojo.com/news/?id=3063.

35 **Profits of the largest film studios declined:** "Between 2007 and 2011, pre-tax profits of the five studios controlled by large media conglomerates (Disney, Universal, Paramount, Twentieth Century Fox and Warner Bros) fell by around 40%, says Benjamin Swinburne of Morgan Stanley." In "Hollywood: Split Screens," *Economist,* February 23, 2013, http:// www.economist.com/news/business/21572218-tale-two-tinseltowns-split-screens.

35 **ticket sales have declined:** Statistics from http://pro.boxoffice.com/statistics/yearly and http://www.the-numbers.com/market/. See also Max Willens, "Box Office Ticket Sales 2014: Revenues Plunge to Lowest in Three Years," *International Business Times,* January 5, 2015.

35 **"Squeezed between rising costs":** "Hollywood: Split Screens," *Economist,* February 23, 2013, http://www.economist.com/news/business/21572218-tale-two-tinseltowns-split -screens.

36 **"the ultimate instrument of intellectual sabotage":** Whittle's comment on the difficulty of bandit problems appears in his discussion of Gittins, "Bandit Processes and Dynamic Allocation Indices."

36 **Robbins proved in 1952:** Robbins, "Some Aspects of the Sequential Design of Experiments" introduces the Win-Stay, Lose-Shift algorithm.

36 **Following Robbins, a series of papers:** Bradt, Johnson, and Karlin, "On Sequential Designs for Maximizing the Sum of N Observations," showed that "stay on a winner" is

always true where the probability of a payoff is unknown for one arm but known for the other. Berry, "A Bernoulli Two-Armed Bandit," proved that the principle is always true for a two-armed bandit. Generalizations of this result (and a characterization of the cases where it doesn't apply) appear in Berry and Fristed, *Bandit Problems*.

37 **exactly how many options and opportunities:** This solution to the "finite horizon" version of the multi-armed bandit problem is presented in Bellman's magnum opus *Dynamic Programming*, a book that is impressive as the starting point (and sometimes endpoint) of a number of topics in optimization and machine learning. Among other uses, dynamic programming can efficiently solve problems that require backward induction— which we also encountered briefly in chapter 1 in the context of the full-information game.

37 **"a byword for intransigence":** Introduction to Gittins, "Bandit Processes and Dynamic Allocation Indices."

38 **"would be a pretty good approximation":** John Gittins, personal interview, August 27, 2013.

38 *Deal or No Deal:* The many worldwide incarnations of this game show began with the Dutch show *Miljoenenjacht*, which first aired in 2000.

39 **the multi-armed bandit problem is no different:** Previous researchers had also found solutions for this "one-armed bandit" problem over a fixed interval (Bellman, "A Problem in the Sequential Design of Experiments"; Bradt, Johnson, and Karlin, "On Sequential Designs for Maximizing the Sum of N Observations").

39 **maximizing a single quantity that accounts for both:** The ideas behind the Gittins index were first presented at a conference in 1972 and appeared in the proceedings as Gittins and Jones, "A Dynamic Allocation Index for the Sequential Design of Experiments," but the canonical presentation is Gittins, "Bandit Processes and Dynamic Allocation Indices."

39 **we provide the Gittins index values:** The table of Gittins index scores for the Bernoulli bandit was taken from Gittins, Glazebrook, and Weber, *Multi-Armed Bandit Allocation Indices*, which is a comprehensive guide to the topic. It assumes complete ignorance about the probability of a payoff.

41 **drives us toward novelty:** Taking this to an extreme results in one simple strategy called the **Least Failures Rule**: always choose the option that's failed the fewest number of times. So, landing in a new city, pick a restaurant at random. If it is good, stick with it. As soon as it fails to satisfy, choose at random from the other restaurants. Continue this process until all restaurants have failed to satisfy once, then go back to the restaurant with the most nights of successful dining and repeat. This strategy builds on the win-stay principle, and it's precisely what the Gittins index yields if you're the patient sort who values tomorrow's payoff as being essentially as good as today's. (The rule appears in Kelly, "Multi-Armed Bandits with Discount Factor Near One"; formally, it is optimal under geometric discounting in the limit as the discount rate approaches 1.) In a big city with plenty of new restaurants opening all the time, a Least Failures policy says quite simply that if you're ever let down, there's too much else out there; don't go back.

41 **a variety of experiments in behavioral economics:** See, for example, Kirby, "Bidding on the Future."

42 **if there's a cost to switching:** This case is analyzed in Banks and Sundaram, "Switching Costs and the Gittins Index."

42 **"Regrets, I've had a few":** Frank Sinatra, "My Way," from *My Way* (1969), lyrics by Paul Anka.

42 **"For myself I am an optimist":** Prime Minister Winston Churchill, speech, Lord

Mayor's Banquet, London, November 9, 1954. Printed in Churchill, *Winston S. Churchill: His Complete Speeches.*

42 **"To try and fail is at least to learn":** Barnard, *The Functions of the Executive.*

43 **"wanted to project myself forward to age 80":** Jeff Bezos, interview with the Academy of Achievement, May 4, 2001, http://www.achievement.org/autodoc/page /bez0int-3.

43 **several key points about regret:** Lai and Robbins, "Asymptotically Efficient Adaptive Allocation Rules."

44 **the guarantee of minimal regret:** Ibid. offered the first such algorithms, which were refined by Katehakis and Robbins, "Sequential Choice from Several Populations"; Agrawal, "Sample Mean Based Index Policies"; and Auer, Cesa-Bianchi, and Fischer, "Finite-Time Analysis of the Multiarmed Bandit Problem," among others. The latter present perhaps the simplest strategy of this kind, which is to assign arm j a score of $\frac{s_j}{n_j} + \sqrt{(2\log n)/n_j}$, where s_j is the number of successes out of n_j plays on that arm, and $n = \Sigma_j n_j$ is the total number of plays of all arms. This is an upper bound on the probability of a successful payoff (which is just $\frac{s_j}{n_j}$). Choosing the arm with the highest score guarantees logarithmic regret (although there are tweaks to this score that result in better performance in practice).

44 **known as the "confidence interval":** Confidence intervals originate with Neyman, "Outline of a Theory of Statistical Estimation."

45 **"optimism in the face of uncertainty":** Kaelbling, Littman, and Moore, "Reinforcement Learning."

45 **"optimistic robots":** Leslie Kaelbling, personal interview, November 22, 2013. See Kaelbling, *Learning in Embedded Systems.*

45 **$57 million of additional donations:** Siroker and Koomen, *A/B Testing.*

45 **A/B testing works as follows:** Christian, "The A/B Test." Also informed by Steve Hanov, personal interview, August 30, 2013, and Noel Welsh, personal interview, August 27, 2013.

46 **In the case of Obama's donation page:** Dan Siroker, "How We Used Data to Win the Presidential Election" (lecture), Stanford University, May 8, 2009, available at https:// www.youtube.com/watch?v=71bH8z6iqSc. See also, Siroker, "How Obama Raised $60 Million," https://blog.optimizely.com/2010/11/29/how-obama-raised-60-million -by-running-a-simple-experiment/.

46 **live A/B tests on their users:** Google's first A/B test was run on February 27, 2000. See, e.g., Christian, "The A/B Test."

46 **Companies A/B test their site navigation:** See, e.g., Siroker and Koomen, *A/B Testing.*

46 **tested forty-one shades of blue:** Laura M. Holson, "Putting a Bolder Face on Google," *New York Times*, February 28, 2009.

46 **"how to make people click ads":** Ashlee Vance, "This Tech Bubble Is Different," *Bloomberg Businessweek*, April 14, 2011. http://www.bloomberg.com/bw/magazine/content/11 _17/b4225060960537.htm.

47 **"destroyed by madness":** Ginsberg, *Howl and Other Poems.*

47 **$50 billion in annual revenue:** Google's finances are detailed in their quarterly shareholder reports. Reported 2013 advertising revenue was $50.6 billion, roughly 91% of total revenue of $55.6 billion. See https://investor.google.com/financial/2013/tables.html.

47 **online commerce comprises hundreds of billions:** Online sales estimated by Forrester Research. See, for instance, "US Online Retail Sales to Reach $370B By 2017; €191B in Europe," *Forbes*, 3/14/2013, http://www.forbes.com/sites/forrester/2013/03/14/us-online -retail-sales-to-reach-370b-by-2017-e191b-in-europe/.

47 **best algorithms to use remain hotly contested:** Chris Stucchio, for instance, penned a cutting article titled "Why Multi-armed Bandit Algorithms Are Superior to A/B Testing," which was then countered by an equally cutting article called "Don't Use Bandit Algorithms—They Probably Won't Work for You"—also written by Chris Stucchio. See https://www.chrisstucchio.com/blog/2012/bandit_algorithms_vs_ab.html and https://www.chrisstucchio.com/blog/2015/dont_use_bandits.html. Stucchio's 2012 post was written partly in reference to an article by Paras Chopra titled "Why Multi-armed Bandit Algorithm Is Not 'Better' than A/B Testing" (https://vwo.com/blog/multi-armed-bandit-algorithm/), which was itself written partly in reference to an article by Steve Hanov titled "20 lines of code that will beat A/B testing every time" (http://stevehanov.ca/blog/index.php?id=132).

48 **it appeared in the *Washington Star*:** Jean Heller, "Syphilis Patients Died Untreated," *Washington Star*, July 25, 1972.

48 **document known as the Belmont Report:** *The Belmont Report: Ethical principles and guidelines for the protection of human subjects of research*, April 18, 1979. Available at http://www.hhs.gov/ohrp/humansubjects/guidance/belmont.html.

49 **proposed conducting "adaptive" trials:** See Zelen, "Play the Winner Rule and the Controlled Clinical Trial." While this was a radical idea, Zelen wasn't the first to propose it. That honor goes to William R. Thompson, an instructor in the School of Pathology at Yale, who formulated the problem of identifying whether one treatment is more effective than another, and proposed his own solution, in 1933 (Thompson, "On the Likelihood That One Unknown Probability Exceeds Another").

The solution that Thompson proposed—randomly sampling options, where the probability of choosing an option corresponds to the probability that it is the best based on the evidence observed so far—is the basis for much recent work on this problem in machine learning (we return to the algorithmic uses of randomness and sampling in chapter 9).

Neither Frederick Mosteller nor Herbert Robbins seemed to be aware of Thompson's work when they started to work on the two-armed bandit problem. Richard Bellman found the "little-known papers" a few years later, noting that "We confess that we found these papers in the standard fashion, namely while thumbing through a journal containing another paper of interest" (Bellman, "A Problem in the Sequential Design of Experiments").

50 **ECMO saved the life of a newborn girl:** University of Michigan Department of Surgery, "'Hope' for ECMO Babies," http://surgery.med.umich.edu/giving/stories/ecmo.shtml.

50 **has now celebrated her fortieth birthday:** University of Michigan Health System, "U-M Health System ECMO team treats its 2,000th patient," March 1, 2011, http://www.uofmhealth.org/news/ECMO%202000th%20patient.

50 **early studies in adults:** Zapol et al., "Extracorporeal Membrane Oxygenation in Severe Acute Respiratory Failure."

50 **a study on newborns:** Bartlett et al., "Extracorporeal Circulation in Neonatal Respiratory Failure."

50 **"did not justify routine use of ECMO":** Quotation from Ware, "Investigating Therapies of Potentially Great Benefit: ECMO," referring to conclusions in Ware and Epstein, "Comments on 'Extracorporeal Circulation in Neonatal Respiratory Failure,'" which is in turn a comment on Bartlett et al., "Extracorporeal Circulation in Neonatal Respiratory Failure."

51 **"difficult to defend further randomization ethically":** Ware, "Investigating Therapies of Potentially Great Benefit: ECMO."

51 **one of the world's leading experts:** It was Berry, in his 1971 PhD dissertation, who

proved that staying on a winner is optimal. The result was published as Berry, "A Bernoulli Two-Armed Bandit."

51 **"Ware study should not have been conducted":** Berry, "Comment: Ethics and ECMO."

51 **nearly two hundred infants in the United Kingdom:** UK Collaborative ECMO Group, "The Collaborative UK ECMO Trial."

52 **clinical trials for a variety of cancer treatments:** Don Berry, personal interview, August 22, 2013.

52 **the FDA released a "guidance" document:** The FDA's "Adaptive Design Clinical Trials for Drugs and Biologics" from February 2010 can be found at http://www.fda.gov /downloads/Drugs/Guidances/ucm201790.pdf.

52 **shown a box with two lights on it:** The study appears in Tversky and Edwards, "Information Versus Reward in Binary Choices."

53 **two airlines:** Meyer and Shi, "Sequential Choice Under Ambiguity."

53 **an experiment with a four-armed bandit:** Steyvers, Lee, and Wagenmakers, "A Bayesian Analysis of Human Decision-Making on Bandit Problems."

54 **what has been termed a "restless bandit":** Restless bandits were introduced by Whittle, "Restless Bandits," which discusses a strategy similar to the Gittins index that can be used in some cases. The computational challenges posed by restless bandits—and the consequent pessimism about efficient optimal solutions—are discussed in Papadimitriou and Tsitsiklis, "The Complexity of Optimal Queuing Network Control."

54 **when the world can change:** Navarro and Newell, "Information Versus Reward in a Changing World," provides recent results supporting the idea that human over-exploration is a result of assuming the world is restless.

54 **"There is in fact a sort of harmony":** Thoreau, "Walking."

54 **"A Coke is a Coke":** Warhol, *The Philosophy of Andy Warhol.*

55 **"a developmental way of solving the exploration/exploitation tradeoff":** Alison Gopnik, personal interview, August 22, 2013. See also Gopnik, *The Scientist in the Crib.*

56 **"a juncture in my reading life":** Lydia Davis, "Someone Reading a Book," *Can't and Won't: Stories.*

56 **challenging our preconceptions about getting older:** Carstensen, "Social and Emotional Patterns in Adulthood" presents the basic "socioemotional selectivity theory" we discuss in this section, as well as some of the evidence for it.

57 **"lifelong selection processes":** Ibid.

57 **about to move across the country:** Fredrickson and Carstensen, "Choosing Social Partners."

57 **their preferences became indistinguishable:** Fung, Carstensen, and Lutz, "Influence of Time on Social Preferences."

58 **older people are generally more satisfied:** Evidence of improvements in emotional well-being with aging are discussed in Charles and Carstensen, "Social and Emotional Aging."

3. SORTING

59 **"Nowe if the word":** Cawdrey, *A Table Alphabeticall,* is the first monolingual dictionary of English. For more on the history of sorting vis-à-vis searching, see Knuth, *The Art of Computer Programming,* §6.2.1. For more on the invention of alphabetical order, see Daly, *Contributions to a History of Alphabetization.*

59 **The roommate pulled a sock out:** Hillis, *The Pattern on the Stone.*

60 **posted to the programming website Stack Overflow:** "Pair socks from a pile efficiently?"

Submitted by user "amit" to Stack Overflow on January 19, 2013, http://stackoverflow
.com/questions/14415881/pair-socks-from-a-pile-efficiently.

As "amit" (real name Amit Gross, a graduate student at the Technion) writes: "Yes-
terday I was pairing the socks from the clean laundry, and figured out the way I was
doing it is not very efficient. I was doing a naive search—picking one sock and 'iterating'
the pile in order to find its pair. This requires iterating over $n/2 \times n/4 = n^2/8$ socks on average. As a computer scientist I was thinking what I could do?"

Amit's question generated a number of answers, but the one that received the most
support from his fellow programmers was to do a **Radix Sort:** identify the dimensions
along which the socks vary (e.g., color, pattern) and sort them into piles on each of these
dimensions. Each sort requires only one pass through all the socks, and the result is a
set of smaller piles. Even if you have to go through all the socks in those piles to find
matches, the amount of time this takes is proportional to the square of the size of the
largest *pile* rather than the square of the total number of socks. (See the endnote below
about sorting a deck of cards for more on Radix Sort.)

But if the reason we are pairing socks is to make it easier to find a pair of socks when
we need them, we can reduce the need for sorting by adopting a better procedure for
searching.

Let's say your socks differ along only one dimension—color—and you have three
different colors of loose, unpaired socks in your sock drawer. Then you are guaranteed
to find a matching pair if you take four socks out of the drawer at random. (To see why,
imagine the worst-case scenario: each of the first three socks that have been pulled out
are a different color. When you go back for a fourth, it has to match one of the three you
have pulled out already.) No matter how many colors you have, taking out one more sock
than the number of colors always guarantees you a matching pair. So don't bother pairing them if you're willing to have your morning run a little slower.

This neat solution to the problem of pairing socks comes courtesy of the Pigeonhole
Principle, a simple but powerful mathematical idea attributed to the nineteenth-century
German mathematician Peter Gustave Lejeune Dirichlet. (Rittaud and Heeffer, "The
Pigeonhole Principle," traces the history of the Pigeonhole Principle, including Dirich-
let as well as what appear to be even earlier references.) The idea is simple: if a group of
pigeons lands in a set of nesting holes, and there are more pigeons than holes, then at
least one hole must contain more than one pigeon. In computer science, the Pigeonhole
Principle is used to establish basic facts about the theoretical properties of algorithms.
For example, it is impossible to make an algorithm that will compress any possible file
without loss of information, because there are more long files than there are short files.

Applying the Pigeonhole Principle suggests a permanent solution to the problem of
sock pairing: only buy one kind of sock. If all your socks are the same, you never need to
pair them, because you can always get a pair by taking two socks out of the drawer. For
many computer scientists (including some of the programmers who responded to Amit's
question) this is the most elegant approach—redefining the problem so it can be solved
efficiently.

One last word of warning, though: when you buy that one kind of sock, be careful
what kind of socks you buy. The reason why Ron Rivest has particular problems with
socks is that he wears socks that are different for left and right feet. This thwarts the
Pigeonhole Principle—to guarantee a match with socks like that, you'll need to pull out
one more sock than the total number of pairs.

60   **"Socks confound me!":** Ronald Rivest, personal interview, July 25, 2013.

60 **"go blind and crazy":** Martin, "Counting a Nation by Electricity."

60 **"unerringly as the mills of the Gods":** Ibid.

60 **"no one will ever use it but governments":** Quoted in Austrian, *Herman Hollerith*.

60 **Hollerith's firm merged with several others:** Austrian, *Herman Hollerith*.

61 **first code ever written for a "stored program" computer:** "Written," here, means literally written out by hand: when the renowned mathematician John von Neumann jotted down the sorting program in 1945, the computer it was meant for was still several years away from completion. Although computer programs in general date back to Ada Lovelace's writing in 1843 on the proposed "Analytical Engine" of Charles Babbage, von Neumann's program was the first one designed to be stored in the memory of the computer itself; earlier computing machines were meant to be guided by punch cards fed into them, or wired for specific calculations. See Knuth, "Von Neumann's First Computer Program."

61 **outsort IBM's dedicated card-sorting machines:** Ibid.

61 **a quarter of the computing resources of the world:** Knuth, *The Art of Computer Programming*, p. 3.

62 **"unit cost of sorting, instead of falling, rises":** Hosken, "Evaluation of Sorting Methods."

63 **the record for sorting a deck of cards:** While we couldn't find a video of Bradáč's performance, there are plenty of videos online of people trying to beat it. They tend to sort cards into the four suits, and then sort the numbers within each suit. "But there is a faster way to do the trick!" urges Donald Knuth in *The Art of Computer Programming*: First, deal out the cards into 13 piles based on their face value (with one pile containing all the 2s, the next all the 3s, etc.). Then, after gathering up all the piles, deal the cards out into the four suits. The result will be one pile for each suit, with the cards ordered within each. This is a Radix Sort, and is related to the Bucket Sort algorithm we discuss later in the chapter. See Knuth, *The Art of Computer Programming*, §5.2.5.

63 **completely sorted by chance:** Sorting things by randomizing them and hoping for the best is actually an algorithm with a name: **Bogosort**, part of computer science's only partly tongue-in-cheek subfield of "pessimal algorithm design." Pessimality is to optimality what pessimism is to optimism; pessimal algorithm designers compete to outdo each other for the *worst* possible computing performance.

   Looking into the matter further, pessimal algorithm designers have concluded that Bogosort is actually far too lean and efficient. Hence their "improvement" **Bogobogosort**, which starts by incrementally Bogosorting the first two elements, then the first three, and so forth. If at any point in time the list gets out of order, Bogobogosort starts over. So the algorithm won't complete a sort of four cards, for instance, until it throws the first two up in the air, sees that they've landed correctly, then throws the first three in the air, sees that *they've* landed correctly, and at last throws the first four in the air and finds them in the correct order too. All in a row. Otherwise it starts over. One of the engineers to first write about Bogobogosort reports running it on his computer overnight and being unable to sort a list of seven items, before he finally turned off the electricity out of mercy.

   Subsequent engineers have suggested that Bogobogosort isn't even the bottom of the well, and have proposed getting even more meta and Bogosorting the *program* rather than the data: randomly flipping bits in the computer memory until it just so happens to take the form of a sorting program that sorts the items. The time bounds of such a monstrosity are still being explored. The quest for pessimality continues.

64 **Computer science has developed a shorthand:** Big-O notation originated in the 1894 book *Die analytische zahlentheorie* by Paul Bachmann. See also Donald Knuth, *The Art of Computer Programming*, §1.2.11.1. Formally, we say that the runtime of an algorithm

is $O(f(n))$ if it is less than or equal to a multiple (with a coefficient that is a positive constant) of $f(n)$. There is also the kindred "Big-Omega" notation, with $\Omega(f(n))$ indicating that the runtime is *greater* than or equal to a multiple of $f(n)$, and "Big-Theta" notation, with $\Theta(f(n))$ meaning the runtime is both $O(f(n))$ and $\Omega(f(n))$.

65 **"He had me at Bubble Sort":** This engineer is Dan Siroker, whom we met earlier in chapter 2. See, e.g., "The A/B Test: Inside the Technology That's Changing the Rules of Business," *Wired*, May 2012.

67 **information processing began in the US censuses:** For more details, see Knuth, *The Art of Computer Programming*, §5.5.

67 **to demonstrate the power of the stored-program computer:** The computer was the EDVAC machine, and at the time von Neumann's program was classified as top-secret military intelligence. See Knuth, "Von Neumann's First Computer Program."

68 **"Mergesort is as important in the history of sorting":** Katajainen and Träff, "A Meticulous Analysis of Mergesort Programs."

68 **large-scale industrial sorting problems:** The current records for sorting are hosted at http://sortbenchmark.org/. As of 2014, a group from Samsung holds the record for sorting the most data in a minute—a whopping 3.7 terabytes of data. That's the equivalent of almost 37 billion playing cards, enough to fill five hundred Boeing 747s to capacity, putting Zdeněk Bradáč's human record for sorting cards in perspective.

70 **167 books a minute:** Says shipping manager Tony Miranda, "We will process—I think our highest is—250 totes in one hour. Our average is about 180 totes in one hour. Keep in mind, each tote has about 40-plus items inside of it." From "KCLS AMH Tour," November 6, 2007, https://www.youtube.com/watch?v=4fq3CWsyde4.

70 **85,000 a day:** "Reducing operating costs," *American Libraries Magazine*, August 31, 2010, http://www.americanlibrariesmagazine.org/aldirect/al-direct-september-1-2010.

70 **"Fuhgeddaboutit":** See Matthew Taub, "Brooklyn & Manhattan Beat Washington State in 4th Annual 'Battle of the Book Sorters,'" *Brooklyn Brief*, October 29, 2014, http://brooklynbrief.com/4th-annual-battle-book-sorters-pits-brooklyn-washington-state/.

70 **the best we can hope to achieve:** A set of n items can have precisely $n!$ distinct orderings, so a sort produces exactly log $n!$ bits of information, which is approximately n log n bits. Recall that $n!$ is $n \times (n-1) \times \ldots \times 2 \times 1$, which is the product of n numbers, of which n is the largest. Consequently, $n! < n^n$, so log $n! <$ log n^n, which then gives us log $n! < n$ log n. This approximation of n log n for log $n!$ is called "Stirling's approximation," named for eighteenth-century Scottish mathematician James Stirling. Because a single pairwise comparison yields at most one bit of information, n log n comparisons are needed to fully resolve our uncertainty about which of the $n!$ possible orders of our n things is the right one. For more detail, see Knuth, *The Art of Computer Programming*, §5.3.1.

71 **"I know from experience":** Jordan Ho, personal interview, October 15, 2013.

73 **a paper on "email overload":** Whittaker and Sidner, "Email Overload."

73 **"sort of wasted a part of their life":** Steve Whittaker, personal interview, November 14, 2013.

74 **"At a Lawn Tennis Tournament":** Dodgson, "Lawn Tennis Tournaments."

75 **an awkward take on triple elimination:** For a computer-scientific critique of Dodgson's tournament proposal, see Donald Knuth's discussion of "minimum-comparison selection" in *The Art of Computer Programming*, §5.3.3.

76 **doesn't produce a full ordering:** An algorithm that, rather than ranking all of the items, identifies one of them as the largest or second-largest or median, etc., is known as a "selection" algorithm, rather than a sorting algorithm.

76 **schedulers for Major League Baseball:** Trick works as part of the Sports Scheduling Group, which he co-founded. From 1981 to 2004, the schedule for Major League Baseball was constructed by hand, by the remarkable husband-and-wife team of Henry and Holly Stephenson. ESPN chronicled the story of the Stephensons in a short film directed by Joseph Garner titled *The Schedule Makers.*

77 **"uncertainty is delayed in its resolution":** Michael Trick, personal interview, November 26, 2013.

77 **"practically no matter who they are":** Ibid.

77 **"A 3:2 score gives the winning team":** Tom Murphy, "Tuning in on Noise?" Published June 22, 2014 on the "Do the Math" blog: http://physics.ucsd.edu/do-the-math/2014/06/tuning-in-on-noise/

78 **recognizing the virtues of robustness in algorithms:** Ackley, "Beyond Efficiency."

78 **"bubble sort has no apparent redeeming features":** Knuth, *The Art of Computer Programming,* §5.5.

78 **The winner of that particular honor:** Dave Ackley, personal interview, November 26, 2013. See Jones and Ackley, "Comparison Criticality in Sorting Algorithms," and Ackley, "Beyond Efficiency." For more about Comparison Counting Sort (also sometimes known as Round-Robin Sort) see Knuth, *The Art of Computer Programming,* §5.2.

79 **"most important skill as a professional poker player":** Isaac Haxton, personal interview, February 20, 2014.

80 **"Imagine two monkeys":** Christof Neumann, personal interview, January 29, 2014.

81 **"aggressive acts per hen increased":** Craig, *Aggressive Behavior of Chickens.*

81 **There's a significant computational burden:** Jessica Flack, personal interview, September 10, 2014. See also DeDeo, Krakauer, and Flack, "Evidence of Strategic Periodicities in Collective Conflict Dynamics"; Daniels, Krakauer, and Flack, "Sparse Code of Conflict in a Primate Society"; Brush, Krakauer, and Flack, "A Family of Algorithms for Computing Consensus About Node State from Network Data." For a broader overview of Flack's work, see Flack, "Life's Information Hierarchy."

82 **This sporting contest is the marathon:** The marathon has an analogue in the world of sorting algorithms. One of the more intriguing (Wikipedia used the word "esoteric" before the article was removed entirely) developments in beyond-comparison sorting theory arose from one of the most unlikely places: the notorious Internet message board 4chan. In early 2011, an anonymous post there proclaimed: "Man, am I a genius. Check out this sorting algorithm I just invented." The poster's "sorting algorithm"—**Sleep Sort**—creates a processing thread for each unsorted item, telling each thread to "sleep" the number of seconds of its value, and then "wake up" and output itself. The final output should, indeed, be sorted. Leaving aside the implementation details that reveal the cracks in Sleep Sort's logic and just taking Sleep Sort on face value, it does seem to promise something rather intoxicating: a sort whose runtime doesn't depend on the *number* of elements at all, but rather on their *size.* (Thus it's still not *quite* as good as a straight-up $O(1)$ constant-time sort.)

82 **"You go to the money":** This is articulated by British entrepreneur Alexander Dean at https://news.ycombinator.com/item?id=8871524.

83 **"the bigger one is the dominant one":** The Law of Gross Tonnage, it seems, really does rule the ocean. This is not to say fish are *entirely* pacifistic. It's worth noting that they will fight—aggressively—when their sizes are similar.

4. CACHING

84 **"In the practical use of our intellect":** James, *Psychology.*

84 **Now you have two problems:** This construction nods to a famous programming joke first coined by Netscape engineer Jamie Zawinski in a Usenet post on August 12, 1997: "Some people, when confronted with a problem, think 'I know, I'll use regular expressions.' Now they have two problems."

84 **"How long have I had it?":** Stewart, *Martha Stewart's Homekeeping Handbook*.

84 **"Hang all your skirts together":** Jay, *The Joy of Less*.

84 **"Items will be sorted by type":** Mellen, *Unstuff Your Life!*

85 **"a very sharp consciousness but almost no memory":** Davis, *Almost No Memory*.

86 **one of the fundamental principles of computing:** Our history of caching is based on that provided by Hennessy and Patterson, *Computer Architecture*, which also has a great treatment of modern caching methods in computer design.

86 **an electrical "memory organ":** Burks, Goldstine, and von Neumann, *Preliminary Discussion of the Logical Design of an Electronic Computing Instrument*.

86 **a supercomputer in Manchester, England, called Atlas:** Kilburn et al., "One-Level Storage System."

86 **"automatically accumulates to itself words":** Wilkes, "Slave Memories and Dynamic Storage Allocation."

87 **implemented in the IBM 360/85 supercomputer:** Conti, Gibson, and Pitkowsky, "Structural Aspects of the System/360 Model 85."

87 **number of transistors in CPUs would double every two years:** Moore's initial 1965 prediction in "Cramming More Components onto Integrated Circuits" was for a doubling every year; in 1975 he then revised this in "Progress in Digital Integrated Electronics" to be a doubling every two years.

88 **six-layer memory hierarchy:** Registers; L1, L2, and L3 caches; RAM; and disk. For more on the "memory wall," see, for instance, Wulf and McKee, "Hitting the Memory Wall."

88 **"not to have useless facts elbowing out the useful ones":** Conan Doyle, "A Study in Scarlet: The Reminiscences of John H. Watson."

88 **"words cannot be preserved in it indefinitely":** Wilkes, "Slave Memories and Dynamic Storage Allocation."

88 **Bélády was born in 1928 in Hungary:** Bélády's personal history is based on an oral history interview he conducted with Philip L. Frana in 2002 (available at https://conservancy .umn.edu/bitstream/107110/1/oh352lab.pdf). His analysis of caching algorithms and results are presented in Bélády, "A Study of Replacement Algorithms for a Virtual-Storage Computer."

88 **the most cited piece of computer science research for fifteen years:** From Bélády himself: "My paper written in 1965 became the Citation Index most-referenced paper in the field of software over a 15-year period." J. A. N. Lee, "Laszlo A. Belady," in *Computer Pioneers*, http://history.computer.org/pioneers/belady.html.

89 **LRU consistently performed the closest to clairvoyance:** A couple of years later, Bélády also showed that FIFO has some curious additional drawbacks—in particular, rare cases where increasing the cache size can actually worsen performance, a phenomenon known as Bélády's Anomaly. Bélády, Nelson, and Shedler, "An Anomaly in Space-Time Characteristics of Certain Programs Running in a Paging Machine."

90 **"the digital equivalent of shuffling papers":** Aza Raskin, "Solving the Alt-Tab Problem," http://www.azarask.in/blog/post/solving-the-alt-tab-problem/.

90 **The literature on eviction policies:** If you're interested in trying a more complex caching algorithm, some popular variants on LRU are the following:

- LRU-K: O'Neil, O'Neil, and Weikum, "The LRU-K Page Replacement Algorithm for Database Disk Buffering," which looks at the time elapsed since the K-th most recent use (which is maximal for items in the cache that have not been used K times). This introduces a frequency bias. LRU-2, which focuses on the penultimate use, is most common.
- 2Q: Johnson and Shasha, "2Q: A Low Overhead High Performance Buffer Management Replacement Algorithm," which organizes items into two separate "queues" to capture a little bit of frequency information. Items start in the first queue, and are promoted to the second queue if they are referred to again while they are in the cache. Items are evicted from this second queue back into the first queue using LRU, which is also used to evict items from the first queue.
- LRFU: Lee et al., "LRFU: A Spectrum of Policies That Subsumes the Least Recently Used and Least Frequently Used Policies," which combines recency and frequency by assigning a numerical score to each item that is incremented when the item is used but decreases gradually over time.
- The Adaptive Replacement Cache (ARC): Megiddo and Modha, "Outperforming LRU with an Adaptive Replacement Cache Algorithm," which uses two queues in a similar fashion to 2Q but adapts the length of the queues based on performance.

All of these algorithms have been shown to outperform LRU in tests of cache-management performance.

90 **overwhelming favorite of computer scientists:** For instance, Pavel Panchekha wrote an article in 2012 for the Dropbox blog where he lays out Dropbox's reasoning for using LRU, at https://tech.dropbox.com/2012/10/caching-in-theory-and-practice/.

90 **Deep within the underground Gardner Stacks:** For those curious to know exactly what UC Berkeley students had been reading when we visited: Thoreau's *Walden*; critical texts on *Song of Myself*, Cormac McCarthy, James Merrill, Thomas Pynchon, Elizabeth Bishop, J. D. Salinger, Anaïs Nin, and Susan Sontag; *Drown* by Junot Díaz; *Telegraph Avenue* and *The Yiddish Policemen's Union* by Michael Chabon; *Bad Dirt* and *Bird Cloud* by Annie Proulx; *Mr. and Mrs. Baby* by Mark Strand; *The Man in the High Castle* by Philip K. Dick; the collected poetry and prose of William Carlos Williams; *Snuff* by Chuck Palahniuk; *Sula* by Toni Morrison; *Tree of Smoke* by Denis Johnson; *The Connection of Everyone with Lungs* by Juliana Spahr; *The Dream of the Unified Field* by Jorie Graham; *Naked*, *Me Talk Pretty One Day*, and *Dress Your Family in Corduroy and Denim* by David Sedaris; *Ariel* by Sylvia Plath and *Oleanna* by David Mamet; D. T. Max's biography of David Foster Wallace; *Like Something Flying Backwards*, *Translations of the Gospel Back into Tongues*, and *Deepstep Come Shining* by C. D. Wright; the prose of T. S. Eliot; *Eureka* by Edgar Allan Poe; *Billy Budd, Sailor* and a collection of short works in poetry and prose by Herman Melville; *The Aspern Papers*, *The Portrait of a Lady*, and *The Turn of the Screw* by Henry James; Harold Bloom on *Billy Budd*, *Benito Cereno*, and "Bartleby the Scrivener"; the plays of Eugene O'Neill; *Stardust* by Neil Gaiman; *Reservation Blues* by Sherman Alexie; *No Country for Old Men* by Cormac McCarthy; and more.

91 **"twelve years, that's the cutoff":** Elizabeth Dupuis, personal interview, September 16, 2014.

92 **"on the scale of a mile to the mile!":** Carroll, *Sylvie and Bruno Concluded*.

92 **A quarter of all Internet traffic:** Stephen Ludin, "Akamai: Why a Quarter of the Internet Is Faster and More Secure than the Rest," lecture, March 19, 2014, International

Computer Science Institute, Berkeley, California. As Akamai claims on their own site, "Akamai delivers between 15–30% of all Web traffic" (http://www.akamai.com/html /about/facts_figures.html).

93 **"distance matters":** Ludin, "Akamai."

93 **eschew any type of human-comprehensible organization:** Amazon's "chaotic storage" system is described here: http://www.ssi-schaefer.de/blog/en/order-picking/chaotic -storage-amazon/.

94 **Amazon was granted a patent:** The patent on preshipping commonly requested items is US Patent No. 8,615,473, granted December 24, 2013, "Method and system for anticipatory package shipping" by Joel R. Spiegel, Michael T. McKenna, Girish S. Lakshman, and Paul G. Nordstrom, on behalf of Amazon Technologies Inc.

94 **which the press seized upon:** See, e.g., Connor Simpson, "Amazon Will Sell You Things Before You Know You Want to Buy Them," *The Wire*, January 20, 2014, http://www .thewire.com/technology/2014/01/amazon-thinks-it-can-predict-your-future/357188/; Chris Matyszczyk, "Amazon to Ship Things Before You've Even Thought of Buying Them?," *CNET*, January 19, 2014, http://www.cnet.com/news/amazon-to-ship-things -before-youve-even-thought-of-buying-them/.

94 **each state's "Local Favorites" from Netflix:** Micah Mertes, "The United States of Netflix Local Favorites," July 10, 2011, http://www.slacktory.com/2011/07/united-states -netflix-local-favorites/.

94 **the enormous files that comprise full-length HD video:** In 2012, Netflix announced that it was tired of paying firms like Akamai and had started building its own global CDN. See Eric Savitz, "Netflix Shifts Traffic to Its Own CDN," *Forbes*, June 5, 2012, http://www.forbes.com/sites/ericsavitz/2012/06/05/netflix-shifts-traffic-to-its-own-cdn -akamai-limelight-shrs-hit/. More information about Netflix's Open Connect CDN can be found at https://www.netflix.com/openconnect.

95 **"Caching is such an obvious thing":** John Hennessy, personal interview, January 9, 2013.

95 **"a crate on the floor of my front coat closet":** Morgenstern, *Organizing from the Inside Out*.

95 **"extra vacuum cleaner bags behind the couch":** Jones, *Keeping Found Things Found*.

96 **search engines from a cognitive perspective:** See Belew, *Finding Out About*.

96 **recommended the use of a valet stand:** Rik Belew, personal interview, October 31, 2013.

96 **"a very fundamental principle in my method":** Yukio Noguchi, personal interview, December 17, 2013.

97 **the "super" filing system was born:** Noguchi's filing system is described in his book *Super Organized Method*, and was initially presented in English by the translator William Lise. The blog article describing the system is no longer available on Lise's site, but it can still be visited via the Internet Archive at https://web.archive.org/web /20031223072329/http://www.lise.jp/honyaku/noguchi.html. Further information comes from Yukio Noguchi, personal interview, December 17, 2013.

98 **The definitive paper on self-organizing lists:** Sleator and Tarjan, "Amortized Efficiency of List Update and Paging Rules," which also provided the clearest results on the theoretical properties of the LRU principle.

98 **"God's algorithm if you will":** Robert Tarjan, personal interview, December 17, 2013.

98 **if you follow the LRU principle:** This application of the LRU principle to self-organizing lists is known as the **Move-to-Front** algorithm.

98 **not merely efficient. It's actually optimal:** This doesn't mean you must entirely give up

on categorization. If you want to make things a bit more gaudy and speed up the search process, Noguchi suggests putting colored tabs on files that fall into different categories. That way if you know you're looking for, say, accounts, you can restrict your linear search to just those items. And they will still be sorted according to the Move-to-Front Rule within each category.

99 **the information retrieval systems of university libraries:** Anderson's findings on human memory are published in Anderson and Milson, "Human Memory," and in the book *The Adaptive Character of Thought*. This book has been influential for laying out a strategy for analyzing everyday cognition in terms of ideal solutions, used by Tom and many others in their research. Anderson and Milson, "Human Memory," in turn, draws from a statistical study of library borrowing that appears in Burrell, "A Simple Stochastic Model for Library Loans."

99 **the missing piece in the study of the mind:** Anderson's initial exploration of connections between information retrieval by computers and the organization of human memory was conducted in an era when most people had never interacted with an information retrieval system, and the systems in use were quite primitive. As search engine research has pushed the boundaries of what information retrieval systems can do, it's created new opportunities for discovering parallels between minds and machines. For example, Tom and his colleagues have shown how ideas behind Google's PageRank algorithm are relevant to understanding human semantic memory. See Griffiths, Steyvers, and Firl, "Google and the Mind."

100 **"I saw that framework laid out before me":** Anderson, *The Adaptive Character of Thought*.

100 **analyzed three human environments:** The analysis of the environment of human memory is presented in Anderson and Schooler, "Reflections of the Environment in Memory."

100 **reality itself has a statistical structure:** "Human memory mirrors, with a remarkable degree of fidelity, the structure that exists in the environment." Ibid.

101 **"fail to appreciate the task before human memory":** Ibid.

102 **"A big book is a big nuisance":** The quotation in Greek is "μέγα βιβλίον μέγα κακόν" (*mega biblion, mega kakon*), which has also been translated as "Big book, big evil." The original reference is intended as a disparagement of epic poetry, but presumably being a scholar at a time when books were in the form of scrolls dozens of feet long meant that big books were a nuisance in more ways than aesthetic. There's a reason why the practice of citation and quotation didn't properly begin until books came in codices with numbered pages. For an excellent recounting of this history, see Boorstin, *The Discoverers*.

102 **"If you make a city bigger":** John Hennessy, personal interview, January 9, 2014.

103 **an unavoidable consequence of the amount of information:** Ramscar et al., "The Myth of Cognitive Decline."

103 **"minds are natural information processing devices:"** Michael Ramscar, "Provider Exclusive: Michael Ramscar on the 'Myth' of Cognitive Decline," interview with Bill Myers, February 19, 2014. http://www.providermagazine.com/news/Pages/0214 /Provider-Exclusive-Michael-Ramscar-On-The-Myth-Of-Cognitive-Decline.aspx.

5. SCHEDULING

105 **"How we spend our days":** Dillard, *The Writing Life*.

105 **"Book-writing, like war-making":** Lawler, "Old Stories."

105 **"We are what we repeatedly do":** In fact, this phrase, frequently attributed to Aristotle

himself, originated with scholar Will Durant, as a summary (in Durant's words) of Aristotle's thinking. See Durant, *The Story of Philosophy*.

106 **any task of two minutes or less:** Allen, *Getting Things Done*.

106 **beginning with the most difficult task:** Tracy, *Eat That Frog!* The book ascribes its titular quotation—"Eat a live frog first thing in the morning and nothing worse will happen to you the rest of the day"—to Mark Twain, although this attribution may be apocryphal. The Quote Investigator website cites eighteenth-century French writer Nicolas Chamfort as the more likely source. See http://quoteinvestigator.com/2013/04/03/eat-frog/ for more.

106 **first scheduling one's social engagements:** Fiore, *The Now Habit*.

106 **"the eternal hanging on of an uncompleted task":** William James, in a letter to Carl Stumpf, January 1, 1886.

106 **deliberately *not* doing things right away:** Partnoy, *Wait*.

106 **developed the Gantt charts:** The role of Taylor and Gantt in the history of scheduling is summarized in Herrmann, "The Perspectives of Taylor, Gantt, and Johnson." Additional biographical details on Taylor are from Kanigel, *The One Best Way*.

106 **firms like Amazon, IKEA, and SpaceX:** Gantt chart software company LiquidPlanner boasts Amazon, IKEA, and SpaceX among its clients at the (counterintuitive) URL http://www.liquidplanner.com/death-to-gantt-charts/.

106 **first hint that this problem even *could* be solved:** Johnson's seminal result (on what is now called "flowshop" scheduling, where jobs flow from one machine to another) appears in "Optimal Two- and Three-Stage Production Schedules with Setup Times Included."

108 **start with the task due soonest:** Earliest Due Date (EDD), also known as Jackson's Rule, was derived in Jackson, *Scheduling a Production Line to Minimize Maximum Tardiness*. James R. Jackson grew up in Los Angeles in the 1930s, and through his work with UCLA's Logistics Research Project spent time visiting machine shops run by various aerospace companies in the area. His thinking about how jobs moved from one machine to another ultimately led him to develop a mathematics for analyzing "network flows"—work that would later be used in the design of algorithms for routing the flow of traffic on the Internet. A brief biography appears in Production and Operations Management Society, "James R. Jackson."

109 **Moore's Algorithm:** Presented in Moore, "An *N* Job, One Machine Sequencing Algorithm for Minimizing the Number of Late Jobs." In the paper, Moore acknowledged a simplification and optimization that had been suggested to him by Thom J. Hodgson. Today the terms "Moore's Algorithm," "Hodgson's Algorithm," and the "Moore–Hodgson Algorithm" are sometimes used interchangeably.

110 **do the quickest task you can:** Shortest Processing Time (SPT), or Smith's Rule, was shown to minimize the sum of completion times in Smith, "Various Optimizers for Single-Stage Production."

111 **shows up in studies of animal foraging:** Stephens and Krebs, *Foraging Theory*.

111 **known as the "debt snowball":** In the popular sphere, author and speaker Dave Ramsey is perhaps the best-known popularizer and advocate of the "debt snowball" strategy, and has garnered many supporters and detractors alike. On the academic side, a 2012 paper by business school researchers at Northwestern, Gal and McShane, "Can Small Victories Help Win the War?" and a 2014 paper by economists at Texas A&M Brown and Lahey, *Small Victories*, for instance, have looked at the impact of "small victories" in helping people get out of consumer debt.

112 **an obsessive-compulsive vampire:** This episode is Season 5, Episode 12, "Bad Blood," which originally aired February 22, 1998.

112 **"a tendency to pre-crastinate"**: Rosenbaum, Gong, and Potts, "Pre-Crastination."

114 **Reeves would blame the bug on "deadline pressures"**: This comes from an email dated December 15, 1997, from Glenn Reeves to his colleagues, subject line "What really happened on Mars?," available online at http://research.microsoft.com/en-us/um/people /mbj/Mars_Pathfinder/Authoritative_Account.html.

115 **"If you're flammable and have legs"**: Hedberg's tale can be found on his 1999 comedy album *Strategic Grill Locations*.

115 **"Things which matter most"**: The first appearance of this quotation in English seems to be in Covey, *How to Succeed with People*, where it's attributed to Goethe without citation.

115 **"That's how I get things done every day"**: Laura Albert McLay, personal interview, September 16, 2014.

115 **"Gene was postponing something"**: Jan Karel Lenstra, personal interview, September 2, 2014; and personal correspondence.

116 **Lawler took an intriguingly circuitous route**: Lawler's biography is drawn from Lawler, "Old Stories," and Lenstra, "The Mystical Power of Twoness."

116 **"the social conscience" of the computer science department**: Richard Karp, "A Personal View of Computer Science at Berkeley," EECS Department, University of California, Berkeley, http://www.eecs.berkeley.edu/BEARS/CS_Anniversary/karp-talk.html.

116 **an award in Lawler's name**: See http://awards.acm.org/lawler/.

116 **build the schedule back to front**: Lawler's analysis of precedence constraints for the maximum lateness problem is in Lawler, "Optimal Sequencing of a Single Machine Subject to Precedence Constraints."

116 **it's what the field calls "intractable"**: This analysis is in Lawler, "Sequencing Jobs to Minimize Total Weighted Completion Time Subject to Precedence Constraints." More precisely, the problem is "*NP*-hard," meaning it has no known efficient solution, and might never have one.

117 **a quest to map the entire landscape of scheduling theory**: The quest emerged one afternoon in 1975, as Lawler, Lenstra, and their colleagues Richard Karp and Ben Lageweg sat around talking scheduling theory in the Mathematisch Centrum in Amsterdam. Perhaps it was the "pungent odors of malt and hops" in the air from the Amstel brewery next door, but something inspired the group to decide that a book containing a list of *all* scheduling problems and whether they had been solved would make a nice gift for their friend and colleague Alexander Rinnooy Kan, who was about to defend his thesis. (This story appears in Lawler, "Old Stories," and Lenstra, "The Mystical Power of Twoness.") Rinnooy Kan would go on to make important contributions not just to academia but also to the Dutch economy, sitting on the board of directors at ING and being named by the newspaper *De Volkskrant* as the most influential person in the Netherlands—three years in a row. See "Rinnooy Kan weer invloedrijkste Nederlander," *De Volkskrant*, December 4, 2009, http://nos.nl/artikel/112743-rinnooy-kan-weer-invloedrijkste-nederlander.html.

Lageweg wrote a computer program that generated the list, enumerating some 4,536 different permutations of the scheduling problem: every possible combination of metrics (maximum lateness, number of late jobs, sum of completion times, etc.) and constraints (weights, precedence, start times, and so on) that they could think of. Over a series of enthralling days, the group "had the pleasure of knocking off one obscure problem type after another in rapid succession."

Their organizational schema for describing the zoo of scheduling problems was a language "laced with shorthand," which they called "Schedulese" (Graham et al., "Optimization and Approximation in Deterministic Sequencing"). The basic idea is that scheduling problems are described by three variables: the nature of the machines involved, the nature of the jobs, and the goal of scheduling. These three variables are specified in that order, with standard codes describing factors such as precedence constraints, preemption, release times, and the goal. For example, $1|r_j|\Sigma C_j$ (pronounced "one-arejay-sum-ceejay") represents a single machine, release times, and the goal of minimizing the sum of completion times. As Eugene Lawler recounts:

> An immediate payoff was the consummate ease with which we could communicate problem types. Visitors to our offices were sometimes baffled to hear exchanges such as: "Since one-arejay-sum-ceejay is NP-hard, does that imply that one-*preemption*-arejay-sum-ceejay is NP-hard, too?" "No, that's easy, remember?" "Well, one-deejay-sum-ceejay is easy and that implies one-*preemption*-deejay-sum-ceejay is easy, so what do we know about one-preemption-arejay-deejay-sum-ceejay?" "Nothing."

(In formal notation: "Since $1|r_j|\Sigma C_j$ is NP-hard, does that imply that $1|pmtn, r_j|\Sigma C_j$ is NP-hard, too?" "No, that's easy, remember?" "Well, $1|d_j|\Sigma C_j$ is easy and that implies $1|pmtn, d_j|\Sigma C_j$ is easy, so what do we know about $1|pmtn, r_j, d_j|\Sigma C_j$?" "Nothing" [Lawler et al., "A Gift for Alexander!"; see also Lawler, "Old Stories"].)

117 **the problem becomes intractable:** In fact, it's equivalent to the "knapsack problem," computer science's most famously intractable problem about how to fill space. The connection between this scheduling problem and the knapsack problem appears in Lawler, *Scheduling a Single Machine to Minimize the Number of Late Jobs*.

117 **a certain time to start some of your tasks:** What we are calling "start times" are referred to in the literature (we think somewhat ambiguously) as "release times." Lenstra, Rinnooy Kan, and Brucker, "Complexity of Machine Scheduling Problems," showed that both minimizing sum of completion times and minimizing maximal lateness with arbitrary release times are *NP*-hard. The case of minimizing the number of late jobs with arbitrary release times is discussed in Lawler, "Scheduling a Single Machine to Minimize the Number of Late Jobs."

117 **A recent survey:** Lawler et al., "Sequencing and Scheduling." The most recent version of this list is available at http://www.informatik.uni-osnabrueck.de/knust/class/.

118 **with a fairly straightforward modification:** The effect of preemption on minimizing maximal lateness with release times is analyzed in Baker et al., "Preemptive Scheduling of a Single Machine." The problem of minimizing the sum of completion times with release times and preemption is analyzed in Schrage, "A Proof of the Optimality of the Shortest Remaining Processing Time Discipline" and Baker, *Introduction to Sequencing and Scheduling*.

118 **still the preemptive version of Earliest Due Date:** The result for minimizing expected maximum lateness by choosing the job with earliest due date is discussed in Pinedo, *Scheduling*.

118 **the preemptive version of Shortest Processing Time:** The effectiveness of choosing the job with the weighted shortest expected processing time for minimizing the sum of weighted completion times in a dynamic setting (provided the estimate of the time to complete a job is nonincreasing in the duration worked on that job) was shown by Sevcik, "Scheduling for Minimum Total Loss Using Service Time Distributions," as part of a more general strategy for dynamic scheduling.

119 **sum of the weighted lateness of those jobs:** Pinedo, "Stochastic Scheduling with Release Dates and Due Dates," showed that this algorithm is optimal for these problems under the (fairly strong) assumption that the times of jobs follow a memoryless distribution,

which means that your estimate of how long they will take remains constant no matter how long you have been doing them. In stochastic scheduling, optimal algorithms won't necessarily be ideal for every possible workload, but rather minimize the *expected* values of their relevant metrics.

119 **"Replace 'plan' with 'guess'"**: Jason Fried, "Let's just call plans what they are: guesses," July 14, 2009, https://signalvnoise.com/posts/1805-lets-just-call-plans-what-they-are -guesses.

119 **"You must not get off the train"**: Ullman, "Out of Time."

120 **can include both delays and errors**: Monsell, "Task Switching."

120 **"I'll just do errands instead"**: Kirk Pruhs, personal interview, September 4, 2014.

121 **"You have part of my attention"**: *The Social Network*, screenplay by Aaron Sorkin; Columbia Pictures, 2010.

121 **"Nobody knew anything about that"**: Peter Denning, personal interview, April 22, 2014.

122 **"caused a complete collapse of service"**: Denning, "Thrashing: Its Causes and Prevention."

122 **"The caches are warm for the current workload"**: Peter Zijlstra, personal interview, April 17, 2014.

122 **any situation where the system grinds to a halt**: Thrashing can also take place in database systems, where the competition between different processes to acquire "locks" to access the database can swamp the system's ability to let the processes currently holding the locks get anything done. Similarly, thrashing can appear in networking contexts, where a cacophony of different signals competing for the network channel can prevent anything at all from getting through. We'll take a closer look at the latter scenario in chapter 10.

124 **replaced their scheduler**: The "$O(n)$ Scheduler" used by Linux starting with version 2.4 in 2001 sorted all processes by priority, which took longer the more processes there were. This was scrapped in favor of the "$O(1)$ Scheduler" starting with Linux 2.6 in 2003, which bucket-sorted all processes into a predetermined number of buckets, regardless of how many processes there were. However, doing this bucket sort required computing complex heuristics, and beginning with Linux 2.6.23 in 2007, the "$O(1)$ Scheduler" was replaced with the even more straightforward "Completely Fair Scheduler."

125 **In Linux this minimum useful slice**: This value is defined in the Linux kernel's "Completely Fair Scheduler" in the variable sysctl_sched_min_granularity.

125 **Methods such as "timeboxing" or "pomodoros"**: Timeboxing has been written about widely in the context of the management of software development teams; the term "timeboxing" appears to originate with Zahniser, "Timeboxing for Top Team Performance." The "Pomodoro Technique," whose name comes from a tomato-shaped kitchen timer (the Italian word for tomato being *pomodoro*), was devised by Francesco Cirillo in the late 1980s and has been taught by Cirillo starting in 1998. See, e.g., Cirillo, *The Pomodoro Technique*.

125 **programmers have turned to psychology**: E.g., Peter Zijlstra, personal interview, April 17, 2014.

126 **Computers themselves do something like this**: Linux added support for timer coalescing in 2007; Microsoft included it in Windows starting with Windows 7 in 2009; and Apple followed suit in OS X Mavericks in 2013.

126 **"just a one-line bug in your algorithm"**: Peter Norvig, personal interview, September 17, 2014.

127 **"I don't swap in and out"**: Shasha and Lazere, *Out of Their Minds*, 101.

127 **"my role is to be on the bottom of things"**: Donald Knuth, "Knuth versus Email," http://www-cs-faculty.stanford.edu/~uno/email.html.

6. BAYES'S RULE

128 **"All human knowledge is uncertain"**: Bertrand Russell, *Human Knowledge: Its Scope and Limits*, 1948, p. 527.

128 **There he saw the Berlin Wall:** Gott, "Implications of the Copernican Principle for Our Future Prospects."

129 **"The Unreasonable Effectiveness of Data"**: The talk was derived from Halevy, Norvig, and Pereira, "The Unreasonable Effectiveness of Data."

129 **"these arguments must be probable only"**: *An Enquiry Concerning Human Understanding*, §IV, "Sceptical Doubts Concerning the Operations of the Understanding."

129 **Bayes's own history:** Our brief biography draws on Dale, *A History of Inverse Probability*, and Bellhouse, "The Reverend Thomas Bayes."

129 **in either 1746, '47, '48, or '49:** Bayes's legendary paper, undated, had been filed between a pair of papers dated 1746 and 1749. See, e.g., McGrayne, *The Theory That Would Not Die*.

130 **defense of Newton's newfangled "calculus"**: *An Introduction to the Doctrine of fluxions, and Defence of the Mathematicians against the Objections of the Author of the analyst, so far as they are assigned to affect their general methods of Reasoning*.

130 **"deserves to be preserved"**: Introduction to Bayes, "An Essay Towards Solving a Problem in the Doctrine of Chances."

130 **"the proportion of *Blanks* to *Prizes*"**: Appendix to ibid.

130 **we need to first reason *forward*:** To be precise, Bayes was arguing that given hypotheses h and some observed data d, we should evaluate those hypotheses by calculating the likelihood $p(d|h)$ for each h. (The notation $p(d|h)$ means the "conditional probability" of d given h—that is, the probability of observing d if h is true.) To convert this back into a probability of each h being true, we then divide by the sum of these likelihoods.

131 **Laplace was born in Normandy:** For more details on Laplace's life and work, see Gillispie, *Pierre-Simon Laplace*.

132 **distilled down to a single estimate:** Laplace's Law is derived by working through the calculation suggested by Bayes—the tricky part is the sum over all hypotheses, which involves a fun application of integration by parts. You can see a full derivation of Laplace's Law in Griffiths, Kemp, and Tenenbaum, "Bayesian Models of Cognition." From the perspective of modern Bayesian statistics, Laplace's Law is the posterior mean of the binomial rate using a uniform prior.

132 **If you try only once and it works out:** You may recall that in our discussion of multiarmed bandits and the explore/exploit dilemma in chapter 2, we also touched on estimates of the success rate of a process—a slot machine—based on a set of experiences. The work of Bayes and Laplace undergirds many of the algorithms we discussed in that chapter, including the Gittins index. Like Laplace's Law, the values of the Gittins index we presented there assumed that any probability of success is equally likely. This implicitly takes the expected overall win rate for a slot machine with a 1–0 record to be two-thirds.

133 **"no more consistent or conceivable than the rest"**: *An Enquiry Concerning Human Understanding*, §IV, "Sceptical Doubts Concerning the Operations of the Understanding."

134 **the real heavy lifting was done by Laplace:** In fairness, an influential 1950 paper (Bailey, *Credibility Procedures*) referred to "Laplace's Generalization of Bayes's Rule," but it didn't quite stick. Discoveries being named after somebody other than their discoverer is a sufficiently common phenomenon that statistician and historian Stephen Stigler has

asserted that it should be considered an empirical law—Stigler's Law of Eponymy. Of course, Stigler wasn't the first person to discover this; he assigns the credit to sociologist Robert K. Merton. See Stigler, "Stigler's Law of Eponymy."

134 **multiply their probabilities together:** For the mathematically inclined, here's the full version of Bayes's Rule. We want to calculate how much probability to assign a hypothesis h given data d. We have prior beliefs about the probability of that hypothesis being true, expressed in a prior distribution $p(h)$. What we want to compute is the "posterior" distribution, $p(h|d)$, indicating how we should update our prior distribution in light of the evidence provided by d. This is given by

$$p(h|d) = \frac{p(d|h)p(h)}{\sum_{h'} p(d|h')p(h')}$$

where h' ranges over the full set of hypotheses under consideration.

134 **"especially about the future":** The uncertain origins of this saying are described in detail in *Quote Investigator*, "It's Difficult to Make Predictions, Especially About the Future," http://quoteinvestigator.com/2013/10/20/no-predict/.

135 **surprising if there were even a New York City:** The *New Yorker* cover is Richard McGuire, "Time Warp," November 24, 2014. For a fascinating and more detailed analysis of the probable life spans of cities and corporations, see the work of Geoffrey West and Luis Bettencourt—e.g., Bettencourt et al., "Growth, Innovation, Scaling, and the Pace of Life in Cities."

136 **a flurry of critical correspondence:** For example, see Garrett and Coles, "Bayesian Inductive Inference and the Anthropic Principles" and Buch, "Future Prospects Discussed."

136 **a raffle where you come in knowing nothing:** The statistician Harold Jeffreys would later suggest, instead of Laplace's $\frac{w+1}{n+2}$, using rather $\frac{w+0.5}{n+1}$, which results from using an "uninformative" prior rather than the "uniform" prior (Jeffreys, *Theory of Probability*; Jeffreys, "An Invariant Form for the Prior Probability in Estimation Problems"). One method for defining more informative priors results in predictions of the form $\frac{w+w'+1}{n+n'+2}$, where w' and n' are the number of wins and attempts for similar processes in your past experience (for details see Griffiths, Kemp, and Tenenbaum, "Bayesian Models of Cognition"). Using this rule, if you had previously seen 100 lottery drawings with only 10 winning tickets ($w = 10$, $n = 100$), your estimate after seeing a single winning draw for this new lottery would be a much more reasonable 12/103 (not far from 10%). Variants on Laplace's Law are used extensively in computational linguistics, where they provide a way to estimate the probabilities of words that have never been seen before (Chen and Goodman, "An Empirical Study of Smoothing Techniques for Language Modeling").

137 **or last for five millennia:** For a quantity like a duration, which ranges from 0 to ∞, the uninformative prior on times t is the probability density $p(t) \propto 1/t$. Changing the scale—defining a new quantity s that is a multiple of t—doesn't change the form of this distribution: if $s = ct$, then $p(s) \propto p(t = s/c) \propto 1/s$. This means that it is scale-invariant. Lots more information on uninformative priors appears in Jeffreys, *Theory of Probability*, and Jeffreys, "An Invariant Form for the Prior Probability in Estimation Problems."

137 **the Copernican Principle emerges:** This was shown by Gott, "Future Prospects Discussed," in responding to Buch, "Future Prospects Discussed."

137 **determining the number of tramcars:** Jeffreys, *Theory of Probability*, §4.8. Jeffreys credits mathematician Max Newman for bringing the problem to his attention.

137 **sought to estimate the number of tanks:** This has come to be known as the "German Tank Problem," and has been documented in a number of sources. See, e.g., Gavyn

Davies, "How a Statistical Formula won the War," the *Guardian*, July 19, 2006, http://www.theguardian.com/world/2006/jul/20/secondworldwar.tvandradio.

138 **fruits in an orchard:** For instance, the 2002 New Zealand Avocado Growers Association Annual Research Report found that "by April, fruit size profiles were normally distributed and remained so for the remainder of the monitored period."

138 **The average population of a town:** This figure comes from Clauset, Shalizi, and Newman, "Power-Law Distributions in Empirical Data," which in turn cites the 2000 US Census.

138 **can plausibly range over many scales:** The general form of a power-law distribution on a quantity t is $p(t) \propto t^{-\gamma}$, where the value of γ describes how quickly the probability of t decreases as t gets larger. As with the uninformative prior, the form of the distribution doesn't change if we take $s = ct$, changing the scale.

139 **a domain full of power laws:** The observation that wealth is distributed according to a power-law function is credited to Pareto, *Cours d'économie politique*. Another good discussion of the power-law distributions of populations and incomes is Simon, "On a Class of Skew Distribution Functions."

139 **The mean income in America:** The mean individual adjusted gross income (AGI), derived from IRS filings, was estimated to be $55,688 for the 2009 tax year, the most recent year for which an estimate was available; see the 2011 working paper "Evaluating the Use of the New Current Population Survey's Annual Social and Economic Supplement Questions in the Census Bureau Tax Model," available at https://www.census.gov/content/dam/Census/library/working-papers/2011/demo/2011_SPM_Tax_Model.pdf, which in turn cites data from the US Census Bureau's 2010 Current Population Survey Annual Social and Economic Supplement.

139 **two-thirds of the US population make less than the mean income:** The cutoff for the top 40% of AGI in 2012 was $47,475, and the cutoff for the top 30% was $63,222, from which we can infer that an AGI of $55,688 lands at approximately the top 33%. See Adrian Dungan, "Individual Income Tax Shares, 2012," *IRS Statistics of Income Bulletin*, Spring 2015, available at https://www.irs.gov/pub/irs-soi/soi-a-ints-id1506.pdf.

139 **the top 1% make almost ten times the mean:** The cutoff for the top 1% was an AGI of $434,682 in 2012, and the cutoff for the top 0.01% was $12,104,014. Ibid.

139 **the process of "preferential attachment":** A good general-audience discussion of the idea of power-law distributions emerging from preferential attachment can be found in Barabási, *Linked*.

139 **"'could go on forever' in a good way?":** Lerner, *The Lichtenberg Figures*.

139 **appropriate prediction strategy is a Multiplicative Rule:** All the prediction rules discussed in this section are derived in Griffiths and Tenenbaum, "Optimal Predictions in Everyday Cognition."

140 **poems follow something closer to a power-law:** Ibid.

141 **formalized the spread of intervals:** Erlang first modeled the rate of phone calls appearing on a network using a Poisson distribution in "The Theory of Probabilities and Telephone Conversations," and in turn developed the eponymous Erlang distribution for modeling the intervals between arriving calls in "Solution of Some Problems in the Theory of Probabilities of Significance in Automatic Telephone Exchanges." For more details on Erlang's life, see Heyde, "Agner Krarup Erlang."

141 **odds of which are about 20 to 1:** To be precise, the odds against being dealt a blackjack hand in the eponymous game are exactly 2,652 to 128, or about 20.7 to 1. To see the derivation of why this leads to an expectation of playing 20.7 hands before getting it, we can

define our expectation recursively: either we land blackjack for a result of 1, or we don't (in which case we're back where we started one hand later). If x is our expectation, $x = 1 + (2524/2652)x$, where 2524/2652 is our chance of *not* getting dealt blackjack. Solving for x gives about 20.7.

141 **known to statisticians as "memoryless"**: Technically, the time to the next blackjack follows a geometric distribution (similar to the exponential distribution for a continuous quantity), which is constantly decreasing, rather than the more wing-like Erlang distribution we describe in the main text. However, both can yield memoryless predictions under the right circumstances. If we encounter a particular phenomenon at some random point in its duration, as Gott assumed regarding the Berlin Wall, then the wing-like Erlang gives us memoryless Additive Rule predictions. And if we continuously observe a phenomenon that has a geometric distribution, as in playing a game of blackjack, the same kind of Additive Rule predictions result.

143 **Kenny Rogers famously advised**: "The Gambler" is best known as sung by Kenny Rogers on his 1978 album of the same name, but it was originally written and performed by Don Schlitz. The Rogers recording of the song would go on to reach the top spot on the Billboard country charts, and win the 1980 Grammy for Best Male Country Vocal Performance.

144 **"I breathed a very long sigh of relief"**: Gould, "The Median Isn't the Message."

144 **asking people to make predictions**: Griffiths and Tenenbaum, "Optimal Predictions in Everyday Cognition."

145 **people's prior distributions across a broad swath**: Studies have examined, for example, how we manage to identify moving shapes from the patterns of light that fall on the retina, infer causal relationships from the interactions between objects, and learn the meaning of new words after seeing them just a few times. See, respectively, Weiss, Simoncelli, and Adelson, "Motion Illusions as Optimal Percepts"; Griffiths et al., "Bayes and Blickets"; Xu and Tenenbaum, "Word Learning as Bayesian Inference."

145 **famous "marshmallow test"**: Mischel, Ebbesen, and Raskoff Zeiss, "Cognitive and Attentional Mechanisms in Delay of Gratification."

146 **all depends on what kind of situation**: McGuire and Kable, "Decision Makers Calibrate Behavioral Persistence on the Basis of Time-Interval Experience," and McGuire and Kable, "Rational Temporal Predictions Can Underlie Apparent Failures to Delay Gratification."

146 **grew into young adults who were more successful**: Mischel, Shoda, and Rodriguez, "Delay of Gratification in Children."

146 **how prior experiences might affect behavior**: Kidd, Palmeri, and Aslin, "Rational Snacking."

148 **Carnegie Hall even half full**: According to figures from the Aviation Safety Network (personal correspondence), the number of fatalities "on board US-owned aircraft that are capable of carrying 12+ passengers, also including corporate jets and military transport planes" during the period 2000–2014 was 1,369, and adding the 2014 figure again to estimate deaths in 2015 yields a total estimate of 1,393 through the end of 2015. Carnegie Hall's famous Isaac Stern Auditorium seats 2,804; see http://www.carnegiehall.org/Information/Stern-Auditorium-Perelman-Stage/.

148 **greater than the entire population of Wyoming**: According to the National Highway Traffic Safety Administration, 543,407 people died in car accidents in the United States in the years 2000–2013. See http://www-fars.nhtsa.dot.gov. Repeating the 2013 figure to estimate deaths in 2014 and 2015 yields an estimate of 608,845 deaths through the end

of 2015. The 2014 population of Wyoming, as estimated by the US Census Bureau, was 584,153. See http://quickfacts.census.gov/qfd/states/56000.html.

148 **gun violence on American news:** Glassner, "Narrative Techniques of Fear Mongering."

7. OVERFITTING

149 **"Marry—Marry—Marry Q.E.D.":** This note by Darwin is dated April 7, 1838; see, e.g., Darwin, *The Correspondence of Charles Darwin, Volume 2: 1837–1843.*

151 **"Moral or Prudential Algebra":** Franklin's letter to Joseph Priestley, London, September 19, 1772.

151 **"Anything you can do I can do better":** "Anything You Can Do," composed by Irving Berlin, in *Annie Get Your Gun,* 1946.

151 **what you know and what you don't:** In the language of machine-learning researchers: the "training" and the "test."

152 **a recent study conducted in Germany:** Lucas et al., "Reexamining Adaptation and the Set Point Model of Happiness."

152 **our job is to figure out the formula:** For math aficionados, we're trying to find the best polynomial function for capturing this relationship. Taking time since marriage to be x and satisfaction to be y, the one-predictor model is $y = ax + b$. The two-predictor model is $y = ax^2 + bx + c$, and the nine-predictor model finds the best coefficients for all values of x up to x^9, estimating a polynomial of degree 9.

153 **through each and every point on the chart:** In fact, it's a mathematical truth that you can always draw a polynomial of degree $n - 1$ through any n points.

153 **people's baseline level of satisfaction:** Lucas et al., "Reexamining Adaptation and the Set Point Model of Happiness."

155 **not always better to use a more complex model:** Statisticians refer to the various factors in the model as "predictors." A model that's too simple, such as a straight line attempting to fit a curve, is said to exhibit "bias." The opposite kind of systemic error, where a model is made too complicated and therefore gyrates wildly because of small changes in the data, is known as "variance."

The surprise is that these two kinds of errors—bias and variance—can be *complementary.* Reducing bias (making the model more flexible and complicated) can increase variance. And increasing bias (simplifying the model and fitting the data less tightly) can sometimes reduce variance.

Like the famous Heisenberg uncertainty principle of particle physics, which says that the more you know about a particle's momentum the less you know about its position, the so-called bias-variance tradeoff expresses a deep and fundamental bound on how good a model can be—on what it's possible to know and to predict. This notion is found in various places in the machine-learning literature. See, for instance, Geman, Bienenstock, and Doursat, "Neural Networks and the Bias/Variance Dilemma," and Grenander, "On Empirical Spectral Analysis of Stochastic Processes."

156 **in the Book of Kings:** The bronze snake, known as Nehushtan, gets destroyed in 2 Kings 18:4.

156 **"pay good money to remove the tattoos":** Gilbert, *Stumbling on Happiness.*

157 **duels less than fifty years ago:** If you're not too fainthearted, you can watch video of a duel fought in 1967 at http://passerelle-production.u-bourgogne.fr/web/atip_insulte /Video/archive_duel_france.swf.

157 **as athletes overfit their tactics:** For an interesting example of very deliberately overfitting fencing, see Harmenberg, *Epee 2.0.*

157 **"Incentive structures work"**: Brent Schlender, "The Lost Steve Jobs Tapes," *Fast Company*, May 2012, http://www.fastcompany.com/1826869/lost-steve-jobs-tapes.

157 **"whatever the CEO decides to measure"**: Sam Altman, "Welcome, and Ideas, Products, Teams and Execution Part I," Stanford CS183B, Fall 2014, "How to Start a Startup," http://startupclass.samaltman.com/courses/lec01/.

157 **Ridgway cataloged a host of such**: Ridgway, "Dysfunctional Consequences of Performance Measurements."

157 **At a job-placement firm**: In this tale, Ridgway is himself citing Blau, *The Dynamics of Bureaucracy*.

158 **"Friends don't let friends measure Page Views"**: Avinash Kaushik, "You Are What You Measure, So Choose Your KPIs (Incentives) Wisely!" http://www.kaushik.net/avinash /measure-choose-smarter-kpis-incentives/.

158 **"dead cops were found"**: Grossman and Christensen, *On Combat*. See http://www .killology.com/on_combat_ch2.htm.

159 **officer instinctively grabbed the gun**: Ibid.

160 **"If you can't explain it simply"**: This quotation is frequently attributed to Albert Einstein, although this attribution is likely to be apocryphal.

161 **Tikhonov proposed one answer**: See, e.g., Tikhonov and Arsenin, *Solution of Ill-Posed Problems*.

161 **invented in 1996 by biostatistician Robert Tibshirani**: Tibshirani, "Regression Shrinkage and Selection via the Lasso."

161 **human brain burns about a fifth**: For more on the human brain's energy consumption see, e.g., Raichle and Gusnard, "Appraising the Brain's Energy Budget," which in turn cites, e.g., Clarke and Sokoloff, "Circulation and Energy Metabolism of the Brain."

162 **brains try to minimize the number of neurons**: Using this neurally inspired strategy (known as "sparse coding"), researchers have developed artificial neurons that have properties similar to those found in the visual cortex. See Olshausen and Field, "Emergence of Simple-Cell Receptive Field Properties."

162 **groundbreaking "mean-variance portfolio optimization"**: The work for which Markowitz was awarded the Nobel Prize appears in his paper "Portfolio Selection" and his book *Portfolio Selection: Efficient Diversification of Investments*.

162 **"I split my contributions fifty-fifty"**: Harry Markowitz, as quoted in Jason Zweig, "How the Big Brains Invest at TIAA–CREF," *Money* 27(1): 114, January 1998.

163 **"less information, computation, and time"**: Gigerenzer and Brighton, "Homo Heuristicus."

163 **more than quadrupled from the mid-1990s to 2013**: From Soyfoods Association of North America, "Sales and Trends," http://www.soyfoods.org/soy-products/sales-and -trends, which in turn cites research "conducted by Katahdin Ventures."

164 **"Nuts are trendy now"**: Vanessa Wong, "Drinkable Almonds," *Bloomberg Businessweek*, August 21, 2013.

164 **an astounding three-hundred-fold since 2004**: Lisa Roolant, "Why Coconut Water Is Now a \$1 Billion Industry," TransferWise, https://transferwise.com/blog/2014-05/why -coconut-water-is-now-a-1-billion-industry/.

164 **"jumped from invisible to unavoidable"**: David Segal, "For Coconut Waters, a Street Fight for Shelf Space," *New York Times*, July 26, 2014.

164 **the kale market grew by 40%**: "Sales of Kale Soar as Celebrity Chefs Highlight Health Benefits," *The Telegraph*, March 25, 2013

164 **Pizza Hut, which put it in their salad bars:** Ayla Withee, "Kale: One Easy Way to Add More Superfoods to Your Diet," *Boston Magazine*, May 31, 2012.

164 **early vertebrates' bodies twisted 180 degrees:** Kinsbourne, "Somatic Twist." Further discussion of body and organ structure in primitive vertebrates can be found in Lowe et al., "Dorsoventral Patterning in Hemichordates." A more approachable overview is Kelly Zalocusky, "Ask a Neuroscientist: Why Does the Nervous System Decussate?," *Stanford Neuroblog*, December 12, 2013, https://neuroscience.stanford.edu/news/ask-neuroscientist-why-does-nervous-system-decussate.

165 **jawbones were apparently repurposed:** See, for example, "Jaws to Ears in the Ancestors of Mammals," Understanding Evolution, http://evolution.berkeley.edu/evolibrary/article/evograms_05.

167 **"the premise that we can't measure what matters":** "The Scary World of Mr Mintzberg," interview with Simon Caulkin, *Guardian*, January 25, 2003, http://www.theguardian.com/business/2003/jan/26/theobserver.observerbusiness11.

168 **"one's whole life like a neuter bee":** Darwin, *The Correspondence of Charles Darwin, Volume 2: 1837–1843*.

168 **"When? Soon or Late":** Ibid.

8. RELAXATION

169 **"successfully design a peptidic inhibitor":** Meghan Peterson (née Bellows), personal interview, September 23, 2014.

170 **about 11^{107} possible seating plans:** More precisely, there would be 11^{107} possibilities if we were choosing a table assignment for each person independently. The number is a little less once we take into account the constraint that only 10 people can sit at each table. But it's still huge.

170 **Bellows was pleased with the computer's results:** The formal framework that Meghan Bellows used to solve her wedding seating chart is described in Bellows and Peterson, "Finding an Optimal Seating Chart."

171 **Lincoln worked as a "prairie lawyer":** You can read more about Lincoln's circuit in Fraker, "The Real Lincoln Highway."

171 **"the postal messenger problem":** Menger, "Das botenproblem," contains a lecture given by Menger on the subject in Vienna on February 5, 1930. For a fuller history of the traveling salesman problem see Schrijver, "On the History of Combinatorial Optimization," as well as Cook's very readable book *In Pursuit of the Traveling Salesman*.

171 **fellow mathematician Merrill Flood:** Flood, "The Traveling-Salesman Problem."

171 **iconic name first appeared in print:** Robinson, *On the Hamiltonian Game*.

172 **"impossibility results would also be valuable":** Flood, "The Traveling-Salesman Problem."

172 **"no good algorithm for the traveling salesman problem":** Edmonds, "Optimum Branchings."

172 **what makes a problem feasible:** Cobham, "The Intrinsic Computational Difficulty of Functions," explicitly considers the question of what should be considered an "efficient" algorithm. Similarly, Edmonds, "Paths, Trees, and Flowers," explains why a solution to a difficult problem is significant and, in making the case for this particular solution, establishes a general framework for what makes algorithms good.

172 **the field's de facto out-of-bounds marker:** There are, in fact, algorithms that run slower than polynomial time but faster than exponential time; these "superpolynomial" run-times also put them outside the set of efficient algorithms.

173 **either efficiently solvable or not:** The set of efficiently solvable problems in computer science is called *P*, short for "polynomial time." The controversially liminal set of problems, meanwhile, is known as *NP*, for "nondeterministic polynomial." Problems in *NP* can have their solutions verified efficiently once found, but whether every problem that can be easily verified can also be easily solved is unknown. For instance, if someone shows you a route and says that it's less than 1,000 miles, the claim is easy to check—but finding a route less than 1,000 miles, or proving that it's impossible, is another feat entirely. The question of whether *P*=*NP* (i.e., whether it's possible to jump efficiently to the solutions of *NP* problems) is the greatest unsolved mystery in computer science.

The main advance toward a solution has been the demonstration that there are certain problems with a special status: if one of them can be solved efficiently, then *any* problem in *NP* can be solved efficiently and *P*=*NP* (Cook, "The Complexity of Theorem-Proving Procedures"). These are known as "*NP*-hard" problems. In the absence of an answer to whether *P*=*NP*, problems in *NP* cannot be solved efficiently, which is why we refer to them as "intractable." (In "A Terminological Proposal," Donald Knuth suggested this as an appropriate label for *NP*-hard problems, in addition to offering a live turkey to anybody who could prove *P*=*NP*.) The intractable scheduling problems that Eugene Lawler encountered in chapter 5 fall into this category. An *NP*-hard problem that is itself in *NP* is known as "*NP*-complete." See Karp, "Reducibility Among Combinatorial Problems," for the classic result showing that a version of the traveling salesman problem is *NP*-complete, and Fortnow, *The Golden Ticket: P, NP, and the Search for the Impossible*, for an accessible introduction to *P* and *NP*.

173 **most computer scientists believe that there *aren't* any:** In a 2002 survey of one hundred leading theoretical computer scientists, sixty-one thought *P*≠*NP* and only nine thought *P*=*NP* (Gasarch, "The *P*=? *NP* Poll"). While proving *P*=*NP* could be done by exhibiting a polynomial-time algorithm for an *NP*-complete problem, proving *P*≠*NP* requires making complex arguments about the limits of polynomial-time algorithms, and there wasn't much agreement among the people surveyed about exactly what kind of mathematics will be needed to solve this problem. But about half of them did think the issue would be resolved before 2060.

173 **What's more, many other optimization problems:** This includes versions of vertex cover and set cover—two problems identified as belonging to *NP* in Karp, "Reducibility Among Combinatorial Problems," where twenty-one problems were famously shown to be in this set. By the end of the 1970s, computer scientists had identified some *three hundred NP*-complete problems (Garey and Johnson, *Computers and Intractability*), and the list has grown significantly since then. These include some problems that are very familiar to humans. In 2003, Sudoku was shown to be *NP*-complete (Yato and Seta, "Complexity and Completeness"), as was maximizing the number of cleared rows in Tetris, even with perfect knowledge of future pieces (Demaine, Hohenberger, and Liben-Nowell, "Tetris Is Hard, Even to Approximate"). In 2012, determining whether there exists a path to the end of the level in platformer games like Super Mario Brothers was officially added to the list (Aloupis, Demaine, and Guo, "Classic Nintendo Games are (*NP*-) Hard").

173 **"you still have to fight it":** Jan Karel Lenstra, personal interview, September 2, 2014.

173 **"The perfect is the enemy of the good":** Voltaire's couplet *Dans ses écrits, un sage Italien / Dit que le mieux est l'ennemi du bien* ("In his writings, an Italian sage / Says the perfect is the enemy of the good") appears at the start of his poem "La Bégueule." Voltaire had

earlier cited the Italian expression *"Le meglio è l'inimico del bene"* in his 1764 *Diction-naire philosophique.*

173 **their minds also turn to relaxation:** Shaw, *An Introduction to Relaxation Methods*; Henderson, *Discrete Relaxation Techniques. Caveat lector*: the math is intense enough that these make for far-from-relaxing reading.

174 **for Lincoln's judicial circuit:** The towns of Lincoln's judicial circuit are derived from the 1847–1853 map of the 8th Judicial Circuit in the *Journal of the Abraham Lincoln Association.* See http://quod.lib.u mich.edu/j/jala/images/fraker_fig01a.jpg.

174 **essentially no time at all:** Well, okay, a little bit of time—linear in the number of cities if you're lucky, linearithmic if you're not. Pettie and Ramachandran, "An Optimal Minimum Spanning Tree Algorithm."

174 **the spanning tree, with its free backtracking:** Approaching the traveling salesman problem via the minimum spanning tree is discussed in Christofides, *Worst-Case Analysis of a New Heuristic.*

175 **visits every single town on Earth:** For more on the state of the art in the all-world-cities traveling salesman problem (the so-called "World TSP"), an up-to-date report can be found at http://www.math.uwaterloo.ca/tsp/world/. For more on the traveling salesman problem in general, Cook, *In Pursuit of the Traveling Salesman*, is a good general reference, and Lawler et al., *The Traveling Salesman Problem*, will satisfy those who want to go deeper.

176 **finding the minimal set of locations:** This classic discrete optimization problem is known as the "set cover" problem.

176 **"when you can't do half of this":** Laura Albert McLay, personal interview, September 16, 2014.

176 **let you lick the fewest envelopes:** In computer science, this is known as the "vertex cover" problem. It's a kind of cousin to the set cover problem,, where instead of seeking the smallest number of fire stations whose coverage *includes* everyone, the goal is to find the smallest number of people who are *connected* to everyone else.

177 **solving the *continuous* versions of these problems:** There are certain kinds of continuous optimization problems that can be solved in polynomial time; the most prominent example is linear programming problems, in which both the metric to be optimized and the constraints on the solution can be expressed as a linear function of the variables involved. See Khachiyan, "Polynomial Algorithms in Linear Programming," and Karmarkar, "A New Polynomial-Time Algorithm for Linear Programming." However, continuous optimization is no panacea: there are also classes of continuous optimization problems that are intractable. For example, see Pardalos and Schnitger, "Checking Local Optimality in Constrained Quadratic Programming is *NP*-hard."

177 **at most twice as many invitations:** Khot and Regev, "Vertex Cover Might Be Hard to Approximate to Within 2-ε."

177 **quickly get us within a comfortable bound:** For more on these approximations, see Vazirani, *Approximation Algorithms.*

177 **not a magic bullet:** It's still an open question within the field whether Continuous Relaxation even offers the best possible *approximation* for the minimum vertex cover (party invitations) problem, or whether better approximations can be found.

178 **"Inconceivable!":** *The Princess Bride*, screenplay by William Goldman; 20th Century Fox, 1987.

178 **computational technique called Lagrangian Relaxation:** Lagrangian Relaxation (initially spelled "Lagrangean") was given its name by Arthur M. Geoffrion of UCLA in

"Lagrangean Relaxation for Integer Programming." The idea itself is considered to have emerged in the work of Michael Held (of IBM) and Richard Karp (of UC Berkeley) on the traveling salesman problem in 1970—see Held and Karp, "The Traveling-Salesman Problem and Minimum Spanning Trees," and Held and Karp, "The Traveling-Salesman Problem and Minimum Spanning Trees: Part II." Earlier precursors, however, also exist—for instance, Lorie and Savage, "Three Problems in Rationing Capital"; Everett III, "Generalized Lagrange Multiplier Method"; and Gilmore and Gomory, "A Linear Programming Approach to the Cutting Stock Problem, Part II." For an overview and reflections see Fisher, "The Lagrangian Relaxation Method for Solving Integer Programming Problems," as well as Geoffrion, "Lagrangian Relaxation for Integer Programming."

179 **"If you end up with fractional games"**: Michael Trick, personal interview, November 26, 2013.

181 **"make-believe can never be reconciled"**: Christopher Booker, "What Happens When the Great Fantasies, Like Wind Power or European Union, Collide with Reality?," the *Telegraph*, April 9, 2011.

9. RANDOMNESS

182 **"why and how is absolutely mysterious"**: Quoted in Shasha and Rabin, "An Interview with Michael Rabin."

182 **a randomized algorithm uses**: Randomized algorithms are discussed in detail in Motwani and Raghavan, *Randomized Algorithms*, and Mitzenmacher and Upfal, *Probability and Computing*. Shorter but older introductions are provided by Karp, "An Introduction to Randomized Algorithms," and Motwani and Raghavan, "Randomized Algorithms."

183 **an interesting probabilistic analysis**: Buffon, "Essai d'arithmétique morale."

183 **simply by dropping needles onto paper**: Laplace, *Théorie analytique des probabilités*.

183 **Lazzarini supposedly made 3,408 tosses**: Lazzarini, "Un'applicazione del calcolo della probabilità."

183 **makes Lazzarini's report seem suspicious**: For further discussion of Lazzarini's results, see Gridgeman, "Geometric Probability and the Number π," and Badger, "Lazzarini's Lucky Approximation of π."

184 **he had contracted encephalitis**: Ulam's story appears in Ulam, *Adventures of a Mathematician*.

184 **"the test of a first-rate intelligence"**: Fitzgerald, "The Crack-Up." Later collected with other essays in *The Crack-Up*.

184 **"it may be much more practical"**: Ulam, *Adventures of a Mathematician*, pp. 196–197. Calculating the winning odds for Klondike solitaire remains an active area of research to this day, driven chiefly by Monte Carlo simulation. For an example of recent work in the area, see Bjarnason, Fern, and Tadepalli, "Lower Bounding Klondike Solitaire with Monte-Carlo Planning."

185 **Metropolis named this approach**: Metropolis claims the naming rights in a letter that appears in Hurd, "Note on Early Monte Carlo Computations."

186 **descendant of a long line of rabbis**: Shasha and Lazere, *Out of Their Minds*.

186 **multiple paths it might follow**: Rabin's key paper here, coauthored with Dana Scott, was "Finite Automata and Their Decision Problems." We've already encountered one of the ways that this concept became central to theoretical computer science in our discussion of the complexity class of the traveling salesman problem in chapter 8; Rabin's notion of "nondeterministic" computing is the "*N*" of *NP*.

186 **"one of the most obviously useless branches"**: The quote is from Hardy, "Prime Num-

bers"; see also Hardy, *Collected Works*. For more about the influence of prime numbers in cryptography, see, e.g., Schneier, *Applied Cryptography*.

186 **In modern encryption, for instance:** One widely used algorithm that is based on the multiplication of prime numbers is RSA, which stands for the initials of its inventors: Ron Rivest, Adi Shamir, and Leonard Adleman. See Rivest, Shamir, and Adleman, "A Method for Obtaining Digital Signatures and Public-Key Cryptosystems." Other cryptographic systems—e.g., Diffie-Hellman—also use prime numbers; see Diffie and Hellman, "New Directions in Cryptography."

187 **The problem, though, is false positives:** The possible breakthrough—or lack thereof— in Miller's approach would come down to how easily these false positives could be dismissed. How many values of x do you need to check to be sure about a given number n? Miller showed that if the "generalized Riemann hypothesis" were true, the minimum number of potential witnesses that would need to be checked is $O((\log n)^2)$—far less than the $\sqrt{n}$ required by algorithms like the Sieve of Erastothenes. But here was the hitch: the generalized Riemann hypothesis was—and still is—unproven.

(The Riemann hypothesis, first offered by the German mathematician Bernhard Riemann in 1859, concerns the properties of a complex mathematical function called the Riemann zeta function. This function is intimately related to the distribution of prime numbers, and in particular how regularly those numbers appear on the number line. If the hypothesis is true, then primes are well enough behaved as to guarantee the efficiency of Miller's algorithm. But nobody knows if it's true. In fact, the Riemann hypothesis is one of six major open problems in mathematics for whose solutions the Clay Mathematics Institute will award a "Millennium Prize" of $1 million. The question of whether $P = NP$, which we saw in chapter 8, is also a Millennium Prize problem.)

188 **"Michael, this is Vaughan":** Rabin tells this story in Shasha and Lazere, *Out of Their Minds*.

188 **quickly identify even gigantic prime numbers:** Rabin's paper on his primality test, "Probabilistic Algorithm for Testing Primality," appeared a few years later. In parallel, Robert Solovay and Volker Strassen had developed a similar probabilistic algorithm based on a different set of equations that primes need to obey, although their algorithm was less efficient; see Solovay and Strassen, "A Fast Monte-Carlo Test for Primality."

188 **less than one in a million billion billion:** The documentation for OpenSSL specifies a function to "perform a Miller-Rabin probabilistic primality test with . . . a number of iterations used . . . that yields a false positive rate of at most 2^{-80} for random input"; see https://www.openssl.org/docs/crypto/BN_generate_prime.html. Likewise the US Federal Information Processing Standard (FIPS) specifies that its Digital Signature Standard (DSS) accept error probability of 2^{-80} (for 1,024-bit keys, at least); see Gallagher and Kerry, *Digital Signature Standard*. Forty Miller-Rabin tests are sufficient to achieve this bound, and work from the 1990s has suggested that in many cases as few as three Miller-Rabin tests will suffice. See Damgård, Landrock, and Pomerance, "Average Case Error Estimates for the Strong Probable Prime Test"; Burthe Jr., "Further Investigations with the Strong Probable Prime Test"; and Menezes, Van Oorschot, and Vanstone, *Handbook of Applied Cryptography*, as well as more recent discussion at http://security.stackexchange.com/questions/4544/how-many-iterations-of-rabin-miller-should-be-used-to-generate-cryptographic-saf.

188 **for the number of grains of sand:** The number of grains of sand on Earth is estimated from various sources at between 10^{18} and 10^{24}.

188 **whether there would ever be an efficient algorithm:** Here by "efficient" we are using the field's standard definition, which is "polynomial-time," as discussed in chapter 8.

188 **one such method did get discovered:** Agrawal, Kayal, and Saxena, "PRIMES Is in *P*."

189 **generate some random *xs* and plug them in:** One of the key results on the role of randomness in polynomial identity testing is what's called the "Schwartz–Zippel lemma." See Schwartz, "Fast Probabilistic Algorithms for Verification of Polynomial Identities"; Zippel, "Probabilistic Algorithms for Sparse Polynomials"; and DeMillo and Lipton, "A Probabilistic Remark on Algebraic Program Testing."

189 **the only practical one we have:** Will an efficient deterministic algorithm for poly-nomial identity testing ever be found? More broadly, does an efficient deterministic algorithm *have* to exist anyplace we find a good randomized one? Or could there be problems that randomized algorithms can solve efficiently but that deterministic algo-rithms simply cannot? It is an interesting problem in theoretical computer science, and the asnwer to it is still unknown.

One of the approaches that has been used to explore the relationship between ran-domized and deterministic algorithms is called *derandomization*—essentially, taking randomized algorithms and removing the randomness from them. In practice, it's hard for a computer to get access to true randomness—so when people implement a random-ized algorithm, they often use a deterministic procedure to generate numbers that obey certain statistical properties of true randomness. Derandomization makes this explicit, examining what happens when the randomness in randomized algorithms is replaced by the output of some other complex computational process.

The study of derandomization shows that it's possible to turn efficient randomized algorithms into efficient deterministic algorithms—provided you can find a function that is sufficiently complex that its output looks random but sufficiently simple that it can be computed efficiently. For (detailed) details, see Impagliazzo and Wigderson, "$P = BPP$ if E Requires Exponential Circuits," and Impagliazzo and Wigderson, "Ran-domness vs. Time."

190 **he called the "veil of ignorance":** The veil of ignorance is introduced in Rawls, *A Theory of Justice.*

190 **Rawls's philosophical critics:** Most prominent among Rawls's critics was economist John Harsanyi; see, e.g., Harsanyi, "Can the Maximin Principle Serve as a Basis for Morality? A Critique of John Rawls's Theory."

190 **the civilization of Omelas:** Le Guin, "The Ones Who Walk Away from Omelas."

190 **These are worthy critiques:** For more on what is sometimes called "the repugnant con-clusion," see Parfit, *Reasons and Persons*, as well as, for instance, Arrhenius, "An Impos-sibility Theorem in Population Axiology."

191 **"concern of engineers rather than philosophers":** Aaronson, "Why Philosophers Should Care About Computational Complexity."

192 **"noticed something you *don't* often see":** Rebecca Lange, "Why So Few Stories?," GiveDirectly blog, November 12, 2014, https://www.givedirectly.org/blog-post.html ?id =2288694352161893466.

193 **"I mean Negative Capability":** John Keats, letter to George and Thomas Keats, Decem-ber 21, 1817.

193 **"assurance sufficient for the purposes of human life":** John Stuart Mill, *On Liberty* (1859).

193 **"there should be a drinking game":** Michael Mitzenmacher, personal interview. November 22, 2013.

193 **well over a trillion distinct URLs:** "We Knew the Web Was Big . . ." July 25, 2008, http://googleblog.blogspot.com/2008/07/we-knew-web-was-big.html.

193 **weighs in at about seventy-seven characters:** Kelvin Tan, "Average Length of a URL (Part 2)," August 16, 2010, http://www.supermind.org/blog/740/average-length-of-a-url-part-2.

194 **the URL is entered into a set of equations:** Bloom, "Space/Time Trade-offs in Hash Coding with Allowable Errors."

194 **shipped with a number of recent web browsers:** Google Chrome until at least 2012 used a Bloom filter: see http://blog.alexyakunin.com/2010/03/nice-bloom-filter-application.html and https://chromiumcodereview.appspot.com/10896048/.

194 **part of cryptocurrencies like Bitcoin:** Gavin Andresen, "Core Development Status Report #1," November 1, 2012, https://bitcoinfoundation.org/2012/11/core-development-status-report-1/.

194 **"The river meanders":** Richard Kenney, "Hydrology; Lachrymation," in *The One-Strand River: Poems, 1994–2007* (New York: Knopf, 2008).

197 **use this approach when trying to decipher codes:** See Berg-Kirkpatrick and Klein, "Decipherment with a Million Random Restarts."

197 **called the Metropolis Algorithm:** Sometimes also known as the Metropolis-Hastings Algorithm, this technique is described in Metropolis et al., "Equation of State Calculations by Fast Computing Machines," and Hastings, "Monte Carlo Methods Using Markov Chains and Their Applications." The Metropolis Algorithm was developed by Nicholas Metropolis and the two husband-and-wife teams of Marshall and Arianna Rosenbluth and Edward and Augusta Teller in the 1950s. Metropolis was the first author on the paper describing the algorithm, so today it is known as the Metropolis Algorithm—which is doubly ironic. For one thing, Metropolis apparently made little contribution to the development of the algorithm, being listed as an author out of courtesy, as the head of the computing laboratory (see Rosenbluth, *Marshall Rosenbluth, Interviewed by Kai-Henrik Barth*). What's more, Metropolis himself liked giving things illustrative names: he claimed to have named the chemical elements technetium and astatine, as well as the MANIAC computer and the Monte Carlo technique itself (Hurd, "Note on Early Monte Carlo Computations").

198 **"Growing a single crystal from a melt":** Kirkpatrick, Gelatt, and Vecchi, "Optimization by Simulated Annealing."

198 **"The guy who was the best at IBM":** Scott Kirkpatrick, personal interview, September 2, 2014.

199 **Finally we'd start going *only* uphill:** If this idea—starting out being willing to move around between options, then focusing more tightly on the good ones—sounds familiar, it should: optimizing a complex function requires facing the explore/exploit tradeoff. And randomness turns out to be a source of pretty good strategies for solving problems like multi-armed bandits as well as the kind of optimization problems that Kirkpatrick was focused on.

  If you recall, the multi-armed bandit offers us several different options—arms we can pull—that provide different, unknown payoffs. The challenge is to find the balance between trying new options (exploring) and pursuing the best option found so far (exploiting). Being more optimistic and more exploratory early on is best, becoming more discerning and exploiting more later. Pursuing such a strategy of gradually decreasing optimism about the alternatives promises the best outcome you can hope for—accumulating regrets at a decreasing rate, with your total regret rising as a logarithmic function of time.

Randomness provides an alternative strategy to optimism. Intuitively, if the problem is one of balancing exploration and exploitation, why not simply do so explicitly? Spend some amount of your time exploring and some amount exploiting. And that's exactly the strategy that multi-armed bandit experts call **Epsilon Greedy**.

Epsilon Greedy has two parts—Epsilon and Greedy. The Epsilon part is that some small proportion of the time (the letter epsilon is used by mathematicians to denote a small number), you choose *at random* from among your options. The Greedy part is that the rest of the time you take the best option you have found so far. So walk into the restaurant and flip a coin (or roll a die, depending on your value of epsilon) to decide whether to try something new. If it says yes, close your eyes and point at the menu. If not, enjoy your current favorite.

Unfortunately, multi-armed bandit researchers don't particularly like Epsilon Greedy. It seems wasteful—you're guaranteed to spend a proportion of your time trying new things even if the best becomes clear very quickly. If you follow Epsilon Greedy, then your regret increases *linearly* in the number of times you play. Each time you dine, there's a chance that you're going to choose something other than the best, so your average regret increases by the same amount every time. This linear growth is much worse than the logarithmic regret guaranteed by deterministic algorithms based on appropriately calibrated optimism.

But if the simplicity of Epsilon Greedy is appealing, there is good news. There's a simple variant of this algorithm—what we are dubbing **Epsilon-Over-N Greedy**—that does guarantee logarithmic regret, and performs well in practice (see Auer, Cesa-Bianchi, and Fischer, "Finite-Time Analysis of the Multiarmed Bandit Problem"). The trick is to decrease the chance of trying something new over time. The first time you make a choice, you choose at random with probability 1/1 (a.k.a. always). If that option is any good, then the second time you choose at random with probability 1/2 (a.k.a. flip a coin: heads you take the same option, tails you try something new). On visit three, you should pick the best thing with probability 2/3, and try something new with probability 1/3. On the Nth visit to the restaurant, you choose at random with probability $1/N$, otherwise taking the best option discovered so far. By gradually decreasing the probability of trying something new, you hit the sweet spot between exploration and exploitation.

There's also another, more sophisticated algorithm for playing the multi-armed bandit that likewise makes use of randomness. It's called **Thompson Sampling**, named after William R. Thompson, the Yale physician who first posed the problem (back in 1933) of how to choose between two treatments (Thompson, "On the Likelihood That One Unknown Probability Exceeds Another"). Thompson's solution was simple: using Bayes's Rule, calculate the probability that each treatment is the best. Then choose that treatment *with that probability*. To begin with you know nothing, and you are equally likely to choose either treatment. As the data accumulate you come to favor one over the other, but some of the time you still choose the dispreferred treatment and have the chance to change your mind. As you become more certain that one treatment is better, you will end up almost always using that treatment. Thompson Sampling balances exploration and exploitation elegantly, and also guarantees that regret will increase only logarithmically (see Agrawal and Goyal, "Analysis of Thompson Sampling").

The advantage of Thompson Sampling over other algorithms for solving multi-armed bandit problems is its flexibility. Even if the assumptions of the problem change—you have information suggesting one option is better than the others, options depend on one another, options change over time—Thompson's strategy of pursuing options with a probability that reflects your sense that they are the best currently available still works. So rather than having to derive a new algorithm in each of these cases, we can simply apply Bayes's Rule and use the results. In real life, those Bayesian calculations can be hard (it took Thompson himself several pages of intricate mathematics to solve the problem with just two options). But trying to choose the best option and allowing an amount of randomness to your choices that is tempered by your degree of certainty is an algorithm that is unlikely to lead you astray.

199 **cited a whopping thirty-two thousand times:** The predominant AI textbook, *Artificial Intelligence: A Modern Approach*, declares that simulated annealing "is now a field in itself, with hundreds of papers published every year" (p. 155).

199 **one of the most promising approaches to optimization:** Intriguingly, a 2014 paper appears to demonstrate that jellyfish use simulated annealing in searching for food; see Reynolds, "Signatures of Active and Passive Optimized Lévy Searching in Jellyfish."

200 **"Not a gambler myself":** Luria, *A Slot Machine, a Broken Test Tube*, p. 75. Also discussed in Garfield, "Recognizing the Role of Chance."

201 **coined the term "serendipity":** In Horace Walpole, letter to Horace Mann (dated January 28, 1754).

201 **"A remarkable parallel":** James, "Great Men, Great Thoughts, and the Environment."

202 **"A blind-variation-and-selective-retention process":** Campbell, "Blind Variation and Selective Retention."

202 **"Newton, Mozart, Richard Wagner, and others":** Quoted in ibid.

203 **"ways of throwing you out of the frame":** Brian Eno, interviewed by Jools Holland, on *Later . . . with Jools Holland*, May 2001.

203 **"vague and constant desire":** The word is *saudade*, and the quoted definition comes from Bell, *In Portugal*.

204 **"stupid to shake it up any further":** Tim Adams, "Dicing with Life," *Guardian*, August 26, 2000.

10. NETWORKING

205 **"*connection* has a wide variety of meanings":** Cerf and Kahn, "A Protocol for Packet Network Intercommunication."

205 **"Only connect":** Forster, *Howards End*.

206 **"handheld, portable, real cellular phone":** Martin Cooper, "Inventor of Cell Phone: We Knew Someday Everybody Would Have One," interview with Tas Anjarwalla, CNN, July 9, 2010.

206 **The message was "login"—or would have been:** Leonard Kleinrock tells the story in a 2014 video interview conducted by Charles Severence and available at "Len Kleinrock: The First Two Packets on the Internet," https://www.youtube.com/watch?v=uY7dU JT7OsU.

206 **portentous and Old Testament despite himself:** Says UCLA's Leonard Kleinrock, "We didn't plan it, but we couldn't have come up with a better message: short and prophetic." The tiles on the floor of UCLA's Boelter Hall, if their colors are interpreted as binary 0s and 1s and parsed as ASCII characters, spell out the phrase "LO AND BEHOLD!" Credit for this tribute goes to architect Erik Hagen. See, e.g., Alison Hewitt, "Discover the Coded Message Hidden in Campus Floor Tiles," *UCLA Newsroom*, July 3, 2013, http://newsroom.ucla.edu/stories/a-coded-message-hidden-in-floor-247232.

206 **rooted in the Greek *protokollon*:** See, e.g., the Online Etymology Dictionary, http://www.etymonline.com/index.php?term=protocol.

207 **"They go *blast!* and they're quiet"**: Leonard Kleinrock, "Computing Conversations: Len Kleinrock on the Theory of Packets," interview with Charles Severance (2013). See https://www.youtube.com/watch?v=qsgrtrwydjw as well as http://www.computer.org /csdl/mags/co/2013/08/mco2013080006.html.

207 **"utter heresy"**: Jacobson, "A New Way to Look at Networking."

207 **"So little boy went away"**: Kleinrock, "Computing Conversations."

207 **would become known as *packet switching***: The term "packet switching" comes from Donald W. Davies of the National Physical Laboratory, another key contributor to packet switching research at the time.

208 **"a consensual illusion between the two endpoints"**: Stuart Cheshire, personal interview, February 26, 2015.

208 **communications could survive a nuclear attack:** Baran, "On Distributed Communications."

208 **a growing network becomes a virtue:** For elaboration on this point, and a broader reflection on the history of networking (including its current problems), see Jacobson, "A New Way to Look at Networking."

209 **a packet-switching network over "Avian Carriers":** See Waitzman, *A Standard for the Transmission of IP Datagrams on Avian Carriers*, Waitzman, *IP Over Avian Carriers with Quality of Service*, and Carpenter and Hinden, *Adaptation of RFC 1149 for IPv6* for descriptions of the avian protocol, and see http://www.blug.linux.no/rfc1149 for details of the actual implementation performed in Bergen, Norway, on April 28, 2001.

209 **"No transmission can be 100 percent reliable":** Cerf and Kahn, "A Protocol for Packet Network Intercommunication."

209 **the "Byzantine generals problem":** Lamport, Shostak, and Pease, "The Byzantine Generals Problem."

210 **signal that the sequence has been restored:** The process being described here is known as "fast retransmit."

211 **almost 10% of upstream Internet traffic:** Jon Brodkin, "Netflix takes up 9.5% of *upstream* traffic on the North American Internet: ACK packets make Netflix an upload monster during peak viewing hours," *Ars Technica*, November 20, 2014. Brodkin in turn cites data from Sandvine's *Global Internet Phenomena Report*, https://www.sandvine .com/trends/global-internet-phenomena/.

211 **"Did the receiver crash? Are they just slow?":** Tyler Treat, "You Cannot Have Exactly-Once Delivery," *Brave New Geek: Introspections of a software engineer*, March 25, 2015, http://bravenewgeek.com/you-cannot-have-exactly-once-delivery/.

211 **"end-to-end retransmissions to recover":** Vint Cerf, interviewed by Charles Severance, "Computing Conversations: Vint Cerf on the History of Packets," 2012.

211 **"you just say, 'Say that again'":** Ibid.

212 **"The world's most difficult word to translate":** Oliver Conway, "Congo Word 'Most Untranslatable,'" *BBC News*, June 22, 2004.

212 **"If at first you don't succeed":** Thomas H. Palmer, *Teacher's Manual* (1840), attested in *The Oxford Dictionary of Proverbs*, 2009.

213 **trying to link together the university's seven campuses:** Abramson, "The ALOHA System."

213 **above a mere 18.6% average utilization:** Ibid. In fact, this figure is $\frac{1}{2e}$, exactly half of the $\frac{n}{e}$, or "37%," figure given in the discussion of optimal stopping in chapter 1.

214 **"only one scheme has any hope of working":** Jacobson, "Congestion Avoidance and Control."

215 **a pilot program called HOPE:** The HOPE program is evaluated in Hawken and Kleiman, *Managing Drug Involved Probationers*.

216 **"what a crazy way to try to change":** For more information, see, e.g., "A New Probation Program in Hawaii Beats the Statistics," *PBS NewsHour*, February 2, 2014.

216 **"this sudden factor-of-thousand drop":** Jacobson, "Congestion Avoidance and Control."

217 **"then it suddenly fell apart":** Jacobson, "Van Jacobson: The Slow-Start Algorithm," interview with Charles Severance (2012), https://www.youtube.com/watch?v=QP4A6 L7CEqA.

217 **ramp up its transmission rate aggressively:** This initial procedure—a tentative single packet followed by a two-for-one acceleration—is known in TCP as Slow Start. This name is a partial misnomer: Slow Start is "slow" in beginning with just a single tentative first packet, but not in its exponential growth thereafter.

218 **"control without hierarchy":** See, e.g., Gordon, "Control without Hierarchy."

218 **ants' solution is similar:** The findings that link ant foraging to flow control algorithms like Slow Start appear in Prabhakar, Dektar, and Gordon, "The Regulation of Ant Colony Foraging Activity without Spatial Information."

219 **"tends to rise to his level of incompetence":** Peter and Hull, *The Peter Principle*.

219 **"Every public servant should be demoted":** This widely reproduced aphorism, in the original Spanish, reads, *"Todos los empleados públicos deberían descender a su grado inmediato inferior, porque han sido ascendidos hasta volverse incompetentes."*

219 **devised by leading law firm Cravath, Swaine & Moore:** The Cravath System is officially documented at the firm's own website: http://www.cravath.com/cravathsystem/. The "up or out" component of the Cravath System is not explicitly discussed there, but is widely referenced elsewhere, e.g., by the American Bar Association: "In the 1920s Cravath, Swaine & Moore became the first law firm on record to openly recruit from law schools with the express understanding that many of the young lawyers it hired would not make partner. Those associates who did not make partner with the rest of their class were expected to leave the firm. However, those deemed best among the associates, who did the necessary work and stayed on track for the requisite number of years, could expect to become stakeholders, earn lockstep increases in compensation, and enjoy lifetime employment in the firm." (Janet Ellen Raasch, "Making Partner—or Not: Is It In, Up or Over in the Twenty-First Century?," *Law Practice* 33, issue 4, June 2007.)

219 **the US Armed Forces adopted:** See, e.g., Rostker et al., *Defense Officer Personnel Management Act of 1980*.

219 **pursued what they call "manning control":** See, e.g., Michael Smith, "Army Corporals Forced Out 'to Save Pension Cash,'" *Telegraph*, July 29, 2002.

220 **as if all communication were written text:** As Bavelas, Coates, and Johnson, "Listeners as Co-Narrators," puts it, "Listeners have at best a tenuous foothold in most theories. At the extreme, listeners are considered nonexistent or irrelevant because the theory either does not mention them or treats them as peripheral. This omission may be attributed, in part, to the implicit use of written text as the prototype for all language use."

221 **"simultaneously engaged in both speaking and listening":** Yngve, "On Getting a Word in Edgewise."

221 **"Narrators who told close-call stories to distracted listeners":** Bavelas, Coates, and Johnson, "Listeners as Co-Narrators."

221 **regulating the flow of information from speaker to listener:** Tolins and Fox Tree, "Addressee Backchannels Steer Narrative Development."

221 **"'bad storytellers' can at least partly blame their audience":** Jackson Tolins, personal correspondence, January 15, 2015.

222 **"misconceptions about the cause and meaning of queues":** Nichols and Jacobson, "Controlling Queue Delay."

222 **the HTTP specification still in use today:** That is HTTP 1.1, as articulated in the RFC 2616 document from June 1999, available at http://tools.ietf.org/html/rfc2616.

222 **"I happened to be copying, or rsyncing":** Jim Gettys, "Bufferbloat: Dark Buffers in the Internet," Google Tech Talk, April 26, 2011.

222 **"not 'Eureka!' but 'That's funny'":** This quotation has appeared in countless publications with an attribution to Isaac Asimov, but its actual authorhip and provenance remain elusive. It seems to have first shown up—complete with the Asimov attribution—as part of the UNIX "fortune" program, which displays quotes or sayings in the style of a fortune cookie. See http://quoteinvestigator.com/2015/03/02/eureka-funny/. Asimov did write an essay about "The Eureka Phenomenon," but this phrase does not appear there.

224 **when they are routinely zeroed out:** See Nichols and Jacobson, "Controlling Queue Delay."

225 **than her home state of California has people:** The US Census Bureau's 2015 estimate for California's population was 39,144,818. See http://www.census.gov/popest/data /state/totals/2015/index.html.

226 **"no really good way to leave messages for people":** Ray Tomlinson, interviewed by Jesse Hicks, "Ray Tomlinson, the Inventor of Email: 'I See Email Being Used, by and Large, Exactly the Way I Envisioned,'" *Verge*, May 2, 2012, http://www.theverge.com/2012/5/2 /2991486/ray-tomlinson-email-inventor-interview-i-see-email-being-used.

226 **simply rejecting all incoming messages:** One such approach was taken, for instance, by University of Sheffield cognitive scientist Tom Stafford. During his 2015 sabbatical, his automated email response read: "I am now on sabbatical until 12th June. Email sent to t.stafford@shef.ac.uk has been deleted."

227 **Explicit Congestion Notification, or ECN:** The Request for Comments (RFC) document for ECN is Ramakrishnan, Floyd, and Black, *The Addition of Explicit Congestion Notification (ECN) to IP*, which is a revision of Ramakrishnan and Floyd, *A Proposal to Add Explicit Congestion Notification (ECN) to IP*. Though the original proposal dates from the 1990s, ECN remains unimplemented in standard networking hardware today (Stuart Cheshire, personal interview, February 26, 2015).

227 **"This is a long-term swamp":** Jim Gettys, personal interview, July 15, 2014.

228 **"would you say that a Boeing 747 is three times 'faster'":** This comes from Cheshire's famous 1996 "rant" "It's the Latency, Stupid." See http://stuartcheshire.org/rants/Latency .html. Twenty years later, the sentiment is only truer.

11. GAME THEORY

229 **"I believe humans are noble and honorable":** Steve Jobs, interview with Gary Wolf, *Wired*, February 1996.

229 **man vs. nature:** Appropriately, schoolchildren in the twenty-first century increasingly learn about "person vs. nature," "person vs. self," "person vs. person," and "person vs. society."

230 **"a clever man would put the poison into his own goblet":** *The Princess Bride*, screenplay by William Goldman; 20th Century Fox, 1987.

230 **"anticipating the anticipations of others":** Attributed to Keynes in Gregory Bergman, *Isms*, Adams Media, 2006.

231 **it was the halting problem that inspired Turing:** Alan Turing considers the halting problem and proposes the Turing machine in "On Computable Numbers, with an Application to the Entscheidungsproblem" and "On Computable Numbers, with an Application to the Entscheidungsproblem. A Correction."

231 **"poker players call it 'leveling'":** Dan Smith, personal interview, September 11, 2014.

232 **"You don't have deuce–seven":** This took place at the "Full Tilt Poker Durrrr Million Dollar Challenge," held at Les Ambassadeurs Club in London, November 17–19, 2009, and was televised on Sky Sports.

232 **"only want to play one level above your opponent":** Vanessa Rousso, "Leveling Wars," https://www.youtube.com/watch?v=Yt5ALnFrwR4.

232 **"knowing or trying to know what Nash is":** Dan Smith, personal interview, September 11, 2014.

233 **a so-called *equilibrium*:** The concept of a game-theoretic equilibrium—and, for that matter, game theory itself—comes from Princeton's John von Neumann and Oskar Morgenstern in *Theory of Games and Economic Behavior*.

233 **In rock-paper-scissors, for example:** For a colorful look into rock-paper-scissors ("RPS") tournament play, including a glossary of the game's various three-move "gambits"—like the Avalanche (RRR), the Bureaucrat (PPP), and Fistful o' Dollars (RPP)—we recommend http://worldrps.com. For a look into *computer* RPS play, check out the Rock Paper Scissors Programming Competition: http://www.rpscontest.com.

233 **choose one of the eponymous hand gestures completely at random:** A strategy, like this one, that incorporates randomness is called a "mixed" strategy. The alternative is a "pure" strategy, which always involves taking the exact same option; this clearly would not work for long in rock-paper-scissors. Mixed strategies appears as part of the equilibrium in many games, especially in "zero-sum" games, where the interests of the players are pitted directly against one another.

233 *every* **two-player game has at least one equilibrium:** Nash, "Equilibrium Points in *N*-Person Games"; Nash, "Non-Cooperative Games."

233 **the fact that a Nash equilibrium always exists:** To be more precise, ibid. proved that every game with a finite number of players and a finite number of strategies has at least one mixed-strategy equilibrium.

234 **"has had a fundamental and pervasive impact":** Myerson, "Nash Equilibrium and the History of Economic Theory."

234 **"a computer scientist's foremost concern":** Papadimitriou, "Foreword."

234 **"Give us something we can use":** Tim Roughgarden, "Algorithmic Game Theory, Lecture 1 (Introduction)," Autumn 2013, https://www.youtube.com/watch?v=TM_QFmQU_VA.

234 **all been proved to be intractable problems:** Gilboa and Zemel, "Nash and Correlated Equilibria."

234 **simply *finding* Nash equilibria is intractable:** Specifically, finding Nash equilibria was shown to belong to a class of problems called *PPAD*, which (like *NP*) is widely believed to be intractable. The link between Nash equilibria and *PPAD* was established in Daskalakis, Goldberg, and Papadimitriou, "The Complexity of Computing a Nash Equilibrium" and Goldberg and Papadimitriou, "Reducibility Between Equilibrium Problems," which was then extended to two-player games by Chen and Deng, "Settling the Complexity of Two-Player Nash Equilibrium," and then further generalized in Daskalakis, Goldberg, and Papadimitriou, "The Complexity of Computing a Nash Equilibrium." *PPAD* stands for "Polynomial Parity Arguments on Directed graphs"; Papadimitriou, who named this class of problems in "On Complexity as Bounded Rationality," insists

any resemblance to his name is a coincidence. (Christos Papadimitriou, personal interview, September 4, 2014.)

PPAD contains other interesting problems, such as the ham sandwich problem: given n sets of $2n$ points in n dimensions, find a plane that divides each set of points exactly in half. (With $n = 3$, this involves figuring out the path a knife would have to travel to cut three sets of points in half; if those sets of points correspond to two pieces of bread and a piece of ham, the result is a perfectly bisected sandwich.) Finding Nash equilibria is actually PPAD-complete, meaning that if there were an efficient algorithm for solving it then all other problems in the class could also be solved efficiently (including making the world's neatest sandwiches). But being PPAD-complete is not quite so bad as being NP-complete. P, the class of efficiently solvable problems, could be equal to PPAD without being equal to NP. As of this writing the jury is still out: it's theoretically possible that somebody could devise an efficient algorithm for finding Nash equilibria, but most experts aren't holding their breath.

235 **"much of its credibility as a prediction"**: Christos Papadimitriou, "The Complexity of Finding Nash Equilibria," in Nisan et al., *Algorithmic Game Theory*.

235 **"should be considered relevant also"**: Aaronson, "Why Philosophers Should Care About Computational Complexity."

235 **"If your laptop cannot find it"**: In Christos Papadimitriou, "The Complexity of Finding Nash Equilibria," in Nisan et al., *Algorithmic Game Theory*, p. 30.

235 **"the prisoner's dilemma"**: The prisoner's dilemma was first conceived by Merrill Flood (of secretary problem and traveling salesman problem fame) and Melvin Drescher at RAND Corporation. In January 1950, they staged a game between UCLA's Armen Alchian and RAND's John D. Williams that had prisoner's dilemma–like payoffs (Flood, "Some Experimental Games"). Princeton's Albert Tucker was intrigued by this experiment, and in preparing to discuss it that May in a lecture at Stanford, he gave the problem its now famous prison formulation and its name. A detailed history of the origins of game theory and its development in the work of the RAND Corporation can be found in Poundstone, *Prisoner's Dilemma*.

237 **a price of anarchy that's a mere 4/3**: Roughgarden and Tardos, "How Bad Is Selfish Routing?" Roughgarden's 2002 Cornell PhD also addresses the topic of selfish routing.

237 **"the pessimist fears this is true"**: Cabell, *The Silver Stallion*.

238 **picture a "commons" of public lawn**: Hardin, "The Tragedy of the Commons."

238 **"there was this thing called leaded gasoline"**: Avrim Blum, personal interview, December 17, 2014.

238 **headline put the trouble succinctly**: Scott K. Johnson, "Stable Climate Demands Most Fossil Fuels Stay in the Ground, but Whose?," *Ars Technica*, January 8, 2015.

238 **"nowhere is the value of work higher"**: "In Search of Lost Time," *Economist*, December 20, 2014.

238 **15% take no vacation at all**: The study is from Glassdoor and is referenced in ibid.

239 **"People will hesitate to take a vacation"**: Mathias Meyer, "From Open (Unlimited) to Minimum Vacation Policy," December 10, 2014, http://www.paperplanes.de/2014/12/10/from-open-to-minimum-vacation-policy.html.

239 **"Stores are opening earlier than ever before"**: Nicole Massabrook, "Stores Open on Thanksgiving 2014: Walmart, Target, Best Buy and Other Store Hours on Turkey Day," *International Business Times*, November 26, 2014.

240 **"Don't hate the player, hate the game"**: Ice-T, "Don't Hate the Playa," *The Seventh Deadly Sin*, 1999.

240 **"Don't ever take sides with anyone against the family"**: *The Godfather*, screenplay by Mario Puzo and Francis Ford Coppola, Paramount Pictures, 1972.

240 **"loaded against the emergence of cooperation"**: This quotation of Binmore's appears in a number of sources, including Binmore, *Natural Justice*, and Binmore, *Game Theory*. Kant's "categorical imperative" originates in his 1785 *Groundwork of the Metaphysic of Morals* and is discussed in his 1788 *Critique of Practical Reason*.

241 **a thousand dollars cash for taking a vacation**: Libin discusses the motivations for the thousand dollars in, for instance, an interview with Adam Bryant, "The Phones Are Out, but the Robot Is In," *New York Times*, April 7, 2012.

242 **make a certain minimal amount of vacation *compulsory***: Compulsory vacation is already a standard practice in finance, although for reasons of fraud detection rather than morale. For more on compulsory vacation and fraud see, e.g., Philip Delves Broughton, "Take Those Two Weeks Off—or Else," *Wall Street Journal*, August 28, 2012.

242 **without federal requirements for paid vacation**: Rebecca Ray, Milla Sanes, and John Schmitt, "No-Vacation Nation Revisited," *Center for Economic Policy and Research*, May 2013, http://www.cepr.net/index.php/publications/reports/no-vacation-nation-2013.

243 *Things a Computer Scientist Rarely Talks About*: Donald E. Knuth.

243 **"The heart has its reasons"**: As Pascal put it in Pascal, *Pensées sur la religion et sur quelques autres sujets*, §277: *"Le cœur a ses raisons, que la raison ne connaît point."*

243 **"The canopy can be thought of as an aerial meadow"**: Dawkins, *The Evidence for Evolution*.

245 **makes mice permanently lose their fear of cats**: Ingram et al., "Mice Infected with Low-Virulence Strains of *Toxoplasma Gondii*."

245 **"Morality is herd instinct in the individual"**: *The Gay Science*, §116, trans. Walter Kaufmann.

245 **"If people *expect* us to respond irrationally"**: Frank, *Passions within Reason*.

246 **"The worry that people will leave relationships"**: Ibid.

246 **"you need a feeling that makes you not want to separate"**: Robert Frank, personal interview, April 13, 2015. Frank, "If Homo Economicus Could Choose," contains this idea, though as he is quick to acknowledge, it builds on work such as Schelling, *The Strategy of Conflict*; Schelling, "Altruism, Meanness, and Other Potentially Strategic Behaviors"; Akerlof, "Loyalty Filters"; Hirshleifer, "On the Emotions as Guarantors of Threats and Promises"; Sen, "Goals, Commitment, and Identity"; and Gauthier, *Morals by Agreement*. Frank treats the ideas at book length in *Passions within Reason*.

246 **"If the prisoner is happy, why lock him in?"**: Shaw, *Man and Superman*.

248 **makes more than 90% of its revenue from selling ads**: Google's 2014 advertising revenue, as detailed in its shareholder report, was $59.6 billion, roughly 90.3% of its total revenue of $66 billion. See https://investor.google.com/financial/tables.html.

248 **raising tens of billions of dollars in revenue**: The AWS-3 auction that closed on January 29, 2015, resulted in winning bids totaling $44.899 billion. See http://wireless.fcc.gov/auctions/default.htm?job=auction_factsheet&id=97.

248 **they're shading their bids based on their prediction of yours!**: The equilibrium strategy for a sealed-bid first-price auction with two players is to bid exactly half what you think the item is worth. More generally, in this auction format with n players, you should bid exactly $\frac{n-1}{n}$ times what you think the item is worth. Note that this strategy is the Nash equilibrium but is not a dominant strategy; that is to say, nothing is better if everyone else is doing it, too, but isn't necessarily optimal under all circumstances. Caveat emptor. Also, if you don't *know* the number of bidders in the auction, the optimal strategy gets complicated

in a hurry; see, for instance, An, Hu, and Shum, "Estimating First-Price Auctions with an Unknown Number of Bidders: A Misclassification Approach." Actually, even the seemingly clean results— $\frac{n-1}{n}$ —require some serious assumptions, namely that the bidders are "risk neutral" and that their different values for the item are distributed evenly across some given range. The $\frac{n-1}{n}$ result here comes from Vickrey, "Counterspeculation, Auctions, and Competitive Sealed Tenders," who warns, "If the assumption of homogeneity among the bidders be abandoned, the mathematics of a complete treatment become intractable."

248 **the largest flower auction in the world:** For more about the Aalsmeer Flower Auction, see http://www.floraholland.com/en/about-floraholland/visit-the-flower-auction/.

249 **a bunch of people all going over a cliff together:** Sometimes these cliffs are all too literal. The *New York Times*, for instance, reported on the deaths of several experienced backcountry skiers in Washington State. The accounts of the survivors show how a group of extremely skilled skiers ended up doing something that almost all the individual members had a bad feeling about.

"If it was up to me, I would never have gone backcountry skiing with twelve people," said one survivor. "That's just way too many. But there were sort of the social dynamics of that—where I didn't want to be the one to say, you know, 'Hey, this is too big a group and we shouldn't be doing this.'"

"There's no way this entire group can make a decision that isn't smart," another said to himself. "Of course it's fine, if we're all going. It's got to be fine."

"Everything in my mind was going off, wanting to tell them to stop," said a third.

"I thought: Oh yeah, that's a bad place to be," recounted a fourth member of the party. "That's a bad place to be with that many people. But I didn't say anything. I didn't want to be the jerk."

As the *Times* summarized: "All the locals in the group presumed they knew what the others were thinking. They did not." See Branch, "Snow Fall."

249 **known as an "information cascade":** Bikhchandani, Hirshleifer, and Welch, "A Theory of Fads." See also Bikhchandani, Hirshleifer, and Welch, "Learning from the Behavior of Others."

250 **"the public pool of information is no longer growing":** David Hirshleifer, personal interview, August 27, 2014.

250 **a sale price of more than $23 million:** The pricing on this particular Amazon title was noticed and reported on by UC Berkeley biologist Michael Eisen; see "Amazon's $23,698,655.93 book about flies," April 23, 2011 on Eisen's blog *it is NOT junk*, http://www.michaeleisen.org/blog/?p=358.

251 **worsen the irrationality of the market:** See, for instance, the reactions of Columbia University economist Rajiv Sethi in the immediate wake of the flash crash. Sethi, "Algorithmic Trading and Price Volatility."

252 **save the entire herd from disaster:** This can also be thought of in terms of mechanism design and evolution. It is better on average for any particular individual to be a somewhat cautious herd follower, yet everyone benefits from the presence of some group members who are headstrong mavericks. In this way, overconfidence can be thought of as a form of altruism. For more on the "socially optimal proportion" of such group members, see Bernardo and Welch, "On the Evolution of Overconfidence and Entrepreneurs."

252 **a way to rethink mechanism design:** The phrase "algorithmic mechanism design" first entered the technical literature in Nisan and Ronen, "Algorithmic Mechanism Design."

252 **It's called the Vickrey auction:** See Vickrey, "Counterspeculation, Auctions, and Competitive Sealed Tenders."

253 **"strategy-proof," or just "truthful":** "Strategy-proof" games are also known as "incentive-compatible." See Noam Nisan, "Introduction to Mechanism Design (for Computer Scientists)," in Nisan et al., eds., *Algorithmic Game Theory*.

253 **honesty is the dominant strategy:** In game theory terms, this makes the Vickrey auction "dominant-strategy incentive-compatible" (DSIC). And a major result in algorithmic game theory, known as "Myerson's Lemma," asserts that there is only one DSIC payment mechanism possible. This means that the Vickrey auction is not just *a* way to avoid strategic, recursive, or dishonest behavior—it's the *only* way. See Myerson, "Optimal Auction Design."

253 **a game-theoretic principle called "revenue equivalence":** The revenue equivalence theorem originated with Vickrey, "Counterspeculation, Auctions, and Competitive Sealed Tenders" and was generalized in Myerson, "Optimal Auction Design," and Riley and Samuelson, "Optimal Auctions."

253 **the Vickrey auction is "awesome":** Tim Roughgarden, "Algorithmic Game Theory, Lecture 3 (Myerson's Lemma)," published October 2, 2013, https://www.youtube.com/watch?v=9qZwchMuslk.

253 **"I think that's really fantastic":** Noam Nisan, personal interview, April 13, 2015.

253 **"one of the best things you can see":** Paul Milgrom, personal interview, April 21, 2015.

254 **"Hell is other people":** Sartre, *No Exit*.

CONCLUSION

256 **"to learn how to live well together":** Flood, "What Future Is There for Intelligent Machines?"

257 **"define this as the *wisest* act":** Russell, "The Elements of Ethics."

257 **a kind of computational Stoicism:** See, e.g., Baltzly, "Stoicism."

258 **knowing a good song when you hear it:** It also happens to be the difference between P and NP. For more delightful philosophical ruminations of this nature, see Aaronson, "Reasons to Believe," and Wigderson, "Knowledge, Creativity, and P versus NP."

258 **none of them had wanted to see the bullfight:** Scenarios like this one sometimes go by the name of "The Abilene Paradox"; see Harvey, "The Abilene Paradox."

259 **moving the group toward resolution:** This point has also been made by Tim Ferriss, who writes, "Stop asking for suggestions or solutions and start proposing them. Begin with the small things. Rather than asking when someone would like to meet next week, propose your ideal times and second choices. If someone asks, 'Where should we eat?,' 'What movie should we watch?,' 'What should we do tonight?,' or anything similar, do not reflect it back with 'Well, what/when/where do you want to . . . ?' Offer a solution. Stop the back and forth and make a decision." See Ferriss, *The 4-Hour Workweek*.

259 **offering one or two concrete proposals:** Ideally, one would want to know the values that each person in the group assigns to *all* the options, and adopt a reasonable policy for making a decision based on those. One potential approach is to simply select the option that maximizes the product of the values assigned by everyone—which also lets anyone veto an option by assigning it a value of zero. There are arguments from economics that this is a good strategy, going all the way back to John Nash. See Nash, "The Bargaining Problem."

259 **minimize the number of coins:** Shallit, "What This Country Needs Is an 18¢ Piece."

259 **ungainly denominations turn change-making:** Lueker, "Two *NP*-Complete Problems in Nonnegative Integer Programming," showed that under certain assumptions, making change with the fewest number of coins is *NP*-hard. This result holds if the coins

are denominated in binary or the familiar base ten, but not if they are denominated in unary (base one), which does have an efficient solution, as shown in Wright, "The Change-Making Problem." For more on the computational complexity of making change, see also Kozen and Zaks, "Optimal Bounds for the Change-Making Problem."

260 **Consider a large parking lot:** Cassady and Kobza, "A Probabilistic Approach to Evaluate Strategies for Selecting a Parking Space," compares the "Pick a Row, Closest Space (PRCS)" and "Cycling (CYC)" parking space–hunting algorithms. The more complicated CYC includes an optimal stopping rule, while PRCS starts at the destination, pointing away, and simply takes the first space. The more aggressive CYC found better spaces on average, but the simpler PRCS actually won in terms of total time spent. Drivers following the CYC algorithm spent more time finding better spaces than those better spaces saved them in walk time. The authors note that research of this nature might be useful in the design of parking lots. Computational models of parking are also explored in, e.g., Benenson, Martens, and Birfir, "PARKAGENT: An Agent-Based Model of Parking in the City."

261 **"spinning" and "blocking":** For a deeper look at when to spin and when to block, see, for instance, Boguslavsky et al., "Optimal Strategies for Spinning and Blocking." (Note that this is the same Leonid Boguslavsky we encountered briefly in chapter 1 on a water-skiing trip.)

Bibliography

Aaronson, Scott. "Reasons to Believe" *Shtetl-Optimized* (blog), September 4, 2006. http://www.scottaaronson.com/blog/?p=122/.

———. "Why Philosophers Should Care About Computational Complexity." *arXiv preprint arXiv:1108.1791*, 2011.

Abramson, Norman. "The ALOHA System: Another Alternative for Computer Communications." In *Proceedings of the November 17–19, 1970, Fall Joint Computer Conference*, 1970, 281–285.

Ackley, David H. "Beyond Efficiency." *Communications of the ACM* 56, no. 10 (2013): 38–40.

Agrawal, Manindra, Neeraj Kayal, and Nitin Saxena. "PRIMES Is in *P*." *Annals of Mathematics* 160 (2004): 781–793.

Agrawal, Rajeev. "Sample Mean Based Index Policies with $O(\log n)$ Regret for the Multi-Armed Bandit Problem." *Advances in Applied Probability* 27 (1995): 1054–1078.

Agrawal, Shipra, and Navin Goyal. "Analysis of Thompson Sampling for the Multi-armed Bandit Problem." In *Proceedings of the 25th Annual Conference on Learning Theory*, 2012.

Akerlof, George A. "Loyalty Filters." *American Economic Review* 1983, 54–63.

Allen, David. *Getting Things Done: The Art of Stress-Free Productivity*. New York: Penguin, 2002.

Aloupis, Greg, Erik D. Demaine, and Alan Guo. "Classic Nintendo Games Are (*NP*-) Hard." *arXiv preprint arXiv:1203.1895*, 2012.

An, Yonghong, Yingyao Hu, and Matthew Shum. "Estimating First-Price Auctions with an Unknown Number of Bidders: A Misclassification Approach." *Journal of Econometrics* 157, no. 2 (2010): 328–341.

Anderson, John R. *The Adaptive Character of Thought*. Hillsdale, NJ: Erlbaum, 1990.

Anderson, John R., and Robert Milson. "Human Memory: An Adaptive Perspective." *Psychological Review* 96, no. 4 (1989): 703–719.

Anderson, John R., and Lael J. Schooler. "Reflections of the Environment in Memory." *Psychological Science* 2, no. 6 (1991): 396–408.

Ariely, Dan, and Simon Jones. *Predictably Irrational*. New York: HarperCollins, 2008.

Arrhenius, Gustaf. "An Impossibility Theorem in Population Axiology with Weak Ordering Assumptions." *Philosophical Studies* 49 (1999): 11–21.

Auer, Peter, Nicolò Cesa-Bianchi, and Paul Fischer. "Finite-Time Analysis of the Multiarmed Bandit Problem." *Machine Learning* 47 (2002): 235–256.

Austen, Jane. *Emma*. London: John Murray, 1815.

Austrian, Geoffrey D. *Herman Hollerith: Forgotten Giant of Information Processing*. New York: Columbia University Press, 1982.

Bachmann, Paul. *Die analytische zahlentheorie*. Leipzig: Teubner, 1894.

Badger, Lee. "Lazzarini's Lucky Approximation of π." *Mathematics Magazine* 67 (1994): 83–91.

Bailey, Arthur L. *Credibility Procedures: Laplace's Generalization of Bayes' Rule and the Combination of Collateral Knowledge with Observed Data*. New York: New York State Insurance Department, 1950.

Baker, Kenneth R. *Introduction to Sequencing and Scheduling*. New York: Wiley, 1974.

Baker, Kenneth R., Eugene L. Lawler, Jan Karel Lenstra, and Alexander H. G. Rinnooy Kan. "Preemptive Scheduling of a Single Machine to Minimize Maximum Cost Subject to Release Dates and Precedence Constraints." *Operations Research* 31, no. 2 (1983): 381–386.

Baltzly, Dirk. "Stoicism." In *The Stanford Encyclopedia of Philosophy* (spring 2014 edition). Edited by Edward N. Zalta. http://plato.stanford.edu/archives/spr2014/entries/stoicism/.

Banks, Jeffrey S., and Rangarajan K Sundaram. "Switching Costs and the Gittins Index." *Econometrica* 62 (1994): 687–694.

Barabási, Albert-László. *Linked: How Everything Is Connected to Everything Else and What It Means for Business, Science, and Everyday Life*. New York: Penguin, 2002.

Baran, Paul. "On Distributed Communications." *Volumes I–XI, RAND Corporation Research Documents*, August 1964, 637–648.

Barnard, Chester I. *The Functions of the Executive*. Cambridge, MA: Harvard University Press, 1938.

Bartlett, Robert H., Dietrich W. Roloff, Richard G. Cornell, Alice French Andrews, Peter W. Dillon, and Joseph B. Zwischenberger. "Extracorporeal Circulation in Neonatal Respiratory Failure: A Prospective Randomized Study." *Pediatrics* 76, no. 4 (1985): 479–487.

Baumgardt, Carola. *Johannes Kepler: Life and Letters*. New York: Philosophical Library, 1951.

Bavelas, Janet B., Linda Coates, and Trudy Johnson. "Listeners as Co-Narrators." *Journal of Personality and Social Psychology* 79, no. 6 (2000): 941–952.

Bayes, Thomas. "An Essay Towards Solving a Problem in the Doctrine of Chances." *Philosophical Transactions* 53 (1763): 370–418.

Bearden, Neil. "A New Secretary Problem with Rank-Based Selection and Cardinal Payoffs." *Journal of Mathematical Psychology* 50 (2006): 58–59.

Bélády, Laszlo A. "A Study of Replacement Algorithms for a Virtual-Storage Computer." *IBM Systems Journal* 5 (1966): 78–101.

Bélády, Laszlo A., Robert A Nelson, and Gerald S. Shedler. "An Anomaly in Space-Time Characteristics of Certain Programs Running in a Paging Machine." *Communications of the ACM* 12, no. 6 (1969): 349–353.

Belew, Richard K. *Finding Out About: A Cognitive Perspective on Search Engine Technology and the WWW*. Cambridge, UK: Cambridge University Press, 2000.

Bell, Aubrey F. G. *In Portugal*. New York: John Lane, 1912.

Bellhouse, David R. "The Reverend Thomas Bayes, FRS: A Biography to Celebrate the Tercentenary of His Birth." *Statistical Science* 19 (2004): 3–43.

Bellman, Richard. *Dynamic Programming*. Princeton, NJ: Princeton University Press, 1957.

———. "A Problem in the Sequential Design of Experiments." *Sankhyā: The Indian Journal of Statistics* 16 (1956): 221–229.

Bellows, Meghan L., and J. D. Luc Peterson. "Finding an Optimal Seating Chart." *Annals of Improbable Research* (2012).

Benenson, Itzhak, Karel Martens, and Slava Birfir. "PARKAGENT: An Agent-Based Model of Parking in the City." *Computers, Environment and Urban Systems* 32, no. 6 (2008): 431–439.

Berezovsky, Boris, and Alexander V. Gnedin. *Problems of Best Choice* (in Russian). Moscow: Akademia Nauk, 1984.

Berg-Kirkpatrick, Taylor, and Dan Klein. "Decipherment with a Million Random Restarts." In *Proceedings of the Conference on Empirical Methods in Natural Language Processing* (2013): 874–878.

Bernardo, Antonio E., and Ivo Welch. "On the Evolution of Overconfidence and Entrepreneurs." *Journal of Economics & Management Strategy* 10, no. 3 (2001): 301–330.

Berry, Donald A. "A Bernoulli Two-Armed Bandit." *Annals of Mathematical Statistics* 43 (1972): 871–897.

———. "Comment: Ethics and ECMO." *Statistical Science* 4 (1989): 306–310.

Berry, Donald A., and Bert Fristed. *Bandit Problems: Sequential Allocation of Experiments.* New York: Chapman and Hall, 1985.

Bettencourt, Luís M. A., José Lobo, Dirk Helbing, Christian Kühnert, and Geoffrey B. West. "Growth, Innovation, Scaling, and the Pace of Life in Cities." *Proceedings of the National Academy of Sciences* 104, no. 17 (2007): 7301–7306.

Bikhchandani, Sushil, David Hirshleifer, and Ivo Welch. "A Theory of Fads, Fashion, Custom, and Cultural Change as Informational Cascades." *Journal of Political Economy* 100, no. 5 (1992): 992–1026.

———. "Learning from the Behavior of Others: Conformity, Fads, and Informational Cascades." *Journal of Economic Perspectives* 12, no. 3 (1998): 151–170.

Binmore, Ken. *Game Theory: A Very Short Introduction.* New York: Oxford University Press, 2007.

———. *Natural Justice.* New York: Oxford University Press, 2005.

Bjarnason, Ronald, Alan Fern, and Prasad Tadepalli. "Lower Bounding Klondike Solitaire with Monte-Carlo Planning." In *Proceedings of the 19th International Conference on Automated Planning and Scheduling, ICAPS 2009.*

Blau, Peter Michael. *The Dynamics of Bureaucracy: A Study of Interpersonal Relations in Two Government Agencies.* Chicago: University of Chicago Press, 1955.

Bloom, Burton H. "Space/Time Trade-offs in Hash Coding with Allowable Errors." *Communications of the ACM* 13, no. 7 (1970): 422–426.

Boguslavsky, Leonid, Karim Harzallah, A. Kreinen, K. Sevcik, and Alexander Vainshtein. "Optimal Strategies for Spinning and Blocking." *Journal of Parallel and Distributed Computing* 21, no. 2 (1994): 246–254.

Boorstin, Daniel J. *The Discoverers: A History of Man's Search to Know His World and Himself.* New York: Random House, 1983.

Bradt, Russell N., S. M. Johnson, and Samuel Karlin. "On Sequential Designs for Maximizing the Sum of N Observations." *Annals of Mathematical Statistics* 27 (1956): 1060–1074.

Branch, John. "Snow Fall: The Avalanche at Tunnel Creek." *New York Times*, December 20, 2012.

Brown, Alexander L., and Joanna N. Lahey. *Small Victories: Creating Intrinsic Motivation in*

Savings and Debt Reduction. Technical report. Cambridge, MA: National Bureau of Economic Research, 2014.

Brush, Eleanor R., David C. Krakauer, and Jessica C. Flack. "A Family of Algorithms for Computing Consensus About Node State from Network Data." *PLoS Computational Biology* 9, no. 7 (2013).

Bruss, F. Thomas. "A Unified Approach to a Class of Best Choice Problems with an Unknown Number of Options." *Annals of Probability* 12 (1984): 882–889.

Buch, P. "Future Prospects Discussed." *Nature* 368 (1994): 107–108.

Buffon, Georges-Louis Leclerc, Comte de. "Essai d'arithmétique morale." *Supplément à l'Histoire naturelle, générale et particuliére* 4 (1777): 46–148.

Burks, Arthur W., Herman H. Goldstine, and John von Neumann. *Preliminary Discussion of the Logical Design of an Electronic Computing Instrument*. Princeton, NJ: Institute for Advanced Study, 1946.

Burrell, Quentin. "A Simple Stochastic Model for Library Loans." *Journal of Documentation* 36, no. 2 (1980): 115–132.

Burthe Jr., Ronald. "Further Investigations with the Strong Probable Prime Test." *Mathematics of Computation of the American Mathematical Society* 65, no. 213 (1996): 373–381.

Cabell, James Branch. *The Silver Stallion*. New York: Robert M. McBride, 1926.

Campbell, Donald T. "Blind Variation and Selective Retention in Creative Thought as in Other Knowledge Processes." *Psychological Review* 67 (1960): 380–400.

Carpenter, Brian, and Robert Hinden. *Adaptation of RFC 1149 for IPv6*. Technical report. RFC 6214, April 2011.

Carroll, Lewis. *Sylvie and Bruno Concluded*. London: Macmillan, 1893.

Carstensen, Laura L. "Social and Emotional Patterns in Adulthood: Support for Socioemotional Selectivity Theory." *Psychology and Aging* 7 (1992): 331–338.

Cassady, C. Richard, and John E. Kobza. "A Probabilistic Approach to Evaluate Strategies for Selecting a Parking Space." *Transportation Science* 32, no. 1 (1998): 30–42.

Cawdrey, Robert. *A Table Alphabeticall, conteyning and teaching the true writing, and vnderstanding of hard vsuall English wordes, borrowed from the Hebrew, Greeke, Latine, or French, &c. With the interpretation thereof by plaine English words, gathered for the benefit & helpe of ladies, gentlewomen, or any other vnskilfull persons. Whereby they may the more easilie and better vnderstand many hard English wordes, which they shall heare or read in Scriptures, Sermons, or elswhere, and also be made able to vse the same aptly themselues*. London: Edmund Weaver, 1604.

Cayley, Arthur. "Mathematical Questions with Their Solutions." *Educational Times* 23 (1875): 18–19.

———. *The Collected Mathematical Papers of Arthur Cayley* 10: 587–588. Cambridge, UK: Cambridge University Press, 1896.

Cerf, Vinton G., and Robert E. Kahn. "A Protocol for Packet Network Intercommunication." *IEEE Transactions on Communications* 22, no. 5 (1974): 637–648.

Chabert, Jean-Luc, Evelyne Barbin, and Christopher John Weeks. *A History of Algorithms: From the Pebble to the Microchip*. Berlin: Springer, 1999.

Charles, Susan T., and Laura L. Carstensen. "Social and Emotional Aging." *Annual Review of Psychology* 61 (2010): 383–409.

Chen, Stanley F., and Joshua Goodman. "An Empirical Study of Smoothing Techniques for Language Modeling." In *Proceedings of the 34th Annual Meeting of the Association for Computational Linguistics*, 1996, 310–318.

Chen, Xi, and Xiaotie Deng. "Settling the Complexity of Two-Player Nash Equilibrium." In *Foundations of Computer Science*, 2006, 261–272.

Chow, Y. S., and Herbert Robbins. "A Martingale System Theorem and Applications." In *Proceedings of the Fourth Berkeley Symposium on Mathematical Statistics and Probability*. Berkeley: University of California Press, 1961.

———. "On Optimal Stopping Rules." *Probability Theory and Related Fields* 2 (1963): 33–49.

Chow, Y. S., Sigaiti Moriguti, Herbert Robbins, and S. M. Samuels. "Optimal Selection Based on Relative Rank (the 'Secretary Problem')." *Israel Journal of Mathematics* 2 (1964): 81–90.

Christian, Brian. "The A/B Test: Inside the Technology That's Changing the Rules of Business." *Wired Magazine* 20, no. 5 (2012).

Christofides, Nicos. *Worst-Case Analysis of a New Heuristic for the Travelling Salesman Problem*. Technical report 388. Pittsburgh: Graduate School of Industrial Administration, Carnegie Mellon University, 1976.

Churchill, Winston. *Winston S. Churchill: His Complete Speeches, 1897–1963*. Edited by Robert Rhodes James. London: Chelsea House, 1974.

Cirillo, Francesco. *The Pomodoro Technique*. Raleigh, NC: Lulu, 2009.

Clarke, Donald D., and Louis Sokoloff. "Circulation and Energy Metabolism of the Brain." In *Basic Neurochemistry: Molecular, Cellular and Medical Aspects*, 6th ed., edited by George J. Siegel, Bernard W. Agranoff, R. Wayne Albers, Stephen K. Fisher, and Michael D. Uhler. Philadelphia: Lippincott-Raven, 1999, 637–669.

Clauset, Aaron, Cosma Rohilla Shalizi, and Mark E. J. Newman. "Power-Law Distributions in Empirical Data." *SIAM Review* 51, no. 4 (2009): 661–703.

Cobham, Alan. "The Intrinsic Computational Difficulty of Functions." In *Proceedings of the 1964 Congress on Logic, Methodology and Philosophy of Science*. Amsterdam: North Holland, 1964.

Conan Doyle, Arthur. "A Study in Scarlet: The Reminiscences of John H. Watson." In *Beeton's Christmas Annual*, vol. 29. London: Ward, Lock, 1887.

Connor, James A. *Kepler's Witch: An Astronomer's Discovery of Cosmic Order Amid Religious War, Political Intrigue, and the Heresy Trial of His Mother*. New York: HarperCollins, 2004.

Conti, Carl J., Donald H. Gibson, and Stanley H. Pitkowsky. "Structural Aspects of the System/360 Model 85, I: General Organization." *IBM Systems Journal* 7 (1968): 2–14.

Cook, Stephen A. "The Complexity of Theorem-Proving Procedures." In *Proceedings of the Third Annual ACM Symposium on Theory of Computing*, 1971, 151–158.

Cook, William. *In Pursuit of the Traveling Salesman: Mathematics at the Limits of Computation*. Princeton, NJ: Princeton University Press, 2012.

Covey, Stephen R. *How to Succeed with People*. Salt Lake City: Shadow Mountain, 1971.

Craig, J. V. *Aggressive Behavior of Chickens: Some Effects of Social and Physical Environments*. Presented at the 27th Annual National Breeder's Roundtable, May 11, Kansas City, MO, 1978.

Dale, Andrew I. *A History of Inverse Probability: From Thomas Bayes to Karl Pearson*. New York: Springer, 1999.

Daly, Lloyd W. *Contributions to a History of Alphabetization in Antiquity and the Middle Ages*. Brussels: Latomus, 1967.

Damgård, Ivan, Peter Landrock, and Carl Pomerance. "Average Case Error Estimates for the Strong Probable Prime Test." *Mathematics of Computation* 61, no. 203 (1993): 177–194.

Daniels, Bryan C., David C. Krakauer, and Jessica C. Flack. "Sparse Code of Conflict in a Primate Society." *Proceedings of the National Academy of Sciences* 109, no. 35 (2012): 14259–14264.

Darwin, Charles. *The Correspondence of Charles Darwin, Volume 2: 1837–1843.* Edited by Frederick Burkhardt and Sydney Smith. Cambridge, UK: Cambridge University Press, 1987.

Daskalakis, Constantinos, Paul W. Goldberg, and Christos H. Papadimitriou. "The Complexity of Computing a Nash Equilibrium." *ACM Symposium on Theory of Computing,* 2006, 71–78.

———. "The Complexity of Computing a Nash Equilibrium." *SIAM Journal on Computing* 39, no. 1 (2009): 195–259.

Davis, Lydia. *Almost No Memory: Stories.* New York: Farrar, Straus & Giroux, 1997.

Dawkins, Richard. *The Evidence for Evolution, the Greatest Show on Earth.* New York: Free Press, 2009.

DeDeo, Simon, David C. Krakauer, and Jessica C. Flack. "Evidence of Strategic Periodicities in Collective Conflict Dynamics." *Journal of The Royal Society Interface,* 2011.

DeGroot, Morris H. *Optimal Statistical Decisions.* New York: McGraw-Hill, 1970.

Demaine, Erik D., Susan Hohenberger, and David Liben-Nowell. "Tetris Is Hard, Even to Approximate." In *Computing and Combinatorics,* 351–363. New York: Springer, 2003.

DeMillo, Richard A., and Richard J. Lipton. "A Probabilistic Remark on Algebraic Program Testing." *Information Processing Letters* 7, no. 4 (1978): 193–195.

Denning, Peter J. "Thrashing: Its Causes and Prevention." In *Proceedings of the December 9–11, 1968, Fall Joint Computer Conference, Part I,* 1968, 915–922.

Diffie, Whitfield, and Martin E. Hellman. "New Directions in Cryptography." *Information Theory, IEEE Transactions on* 22, no. 6 (1976): 644–654.

Dillard, Annie. *Pilgrim at Tinker Creek.* New York: Harper's Magazine Press, 1974.

———. *The Writing Life.* New York: Harper & Row, 1989.

Dodgson, Charles Lutwidge. "Lawn Tennis Tournaments: The True Method of Assigning Prizes with a Proof of the Fallacy of the Present Method." *St. James's Gazette,* August 1, 1883: 5–6.

Durant, Will. *The Story of Philosophy: The Lives and Opinions of the Greater Philosophers.* New York: Simon & Schuster, 1924.

Edmonds, Jack. "Optimum Branchings." *Journal of Research of the National Bureau of Standards* 71B, no. 4 (1967): 233–240.

———. "Paths, Trees, and Flowers." *Canadian Journal of Mathematics* 17, no. 3 (1965): 449–467.

Erlang, Agner Krarup. "Solution of Some Problems in the Theory of Probabilities of Significance in Automatic Telephone Exchanges." *Elektrotkeknikeren* 13 (1917): 5–13.

———. "The Theory of Probabilities and Telephone Conversations." *Nyt Tidsskrift for Matematik B* 20, nos. 33–39 (1909): 16.

Everett III, Hugh. "Generalized Lagrange Multiplier Method for Solving Problems of Optimum Allocation of Resources." *Operations Research* 11, no. 3 (1963): 399–417.

Feldman, Dorian. "Contributions to the 'Two-Armed Bandit' Problem." *Annals of Mathematical Statistics* 33 (1962): 847–856.

Ferguson, Thomas S. *Optimal Stopping and Applications.* Available at http://www.math.ucla.edu/~tom/Stopping/2008.

———. "Stopping a Sum During a Success Run." *Annals of Statistics* 4 (1976): 252–264.

———. "Who Solved the Secretary Problem?" *Statistical Science* 4 (1989): 282–289.

Ferguson, Thomas S., Janis P. Hardwick, and Mitsushi Tamaki. "Maximizing the Duration of Owning a Relatively Best Object." In *Strategies for Sequential Search and Selection in Real Time*, 37–57. Providence: American Mathematical Society, 1992.

Ferriss, Timothy. *The 4-Hour Workweek*. New York: Crown, 2007.

Fiore, Neil A. *The Now Habit: A Strategic Program for Overcoming Procrastination and Enjoying Guilt-Free Play*. New York: Penguin, 2007.

Fisher, Marshall L. "The Lagrangian Relaxation Method for Solving Integer Programming Problems." *Management Science* 27, no. 1 (1981): 1–18.

Fitzgerald, F. Scott. "The Crack-Up." *Esquire* 5, nos. 2–4 (1936).

———. *The Crack-Up with Other Uncollected Pieces*. New York: New Directions, 1956.

Flood, Merrill M. "Soft News." *Datamation* 30, no. 20 (1984): 15–16.

———. "Some Experimental Games." In *Research Memorandum RM-789*. Santa Monica, CA: RAND, 1952.

———. "The Traveling-Salesman Problem." *Operations Research* 4, no. 1 (1956): 61–75.

———. "What Future Is There for Intelligent Machines?" *Audio Visual Communication Review* 11, no. 6 (1963): 260–270.

Forster, Edward M. *Howards End*. London: Edward Arnold, 1910.

Fortnow, Lance. *The Golden Ticket: P, NP, and the Search for the Impossible*. Princeton, NJ: Princeton University Press, 2013.

Fraker, Guy C. "The Real Lincoln Highway: The Forgotten Lincoln Circuit Markers." *Journal of the Abraham Lincoln Association* 25 (2004): 76–97.

Frank, Robert H. "If Homo Economicus Could Choose His Own Utility Function, Would He Want One with a Conscience?" *American Economic Review* 1987, 593–604.

———. *Passions within Reason: The Strategic Role of the Emotions*. New York: Norton, 1988.

Fredrickson, Barbara L., and Laura L. Carstensen. "Choosing Social Partners: How Old Age and Anticipated Endings Make People More Selective." *Psychology and Aging* 5 (1990): 335–347.

Freeman, P. R. "The Secretary Problem and Its Extensions: A Review." *International Statistical Review* 51 (1983): 189–206.

Fung, Helene H., Laura L. Carstensen, and Amy M. Lutz. "Influence of Time on Social Preferences: Implications for Life-Span Development." *Psychology and Aging* 14 (1999): 595–604.

Gal, David, and Blakeley B. McShane. "Can Small Victories Help Win the War? Evidence from Consumer Debt Management." *Journal of Marketing Research* 49 (2012): 487–501.

Gallagher, P., and C. Kerry. *Digital Signature Standard*. FIPS PUB 186-4, 2013.

Garey, Michael R., and David S. Johnson. *Computers and Intractability: A Guide to NP-Completeness*. New York: W. H. Freeman, 1979.

Garfield, Eugene. "Recognizing the Role of Chance." *Scientist* 2, no. 8 (1988): 10.

Garrett, A. J. M., and P. Coles. "Bayesian Inductive Inference and the Anthropic Cosmological Principle." *Comments on Astrophysics*. 17 (1993): 23–47.

Gasarch, William I. "The $P=?$ NP Poll." *SIGACT News* 33, no. 2 (2002): 34–47.

Gauthier, David P. *Morals by Agreement*. New York: Oxford University Press, 1985.

Geman, Stuart, Elie Bienenstock, and René Doursat. "Neural Networks and the Bias/Variance Dilemma." *Neural Computation* 4, no. 1 (1992): 1–58.

Geoffrion, Arthur M. "Lagrangean Relaxation for Integer Programming." *Mathematical Programming Study* 2 (1974): 82–114.

———. "Lagrangian Relaxation for Integer Programming." In *50 Years of Integer Programming 1958–2008: From Early Years to State of the Art*. Edited by Michael Juenger, Thomas M.

Liebling, Denis Naddef, George L. Nemhauser, William R. Pulleyblank, Gerhard Reinelt, Giovanni Rinaldi, and Laurence A. Wolsey. Berlin: Springer, 2010, 243–281.

Gigerenzer, Gerd, and Henry Brighton. "Homo Heuristicus: Why Biased Minds Make Better Inferences." *Topics in Cognitive Science* 1, no. 1 (2009): 107–143.

Gilbert, Daniel. *Stumbling on Happiness.* New York: Knopf, 2006.

Gilbert, John P. and Frederick Mosteller. "Recognizing the Maximum of a Sequence." *Journal of the American Statistical Association* 61 (1966): 35–75.

Gilboa, Itzhak, and Eitan Zemel. "Nash and Correlated Equilibria: Some Complexity Considerations." *Games and Economic Behavior* 1, no. 1 (1989): 80–93.

Gillispie, Charles Coulston. *Pierre-Simon Laplace, 1749–1827: A Life in Exact Science.* Princeton, NJ: Princeton University Press, 2000.

Gilmore, Paul C., and Ralph E. Gomory. "A Linear Programming Approach to the Cutting Stock Problem, Part II." *Operations Research* 11, no. 6 (1963): 863–888.

Gilovich, Thomas. *How We Know What Isn't So.* New York: Simon & Schuster, 2008.

Ginsberg, Allen. *Howl and Other Poems.* San Francisco: City Lights Books, 1956.

Gittins, John C. "Bandit Processes and Dynamic Allocation Indices." *Journal of the Royal Statistical Society, Series B (Methodological)* 41 (1979): 148–177.

Gittins, John C., Kevin Glazebrook, and Richard Weber. *Multi-Armed Bandit Allocation Indices*, 2nd ed. Chichester, UK: Wiley, 2011.

Gittins, John C., and D. Jones. "A Dynamic Allocation Index for the Sequential Design of Experiments." In *Progress in Statistics.* Amsterdam: North Holland, 1974, 241–266.

Glassner, Barry. "Narrative Techniques of Fear Mongering." *Social Research* 71 (2004): 819–826.

Goldberg, Paul W., and Christos H. Papadimitriou. "Reducibility Between Equilibrium Problems." *ACM Symposium on Theory of Computing* 2006, 62–70.

Good, Irving John. *Good Thinking: The Foundations of Probability and Its Applications.* Minneapolis, MN: University of Minnesota Press, 1983.

Gopnik, Alison, Andrew N. Meltzoff, and Patricia K. Kuhl. *The Scientist in the Crib.* New York: Morrow, 1999.

Gordon, Deborah M. "Control Without Hierarchy." *Nature* 446, no. 7132 (2007): 143.

Gott, J. R. "Future Prospects Discussed." *Nature* 368 (1994): 108.

———. "Implications of the Copernican Principle for Our Future Prospects." *Nature* 363 (1993): 315–319.

Gould, Stephen Jay. "The Median Isn't the Message." *Discover* 6, no. 6 (1985): 40–42.

Graham, Ronald L., Eugene L. Lawler, Jan Karel Lenstra, and Alexander H. G. Rinnooy Kan. "Optimization and Approximation in Deterministic Sequencing and Scheduling: A Survey." *Annals of Discrete Mathematics* 5 (1979): 287–326.

Grenander, Ulf. "On Empirical Spectral Analysis of Stochastic Processes." *Arkiv för Matematik* 1, no. 6 (1952): 503–531.

Gridgeman, T. "Geometric Probability and the Number π." *Scripta Mathematika* 25, no. 3 (1960): 183–195.

Griffiths, Thomas L., Charles Kemp, and Joshua B. Tenenbaum. "Bayesian Models of Cognition." In *The Cambridge Handbook of Computational Cognitive Modeling.* Edited by Ron Sun. Cambridge, UK: Cambridge University Press, 2008.

Griffiths, Thomas L., Falk Lieder, and Noah D. Goodman. "Rational Use of Cognitive Resources: Levels of Analysis Between the Computational and the Algorithmic." *Topics in Cognitive Science* 7 (2015): 217–229.

Griffiths, Thomas L., David M. Sobel, Joshua B. Tenenbaum, and Alison Gopnik. "Bayes and

Blickets: Effects of Knowledge on Causal Induction in Children and Adults." *Cognitive Science* 35 (2011): 1407–1455.

Griffiths, Thomas L., Mark Steyvers, and Alana Firl. "Google and the Mind: Predicting Fluency with PageRank." *Psychological Science* 18 (2007): 1069–1076.

Griffiths, Thomas L., and Joshua B. Tenenbaum. "Optimal Predictions in Everyday Cognition." *Psychological Science* 17 (2006): 767–773.

Grossman, Dave, and L. W. Christensen. *On Combat.* Belleville, IL: PPCT Research Publications, 2004.

Haggstrom, Gus W. "Optimal Sequential Procedures When More Than One Stop Is Required." *Annals of Mathematical Statistics* 38 (1967): 1618–1626.

Halevy, Alon, Peter Norvig, and Fernando Pereira. "The Unreasonable Effectiveness of Data." *Intelligent Systems, IEEE* 24, no. 2 (2009): 8–12.

Hardin, Garrett. "The Tragedy of the Commons." *Science* 162, no. 3859 (1968): 1243–1248.

Hardy, G. H. *Collected Works.* Vol. II. Oxford, UK: Oxford University Press, 1967.

———. "Prime Numbers." *British Association Report* 10 (1915): 350–354.

Harmenberg, J. *Epee 2.0: The Birth of the New Fencing Paradigm.* New York: SKA Swordplay Books, 2007.

Harsanyi, John C. "Can the Maximin Principle Serve as a Basis for Morality? A Critique of John Rawls's Theory." *The American Political Science Review* 69, no. 2 (1975): 594–606.

Harvey, Jerry B. "The Abilene Paradox: The Management of Agreement." *Organizational Dynamics* 3, no. 1 (1974): 63–80.

Hastings, W. K. "Monte Carlo Methods Using Markov Chains and Their Applications." *Biometrika* 57 (1970): 97–109.

Hawken, Angela, and Mark Kleiman. *Managing Drug Involved Probationers with Swift and Certain Sanctions: Evaluating Hawaii's HOPE.* Report submitted to the National Institute of Justice. 2009. http://www.ncjrs.gov/pdffiles1/nij/grants/229023.pdf.

Held, Michael, and Richard M. Karp. "The Traveling-Salesman Problem and Minimum Spanning Trees." *Operations Research* 18, no. 6 (1970): 1138–1162.

———. "The Traveling-Salesman Problem and Minimum Spanning Trees: Part II." *Mathematical Programming* 1, no. 1 (1971): 6–25.

Henderson, T. *Discrete Relaxation Techniques.* Oxford, UK: Oxford University Press, 1989.

Hennessy, John L., and David A. Patterson. *Computer Architecture: A Quantitative Approach.* New York: Elsevier, 2012.

Herrmann, Jeffrey W. "The Perspectives of Taylor, Gantt, and Johnson: How to Improve Production Scheduling." *International Journal of Operations and Quality Management* 16 (2010): 243–254.

Heyde, C. C. "Agner Krarup Erlang." In *Statisticians of the Centuries.* Edited by C. C. Heyde, E. Seneta, P. Crepel, S. E. Fienberg, and J. Gani, 328–330. New York: Springer, 2001.

Hill, Theodore. "Knowing When to Stop." *American Scientist* 97 (2009): 126–131.

Hillis, W. Daniel. *The Pattern on the Stone: The Simple Ideas That Make Computers Work.* New York: Basic Books, 1998.

Hirshleifer, Jack. "On the Emotions as Guarantors of Threats and Promises." In *The Latest on the Best: Essays in Evolution and Optimality.* Edited by John Dupre, 307–326. Cambridge, MA: MIT Press, 1987.

Hoffman, David. *The Oligarchs: Wealth and Power in the New Russia.* New York: PublicAffairs, 2003.

Horvitz, Eric, and Shlomo Zilberstein. "Computational Tradeoffs Under Bounded Resources." *Artificial Intelligence* 126 (2001): 1–4.

Hosken, James C. "Evaluation of Sorting Methods." In *Papers and Discussions Presented at the November 7–9, 1955, Eastern Joint AIEE-IRE Computer Conference: Computers in Business and Industrial Systems*, 39–55.

Hurd, Cuthbert C. "A Note on Early Monte Carlo Computations and Scientific Meetings." *IEEE Annals of the History of Computing* 7, no. 2 (1985): 141–155.

Impagliazzo, Russell, and Avi Wigderson. "$P = BPP$ if E Requires Exponential Circuits: Derandomizing the *XOR* Lemma." In *Proceedings of the Twenty-Ninth Annual ACM Symposium on Theory of Computing*, 1997, 220–229.

———. "Randomness vs. Time: De-Randomization Under a Uniform Assumption." In *Proceedings of the 39th Annual Symposium on Foundations of Computer Science*, 1998, 734–743.

Ingram, Wendy Marie, Leeanne M. Goodrich, Ellen A. Robey, and Michael B. Eisen. "Mice Infected with Low-Virulence Strains of Toxoplasma Gondii Lose Their Innate Aversion to Cat Urine, Even After Extensive Parasite Clearance." *PLOS ONE*, no. 9 (2013): e75246.

Jackson, James R. *Scheduling a Production Line to Minimize Maximum Tardiness*. Technical report 43. Management Science Research Project, University of California, Los Angeles, 1955.

Jacobson, Van. "Congestion Avoidance and Control." In *ACM SIGCOMM Computer Communication Review* 18, no. 4 (1988): 314–329.

———. "A New Way to Look at Networking." Lecture at Google, Mountain View, CA, August 2006. https://www.youtube.com/watch?v=oCZMoY3q2uM.

James, William. "Great Men, Great Thoughts, and the Environment." *Atlantic Monthly* 46 (1880): 441–459.

———. *Psychology: Briefer Course*. New York: Holt, 1892.

Jay, Francine. *The Joy of Less: A Minimalist Living Guide: How to Declutter, Organize, and Simplify Your Life*. Medford, NJ: Anja Press, 2010.

Jeffreys, Harold. "An Invariant Form for the Prior Probability in Estimation Problems." *Proceedings of the Royal Society of London. Series A. Mathematical and Physical Sciences* 186 (1946): 453–461.

———. *Theory of Probability*, 3rd ed. Oxford, UK: Oxford University Press, 1961.

Johnson, Selmer Martin. "Optimal Two- and Three-Stage Production Schedules with Setup Times Included." *Naval Research Logistics Quarterly* 1, no. 1 (1954): 61–68.

Johnson, Theodore, and Dennis Shasha. "2Q: A Low Overhead High Performance Buffer Management Replacement Algorithm." *VLDB '94 Proceedings of the 20th International Conference on Very Large Data Bases*, 1994, 439–450.

Jones, Thomas B., and David H. Ackley. "Comparison Criticality in Sorting Algorithms." In *2014 44th Annual IEEE/IFIP International Conference on Dependable Systems and Networks (DSN)*, June 2014, 726–731.

Jones, William. *Keeping Found Things Found: The Study and Practice of Personal Information Management*. Burlington, MA: Morgan Kaufmann, 2007.

Kaelbling, Leslie Pack. *Learning in Embedded Systems*. Cambridge, MA: MIT Press, 1993.

Kaelbling, Leslie Pack, Michael L. Littman, and Andrew W. Moore. "Reinforcement Learning: A Survey." *Journal of Artificial Intelligence Research* 4 (1996): 237–285.

Kanigel, Robert. *The One Best Way: Frederick Winslow Taylor and the Enigma of Efficiency*. New York: Viking Penguin, 1997.

Kant, Immanuel. *Grundlegung zur Metaphysik der Sitten*. Riga: Johann Friedrich Hartknoch, 1785.

———. *Kritik der praktischen Vernunft*. Riga: Johann Friedrich Hartknoch, 1788.

Karmarkar, Narendra. "A New Polynomial-Time Algorithm for Linear Programming." In *Proceedings of the Sixteenth Annual ACM Symposium on Theory of Computing*, 1984, 302–311.

Karp, Richard M. "An Introduction to Randomized Algorithms." *Discrete Applied Mathematics* 34, no. 1 (1991): 165–201.

———. "Reducibility Among Combinatorial Problems." In *Complexity of Computer Computations*, 85–103. New York: Plenum, 1972.

Katajainen, Jyrki, and Jesper Larsson Träff. "A Meticulous Analysis of Mergesort Programs." In *Algorithms and Complexity: Third Italian Conference CIAC '97*. Berlin: Springer, 1997.

Katehakis, Michael N., and Herbert Robbins. "Sequential Choice from Several Populations." *Proceedings of the National Academy of Sciences* 92 (1995): 8584–8585.

Kelly, F. P. "Multi-Armed Bandits with Discount Factor Near One: The Bernoulli Case." *Annals of Statistics* 9 (1981): 987–1001.

Kelly, John L. "A New Interpretation of Information Rate." *Information Theory, IRE Transactions on* 2, no. 3 (1956): 185–189.

Khachiyan, Leonid G. "Polynomial Algorithms in Linear Programming." *USSR Computational Mathematics and Mathematical Physics* 20, no. 1 (1980): 53–72.

Khot, Subhash, and Oded Regev. "Vertex Cover Might Be Hard to Approximate to Within $2-\varepsilon$." *Journal of Computer and System Sciences* 74, no. 3 (2008): 335–349.

Kidd, Celeste, Holly Palmeri, and Richard N. Aslin. "Rational Snacking: Young Children's Decision-Making on the Marshmallow Task Is Moderated by Beliefs About Environmental Reliability." *Cognition* 126, no. 1 (2013): 109–114.

Kilburn, Tom, David B. G. Edwards, M. J. Lanigan, and Frank H. Sumner. "One-Level Storage System." *IRE Transactions on Electronic Computers* (1962): 223–235.

Kinsbourne, Marcel. "Somatic Twist: A Model for the Evolution of Decussation." *Neuropsychology* 27, no. 5 (2013): 511.

Kirby, Kris N. "Bidding on the Future: Evidence Against Normative Discounting of Delayed Rewards." *Journal of Experimental Psychology: General* 126, no. 1 (1997): 54–70.

Kirkpatrick, Scott, C. D. Gelatt, and M. P. Vecchi. "Optimization by Simulated Annealing." *Science* 220, no. 4598 (1983): 671–680.

Knuth, Donald E. "Ancient Babylonian Algorithms." *Communications of the ACM* 15, no. 7 (1972): 671–677.

———. *The Art of Computer Programming, Volume 1: Fundamental Algorithms*, 3rd ed. Boston: Addison-Wesley, 1997.

———. *The Art of Computer Programming, Volume 3: Sorting and Searching*, 3rd ed. Boston: Addison-Wesley, 1997.

———. "A Terminological Proposal." *ACM SIGACT News* 6, no. 1 (1974): 12–18.

———. "The TeX Tuneup of 2014." *TUGboat* 35, no. 1 (2014).

———. *Things a Computer Scientist Rarely Talks About*. Stanford, CA: Center for the Study of Language/Information, 2001.

———. "Von Neumann's First Computer Program." *ACM Computing Surveys (CSUR)* 2, no. 4 (December 1970): 247–260.

Koestler, Arthur. *The Watershed: A Biography of Johannes Kepler*. Garden City, NY: Doubleday, 1960.

Kozen, Dexter, and Shmuel Zaks. "Optimal Bounds for the Change-Making Problem." In *Automata, Languages and Programming*, 700: 150–161. Edited by Andrzej Lingas, Rolf Karlsson, and Svante Carlsson. Berlin: Springer, 1993.

Lai, Tze Leung, and Herbert Robbins. "Asymptotically Efficient Adaptive Allocation Rules." *Advances in Applied Mathematics* 6 (1985): 4–22.

Lamport, Leslie, Robert Shostak, and Marshall Pease. "The Byzantine Generals Problem." *ACM Transactions on Programming Languages and Systems (TOPLAS)* 4, no. 3 (1982): 382–401.

Laplace, Pierre-Simon. *A Philosophical Essay on Probabilities*. 1812. Reprint, New York: Dover, 1951.

———. "Memoir on the Probability of the Causes of Events." *Statistical Science* 1 (1774/1986): 364–378.

———. *Théorie analytique des probabilités*. Paris: Mme Ve Courcier, 1812.

Lawler, Eugene L. "Old Stories." In *History of Mathematical Programming. A Collection of Personal Reminiscences*, 97–106. Amsterdam: CWI/North-Holland, 1991.

———. "Optimal Sequencing of a Single Machine Subject to Precedence Constraints." *Management Science* 19, no. 5 (1973): 544–546.

———. *Scheduling a Single Machine to Minimize the Number of Late Jobs*. Technical report. Berkeley: University of California, 1983.

———. "Scheduling a Single Machine to Minimize the Number of Late Jobs," no. UCB/CSD-83-139 (1983). http://www.eecs.berkeley.edu/Pubs/TechRpts/1983/6344.html.

———. "Sequencing Jobs to Minimize Total Weighted Completion Time Subject to Precedence Constraints." *Annals of Discrete Mathematics* 2 (1978): 75–90.

Lawler, Eugene L., Jan Karel Lenstra, and Alexander H. G. Rinnooy Kan. "A Gift for Alexander!: At Play in the Fields of Scheduling Theory." *Optima* 7 (1982): 1–3.

Lawler, Eugene L., Jan Karel Lenstra, Alexander H. G. Rinnooy Kan, and David B. Shmoys. "Sequencing and Scheduling: Algorithms and Complexity." In *Handbooks in Operations Research and Management Science, Volume 4: Logistics of Production and Inventory*, edited by S. S. Graves, A. H. G. Rinnooy Kan, and P. Zipkin, 445–522. Amsterdam: North Holland, 1993.

———. *The Traveling Salesman Problem: A Guided Tour of Combinatorial Optimization*. New York: Wiley, 1985.

Lazzarini, Mario. "Un'applicazione del calcolo della probabilità alla ricerca sperimentale di un valore approssimato di π." *Periodico di Matematica* 4 (1901): 140–143.

Lee, Donghee, S. H. Noh, S. L. Min, J. Choi, J. H. Kim, Yookun Cho, and Chong Sang Kim. "LRFU: A Spectrum of Policies That Subsumes the Least Recently Used and Least Frequently Used Policies." *IEEE Transactions on Computers* 50 (2001): 1352–1361.

Le Guin, Ursula K. "The Ones Who Walk Away from Omelas." In *New Dimensions 3*. Edited by Robert Silverberg. New York: Signet, 1973.

Lenstra, Jan Karel. "The Mystical Power of Twoness: In Memoriam Eugene L. Lawler." *Journal of Scheduling* 1, no. 1 (1998): 3–14.

Lenstra, Jan Karel, Alexander H. G. Rinnooy Kan, and Peter Brucker. "Complexity of Machine Scheduling Problems." *Annals of Discrete Mathematics* 1 (1977): 343–362.

Lerner, Ben. *The Lichtenberg Figures*. Port Townsend, WA: Copper Canyon Press, 2004.

Lindley, Denis V. "Dynamic Programming and Decision Theory." *Applied Statistics* 10 (1961): 39–51.

Lippman, Steven A., and John J. McCall. "The Economics of Job Search: A Survey." *Economic Inquiry* 14 (1976): 155–189.

Lorie, James H., and Leonard J. Savage. "Three Problems in Rationing Capital." *Journal of Business* 28, no. 4 (1955): 229–239.

Lowe, Christopher J., Mark Terasaki, Michael Wu, Robert M. Freeman Jr., Linda Runft, Kristen Kwan, Saori Haigo, Jochanan Aronowicz, Eric Lander, Chris Gruber, et al. "Dorsoventral Patterning in Hemichordates: Insights into Early Chordate Evolution." *PLoS Biology* 4, no. 9 (2006): e291.

Lucas, Richard E., Andrew E. Clark, Yannis Georgellis, and Ed Diener. "Reexamining Adaptation and the Set Point Model of Happiness: Reactions to Changes in Marital Status." *Journal of Personality and Social Psychology* 84, no. 3 (2003): 527–539.

Lueker, George S. "Two NP-Complete Problems in Nonnegative Integer Programming." *Technical Report TR-178*, Computer Science Laboratory, Princeton University, 1975.

Luria, Salvador E. *A Slot Machine, a Broken Test Tube: An Autobiography.* New York: Harper & Row, 1984.

MacQueen, J., and R. G. Miller. "Optimal Persistence Policies." *Operations Research* 8 (1960): 362–380.

Malthus, Thomas Robert. *An Essay on the Principle of Population.* London: J. Johnson, 1798.

Marcus, Gary. *Kluge: The Haphazard Evolution of the Human Mind.* New York: Houghton Mifflin Harcourt, 2009.

Markowitz, Harry. "Portfolio Selection." *Journal of Finance* 7, no. 1 (1952): 77–91.

———. *Portfolio Selection: Efficient Diversification of Investments.* New York: Wiley, 1959.

Martin, Thomas Commerford. "Counting a Nation by Electricity." *Electrical Engineer* 12, no. 184 (1891): 521–530.

McCall, John. "Economics of Information and Job Search." *Quarterly Journal of Economics* 84 (1970): 113–126.

McGrayne, Sharon Bertsch. *The Theory That Would Not Die: How Bayes' Rule Cracked the Enigma Code, Hunted Down Russian Submarines, & Emerged Triumphant from Two Centuries of Controversy.* New Haven, CT: Yale University Press, 2011.

McGuire, Joseph T., and Joseph W. Kable. "Decision Makers Calibrate Behavioral Persistence on the Basis of Time-Interval Experience." *Cognition* 124, no. 2 (2012): 216–226.

———. "Rational Temporal Predictions Can Underlie Apparent Failures to Delay Gratification." *Psychological Review* 120, no. 2 (2013): 395.

Megiddo, Nimrod, and Dharmendra S. Modha. "Outperforming LRU with an Adaptive Replacement Cache Algorithm." *Computer* 37, no. 4 (2004): 58–65.

Mellen, Andrew. *Unstuff Your Life! Kick the Clutter Habit and Completely Organize Your Life for Good.* New York: Avery, 2010.

Menezes, Alfred J., Paul C. Van Oorschot, and Scott A Vanstone. *Handbook of Applied Cryptography.* Boca Raton, FL: CRC Press, 1996.

Menger, Karl. "Das botenproblem." *Ergebnisse eines mathematischen kolloquiums* 2 (1932): 11–12.

Metropolis, Nicholas, Arianna W. Rosenbluth, Marshall N. Rosenbluth, Augusta H. Teller, and Edward Teller. "Equation of State Calculations by Fast Computing Machines." *Journal of Chemical Physics* 21, no. 6 (1953): 1087–1092.

Meyer, Robert J., and Yong Shi. "Sequential Choice Under Ambiguity: Intuitive Solutions to the Armed-Bandit Problem." *Management Science* 41 (1995): 817–834.

Millard-Ball, Adam, Rachel R. Weinberger, and Robert C. Hampshire. "Is the Curb 80% Full or 20% Empty? Assessing the Impacts of San Francisco's Parking Pricing Experiment." *Transportation Research Part A: Policy and Practice* 63 (2014): 76–92.

Mischel, Walter, Ebbe B. Ebbesen, and Antonette Raskoff Zeiss. "Cognitive and Attentional Mechanisms in Delay of Gratification." *Journal of Personality and Social Psychology* 21, no. 2 (1972): 204.

Mischel, Walter, Yuichi Shoda, and Monica I. Rodriguez. "Delay of Gratification in Children." *Science* 244, no. 4907 (1989): 933–938.

Mitzenmacher, Michael, and Eli Upfal. *Probability and Computing: Randomized Algorithms and Probabilistic Analysis.* Cambridge, UK: Cambridge University Press, 2005.

Monsell, Stephen. "Task Switching." *Trends in Cognitive Sciences* 7, no. 3 (2003): 134–140.

Moore, Gordon E. "Cramming More Components onto Integrated Circuits." *Electronics Magazine* 38 (1965): 114–117.

———. "Progress in Digital Integrated Electronics." In *International Electronic Devices Meeting 1975 Technical Digest*, 1975, 11–13.

Moore, J. Michael. "An *N* Job, One Machine Sequencing Algorithm for Minimizing the Number of Late Jobs." *Management Science* 15, no. 1 (1968): 102–109.

Morgenstern, Julie. *Organizing from the Inside Out: The Foolproof System for Organizing Your Home, Your Office and Your Life.* New York: Macmillan, 2004.

Moser, L. "On a Problem of Cayley." *Scripta Mathematica* 22 (1956): 289–292.

Motwani, Rajeev, and Prabhakar Raghavan. *Randomized Algorithms.* Cambridge, UK: Cambridge University Press, 1995.

———. "Randomized Algorithms." *ACM Computing Surveys (CSUR)* 28, no. 1 (1996): 33–37.

Mucci, A. G. "On a Class of Secretary Problems." *Annals of Probability* 1 (1973): 417–427.

Murray, David. *Chapters in the History of Bookkeeping, Accountancy and Commercial Arithmetic.* Glasgow, UK: Jackson, Wylie, 1930.

Myerson, Roger B. "Nash Equilibrium and the History of Economic Theory." *Journal of Economic Literature* 1999, 1067–1082.

———. "Optimal Auction Design." *Mathematics of Operations Research* 6, no. 1 (1981): 58–73.

Nash, John F. "Equilibrium Points in *N*-Person Games." *Proceedings of the National Academy of Sciences* 36, no. 1 (1950): 48–49.

———. "Non-Cooperative Games." *Annals of Mathematics* 54, no. 2 (1951): 286–295.

———. "The Bargaining Problem." *Econometrica* 18, no. 2 (1950): 155–162.

Navarro, Daniel J., and Ben R. Newell. "Information Versus Reward in a Changing World." In *Proceedings of the 36th Annual Conference of the Cognitive Science Society*, 2014, 1054–1059.

Neumann, John von, and Oskar Morgenstern. *Theory of Games and Economic Behavior.* Princeton, NJ: Princeton University Press, 1944.

Neyman, Jerzy. "Outline of a Theory of Statistical Estimation Based on the Classical Theory of Probability." *Philosophical Transactions of the Royal Society of London. Series A, Mathematical and Physical Sciences* 236, no. 767 (1937): 333–380.

Nichols, Kathleen, and Van Jacobson. "Controlling Queue Delay: A Modern AQM Is Just One Piece of the Solution to Bufferbloat." *ACM Queue Networks* 10, no. 5 (2012): 20–34.

Nisan, Noam, and Amir Ronen. "Algorithmic Mechanism Design." In *Proceedings of the Thirty-First Annual ACM Symposium on Theory of Computing*, 1999, 129–140.

Olshausen, Bruno A., and David J. Field. "Emergence of Simple-Cell Receptive Field Properties by Learning a Sparse Code for Natural Images." *Nature* 381 (1996): 607–609.

O'Neil, Elizabeth J., Patrick E. O'Neil, and Gerhard Weikum. "The LRU-*K* Page Replacement Algorithm for Database Disk Buffering," *ACM SIGMOD Record* 22, no. 2 (1993): 297–306.

Papadimitriou, Christos. "Foreword." In *Algorithmic Game Theory.* Edited by Noam Nisan, Tim Roughgarden, Éva Tardos, and Vijay V. Vazirani. Cambridge, UK: Cambridge University Press, 2007.

Papadimitriou, Christos H., and John N. Tsitsiklis. "The Complexity of Optimal Queuing Network Control." *Mathematics of Operations Research* 24 (1999): 293–305.

Papadimitriou, Christos H., and Mihalis Yannakakis. "On Complexity as Bounded Rationality." In *Proceedings of the Twenty-Sixth Annual ACM Symposium on Theory of Computing*, 1994, 726–733.

Pardalos, Panos M., and Georg Schnitger. "Checking Local Optimality in Constrained Quadratic Programming is *NP*-hard." *Operations Research Letters* 7 (1988): 33–35.

Pareto, Vilfredo. *Cours d'économie politique.* Lausanne: F. Rouge, 1896.

Parfit, Derek. *Reasons and Persons.* Oxford, UK: Oxford University Press, 1984.

Partnoy, Frank. *Wait: The Art and Science of Delay.* New York: PublicAffairs, 2012.

Pascal, Blaise. *Pensées sur la religion et sur quelques autres sujets.* Paris: Guillaume Desprez, 1670.

Peter, Laurence J., and Raymond Hull. *The Peter Principle: Why Things Always Go Wrong.* New York: Morrow, 1969.

Petruccelli, Joseph D. "Best-Choice Problems Involving Uncertainty of Selection and Recall of Observations." *Journal of Applied Probability* 18 (1981): 415–425.

Pettie, Seth, and Vijaya Ramachandran. "An Optimal Minimum Spanning Tree Algorithm." *Journal of the ACM* 49, no. 1 (2002): 16–34.

Pinedo, Michael. *Scheduling: Theory, Algorithms, and Systems.* New York: Springer, 2012.

———. "Stochastic Scheduling with Release Dates and Due Dates." *Operations Research* 31, no. 3 (1983): 559–572.

Pirsig, Robert M. *Zen and the Art of Motorcycle Maintenance.* New York: Morrow, 1974.

Poundstone, William. *Fortune's Formula: The Untold Story of the Scientific Betting System That Beat the Casinos and Wall Street.* New York: Macmillan, 2005.

———. *Prisoner's Dilemma: John von Neumann, Game Theory, and the Puzzle of the Bomb.* New York: Doubleday, 1992.

Prabhakar, Balaji, Katherine N. Dektar, and Deborah M. Gordon. "The Regulation of Ant Colony Foraging Activity Without Spatial Information." *PLoS Computational Biology* 8, no. 8 (2012): e1002670.

Presman, Ernst L'vovich, and Isaac Mikhailovich Sonin. "The Best Choice Problem for a Random Number of Objects." *Teoriya Veroyatnostei i ee Primeneniya* 17 (1972): 695–706.

Production and Operations Management Society. "James R. Jackson." *Production and Operations Management* 17, no. 6 (2008): i–ii.

Rabin, Michael O. "Probabilistic Algorithm for Testing Primality." *Journal of Number Theory* 12, no. 1 (1980): 128–138.

Rabin, Michael O., and Dana Scott. "Finite Automata and Their Decision Problems." *IBM Journal of Research and Development* 3 (1959): 114–125.

Raichle, Marcus E., and Debra A. Gusnard. "Appraising the Brain's Energy Budget." *Proceedings of the National Academy of Sciences* 99, no. 16 (2002): 10237–10239.

Ramakrishnan, Kadangode, and Sally Floyd. *A Proposal to Add Explicit Congestion Notification (ECN) to IP.* Technical report. RFC 2481, January 1999.

Ramakrishnan, Kadangode, Sally Floyd, and David Black. *The Addition of Explicit Congestion Notification (ECN) to IP.* Technical report. RFC 3168, September 2001.

Ramscar, Michael, Peter Hendrix, Cyrus Shaoul, Petar Milin, and Harald Baayen. "The Myth of Cognitive Decline: Non-Linear Dynamics of Lifelong Learning." *Topics in Cognitive Science* 6, no. 1 (2014): 5–42.

Rasmussen, Willis T., and Stanley R. Pliska. "Choosing the Maximum from a Sequence with a Discount Function." *Applied Mathematics and Optimization* 2 (1975): 279–289.

Rawls, John. *A Theory of Justice.* Cambridge, MA: Harvard University Press, 1971.

Revusky, Samuel H., and Erwin W. Bedarf. "Association of Illness with Prior Ingestion of Novel Foods." *Science* 155, no. 3759 (1967): 219–220.

Reynolds, Andy M. "Signatures of Active and Passive Optimized Lévy Searching in Jellyfish." *Journal of the Royal Society Interface* 11, no. 99 (2014): 20140665.

Ridgway, Valentine F. "Dysfunctional Consequences of Performance Measurements." *Administrative Science Quarterly* 1, no. 2 (1956): 240–247.

Riley, John G., and William F. Samuelson. "Optimal Auctions." *American Economic Review* 71, no. 3 (1981): 381–392.

Rittaud, Benoît, and Albrecht Heeffer. "The Pigeonhole Principle, Two Centuries Before Dirichlet." *Mathematical Intelligencer* 36, no. 2 (2014): 27–29.

Rivest, Ronald L., Adi Shamir, and Leonard Adleman. "A Method for Obtaining Digital Signatures and Public-Key Cryptosystems." *Communications of the ACM* 21, no. 2 (1978): 120–126.

Robbins, Herbert. "Some Aspects of the Sequential Design of Experiments." *Bulletin of the American Mathematical Society* 58 (1952): 527–535.

Robinson, Julia. *On the Hamiltonian Game (a Traveling Salesman Problem)*. Technical report RAND/RM-303. Santa Monica, CA: RAND, 1949.

Rogerson, Richard, Robert Shimer, and Randall Wright. *Search-Theoretic Models of the Labor Market: A Survey*. Technical report. Cambridge, MA: National Bureau of Economic Research, 2004.

Rose, John S. "A Problem of Optimal Choice and Assignment." *Operations Research* 30 (1982): 172–181.

Rosenbaum, David A., Lanyun Gong, and Cory Adam Potts. "Pre-Crastination: Hastening Subgoal Completion at the Expense of Extra Physical Effort." *Psychological Science* 25, no. 7 (2014): 1487–1496.

Rosenbluth, Marshall. *Marshall Rosenbluth, interviewed by Kai-Henrik Barth*. August 11, 2003, College Park, MD.

Rostker, Bernard D., Harry J. Thie, James L. Lacy, Jennifer H. Kawata, and Susanna W. Purnell. *The Defense Officer Personnel Management Act of 1980: A Retrospective Assessment*. Santa Monica, CA: RAND, 1993.

Roughgarden, Tim, and Éva Tardos. "How Bad Is Selfish Routing?" *Journal of the ACM* 49, no. 2 (2002): 236–259.

Russell, Bertrand. "The Elements of Ethics." In *Philosophical Essays*, 13–59. London: Longmans, Green, 1910.

Russell, Stuart, and Peter Norvig. *Artificial Intelligence: A Modern Approach*, 3rd ed. Upper Saddle River, NJ: Pearson, 2009.

Russell, Stuart, and Eric Wefald. *Do the Right Thing*. Cambridge, MA: MIT Press, 1991.

Sagan, Carl. *Broca's Brain: Reflections on the Romance of Science*. New York: Random House, 1979.

Sakaguchi, Minoru. "Bilateral Sequential Games Related to the No-Information Secretary Problem." *Mathematica Japonica* 29 (1984): 961–974.

———. "Dynamic Programming of Some Sequential Sampling Design." *Journal of Mathematical Analysis and Applications* 2 (1961): 446–466.

Sakaguchi, Minoru, and Mitsushi Tamaki. "On the Optimal Parking Problem in Which Spaces Appear Randomly." *Bulletin of Informatics and Cybernetics* 20 (1982): 1–10.

Sartre, Jean-Paul. *No Exit: A Play in One Act*. New York: Samuel French, 1958.

Schelling, Thomas C. "Altruism, Meanness, and Other Potentially Strategic Behaviors." *American Economic Review* 68, no. 2 (1978): 229–230.

———. *The Strategy of Conflict*. Cambridge, MA: Harvard University Press, 1960.

Schneier, Bruce. *Applied Cryptography*. New York: Wiley, 1994.

Schrage, Linus. "A Proof of the Optimality of the Shortest Remaining Processing Time Discipline." *Operations Research* 16, no. 3 (1968): 687–690.

Schrijver, Alexander. "On the History of Combinatorial Optimization (Till 1960)." In *Handbooks in Operations Research and Management Science: Discrete Optimization*. Edited by Karen Aardal, George L. Nemhauser, and Robert Weismantel. Amsterdam: Elsevier, 2005, 1–68.

Schwartz, Jacob T. "Fast Probabilistic Algorithms for Verification of Polynomial Identities." *Journal of the ACM* 27, no. 4 (1980): 701–717.

Seale, Darryl A., and Amnon Rapoport. "Sequential Decision Making with Relative Ranks: An Experimental Investigation of the 'Secretary Problem.'" *Organizational Behavior and Human Decision Processes* 69 (1997): 221–236.

Sen, Amartya. "Goals, Commitment, and Identity." *Journal of Law, Economics, and Organization* 1 (1985): 341–355.

Sethi, Rajiv. "Algorithmic Trading and Price Volatility." *Rajiv Sethi* (blog), May 7, 2010, http://rajivsethi.blogspot.com/2010/05/algorithmic-trading-and-price.html.

Sevcik, Kenneth C. "Scheduling for Minimum Total Loss Using Service Time Distributions." *Journal of the ACM* 21, no. 1 (1974): 66–75.

Shallit, Jeffrey. "What This Country Needs Is an 18¢ Piece." *Mathematical Intelligencer* 25, no. 2 (2003): 20–23.

Shasha, Dennis, and Cathy Lazere. *Out of Their Minds: The Lives and Discoveries of 15 Great Computer Scientists*. New York: Springer, 1998.

Shasha, Dennis, and Michael Rabin. "An Interview with Michael Rabin." *Communications of the ACM* 53, no. 2 (2010): 37–42.

Shaw, Frederick S. *An Introduction to Relaxation Methods*. New York: Dover, 1953.

Shaw, George Bernard. *Man and Superman: A Comedy and a Philosophy*. Cambridge, MA: Harvard University Press, 1903.

Shoup, Donald. *The High Cost of Free Parking*. Chicago: APA Planners Press, 2005.

Simon, Herbert A. "A Behavioral Model of Rational Choice." *Quarterly Journal of Economics* 69, no. 1 (1955): 99–118.

———. *Models of Man*. New York: Wiley, 1957.

———. "On a Class of Skew Distribution Functions." *Biometrika*, 1955, 425–440.

Siroker, Dan. "How Obama Raised $60 Million by Running a Simple Experiment." *The Optimizely Blog: A/B Testing You'll Actually Use* (blog), November 29, 2010, https://blog.optimizely.com/2010/11/29/how-obama-raised-60-million-by-running-a-simple-experiment/.

Siroker, Dan, and Pete Koomen. *A/B Testing: The Most Powerful Way to Turn Clicks into Customers*. New York: Wiley, 2013.

Sleator, Daniel D., and Robert E. Tarjan. "Amortized Efficiency of List Update and Paging Rules." *Communications of the ACM* 28 (1985): 202–208.

Smith, Adam. *The Theory of Moral Sentiments*. Printed for A. Millar, in the Strand; and A. Kincaid and J. Bell, in Edinburgh, 1759.

Smith, M. H. "A Secretary Problem with Uncertain Employment." *Journal of Applied Probability* 12, no. 3 (1975): 620–624.

Smith, Wayne E. "Various Optimizers for Single-Stage Production." *Naval Research Logistics Quarterly* 3, nos. 1–2 (1956): 59–66.

Solovay, Robert, and Volker Strassen. "A Fast Monte-Carlo Test for Primality." *SIAM Journal on Computing* 6 (1977): 84–85.

Starr, Norman. "How to Win a War if You Must: Optimal Stopping Based on Success Runs." *Annals of Mathematical Statistics* 43, no. 6 (1972): 1884–1893.

Stephens, David W., and John R. Krebs. *Foraging Theory*. Princeton, NJ: Princeton University Press, 1986.

Stewart, Martha. *Martha Stewart's Homekeeping Handbook: The Essential Guide to Caring for Everything in Your Home*. New York: Clarkson Potter, 2006.

Steyvers, Mark, Michael D. Lee, and Eric-Jan Wagenmakers. "A Bayesian Analysis of Human Decision-Making on Bandit Problems." *Journal of Mathematical Psychology* 53 (2009): 168–179.

Stigler, George J. "The Economics of Information." *Journal of Political Economy* 69 (1961): 213–225.

———. "Information in the Labor Market." *Journal of Political Economy* 70 (1962): 94–105.

Stigler, Stephen M. "Stigler's Law of Eponymy." *Transactions of the New York Academy of Sciences* 39 (1980): 147–157.

Tamaki, Mitsushi. "Adaptive Approach to Some Stopping Problems." *Journal of Applied Probability* 22 (1985): 644–652.

———. "An Optimal Parking Problem." *Journal of Applied Probability* 19 (1982): 803–814.

———. "Optimal Stopping in the Parking Problem with U-Turn." *Journal of Applied Probability* 25 (1988): 363–374.

Thomas, Helen. *Front Row at the White House: My Life and Times*. New York: Simon & Schuster, 2000.

Thompson, William R. "On the Likelihood That One Unknown Probability Exceeds Another in View of the Evidence of Two Samples." *Biometrika* 25 (1933): 285–294.

Thoreau, Henry David. "Walking." *Atlantic Monthly* 9 (1862): 657–674.

Tibshirani, Robert. "Regression Shrinkage and Selection via the Lasso." *Journal of the Royal Statistical Society. Series B (Methodological)* 58, no. 1 (1996): 267–288.

Tikhonov, A. N., and V. Y. Arsenin. *Solution of Ill-Posed Problems*. Washington, DC: Winston, 1977.

Todd, Peter M. "Coevolved Cognitive Mechanisms in Mate Search." *Evolution and the Social Mind: Evolutionary Psychology and Social Cognition* (New York) 9 (2007): 145–159.

Todd, Peter M., and G. F. Miller. "From Pride and Prejudice to Persuasion: Satisficing in Mate Search." In *Simple Heuristics That Make Us Smart*. Edited by G. Gigerenzer and P. M. Todd. New York: Oxford University Press, 1999, 287–308.

Tolins, Jackson, and Jean E. Fox Tree. "Addressee Backchannels Steer Narrative Development." *Journal of Pragmatics* 70 (2014): 152–164.

Tracy, Brian. *Eat That Frog! 21 Great Ways to Stop Procrastinating and Get More Done in Less Time*. Oakland, CA: Berrett-Koehler, 2007.

Turing, Alan M. "On Computable Numbers, with an Application to the Entscheidungsproblem." Read November 12, 1936. *Proceedings of the London Mathematical Society* s2-42, no. 1 (1937): 230–265.

———. "On Computable Numbers, with an Application to the Entscheidungsproblem: A Correction." *Proceedings of the London Mathematical Society* s2-43, no. 1 (1938): 544–546.

Tversky, Amos, and Ward Edwards. "Information Versus Reward in Binary Choices." *Journal of Experimental Psychology* 71 (1966): 680–683.

Ulam, Stanislaw M. *Adventures of a Mathematician*. New York: Scribner, 1976.

Ullman, Ellen. "Out of Time: Reflections on the Programming Life." *Educom Review* 31 (1996): 53–59.

UK Collaborative ECMO Group. "The Collaborative UK ECMO Trial: Follow-up to 1 Year of Age." *Pediatrics* 101, no. 4 (1998): e1.

Vazirani, Vijay V. *Approximation Algorithms*. New York: Springer, 2001.

Vickrey, William. "Counterspeculation, Auctions, and Competitive Sealed Tenders." *Journal of Finance* 16, no. 1 (1961): 8–37.

Waitzman, David. *A Standard for the Transmission of IP Datagrams on Avian Carriers.* Technical report. RFC 1149, April 1990.

———. *IP Over Avian Carriers with Quality of Service.* Technical report. RFC 2549, April 1999.

Ware, James H. "Investigating Therapies of Potentially Great Benefit: ECMO." *Statistical Science* 4 (1989): 298–306.

Ware, James H., and Michael F. Epstein. "Comments on 'Extracorporeal Circulation in Neonatal Respiratory Failure: A Prospective Randomized Study' by R. H. Bartlett et al." *Pediatrics* 76, no. 5 (1985): 849–851.

Warhol, Andy. *The Philosophy of Andy Warhol (from A to B and Back Again).* New York: Harcourt Brace Jovanovich, 1975.

Weiss, Yair, Eero P. Simoncelli, and Edward H. Adelson. "Motion Illusions as Optimal Percepts." *Nature Neuroscience* 5 (2002): 598–604.

Whittaker, Steve, and Candace Sidner. "Email Overload: Exploring Personal Information Management of Email." In *Proceedings of the SIGCHI Conference on Human Factors in Computing Systems,* 1996, 276–283.

Whittaker, Steve, Tara Matthews, Julian Cerruti, Hernan Badenes, and John Tang. "Am I Wasting My Time Organizing Email? A Study of Email Refinding." In *Proceedings of the SIGCHI Conference on Human Factors in Computing Systems,* 2011, 3449–3458.

Whittle, Peter. *Optimization over Time: Dynamic Programming and Stochastic Control.* New York: Wiley, 1982.

———. "Restless Bandits: Activity Allocation in a Changing World." *Journal of Applied Probability* 25 (1988): 287–298.

Wigderson, Avi. "Knowledge, Creativity, and *P* versus *NP.*" http://www.math.ias.edu/~avi /PUBLICATIONS/MYPAPERS/AW09/AW09.pdf, 2009.

Wilkes, Maurice V. "Slave Memories and Dynamic Storage Allocation." *IEEE Transactions on Electronic Computers* 14 (1965): 270–271.

Wright, J. W. "The Change-Making Problem." *Journal of the Association of Computing Machinery* 22 (1975): 125–128.

Wulf, William Allan, and Sally A. McKee. "Hitting the Memory Wall: Implications of the Obvious." *ACM SIGARCH Computer Architecture News* 23, no. 1 (1995): 20–24.

Xu, Fei, and Joshua B. Tenenbaum. "Word Learning as Bayesian Inference." *Psychological Review* 114 (2007): 245–272.

Yang, Mark C. K. "Recognizing the Maximum of a Random Sequence Based on Relative Rank with Backward Solicitation." *Journal of Applied Probability* 11 (1974): 504–512.

Yato, Takayuki, and Takahiro Seta. "Complexity and Completeness of Finding Another Solution and Its Application to Puzzles." *IEICE Transactions on Fundamentals of Electronics, Communications and Computer Sciences* 86, no. 5 (2003): 1052–1060.

Yngve, Victor H. "On Getting a Word in Edgewise." In *Chicago Linguistics Society, 6th Meeting,* 1970, 567–578.

Zahniser, Rick. "Timeboxing for Top Team Performance." *Software Development* 3, no. 3 (1995): 34–38.

Zapol, Warren M., Michael T. Snider, J. Donald Hill, Robert J. Fallat, Robert H. Bartlett, L. Henry Edmunds, Alan H. Morris, E. Converse Peirce, Arthur N. Thomas, Herbert J. Proctor, et al. "Extracorporeal Membrane Oxygenation in Severe Acute Respiratory Failure: A Randomized Prospective Study." *Journal of the American Medical Association* 242, no. 20 (1979): 2193–2196.

Zelen, Marvin. "Play the Winner Rule and the Controlled Clinical Trial." *Journal of the American Statistical Association* 64, no. 325 (1969): 131–146.

Zippel, Richard. "Probabilistic Algorithms for Sparse Polynomials." In *EUROSAM '79 Proceedings of the International Symposium on Symbolic and Algebraic Computation*. London: Springer, 1979, 216–226.

Acknowledgments

Thank you, first, to the researchers, practitioners, and experts who made time to sit down with us and discuss their work and broader perspectives: to Dave Ackley, Steve Albert, John Anderson, Jeff Atwood, Neil Bearden, Rik Belew, Donald Berry, Avrim Blum, Laura Carstensen, Nick Chater, Stuart Cheshire, Paras Chopra, Herbert Clark, Ruth Corbin, Robert X. Cringely, Peter Denning, Raymond Dong, Elizabeth Dupuis, Joseph Dwyer, David Estlund, Christina Fang, Thomas Ferguson, Jessica Flack, James Fogarty, Jean E. Fox Tree, Robert Frank, Stuart Geman, Jim Gettys, John Gittins, Alison Gopnik, Deborah Gordon, Michael Gottlieb, Steve Hanov, Andrew Harbison, Isaac Haxton, John Hennessy, Geoff Hinton, David Hirshliefer, Jordan Ho, Tony Hoare, Kamal Jain, Chris Jones, William Jones, Leslie Kaelbling, David Karger, Richard Karp, Scott Kirkpatrick, Byron Knoll, Con Kolivas, Michael Lee, Jan Karel Lenstra, Paul Lynch, Preston McAfee, Jay McClelland, Laura Albert McLay, Paul Milgrom, Anthony Miranda, Michael Mitzenmacher, Rosemarie Nagel, Christof Neumann, Noam Nisan, Yukio Noguchi, Peter Norvig, Christos Papadimitriou, Meghan Peterson, Scott Plagenhoef, Anita Pomerantz, Balaji Prabhakar, Kirk Pruhs, Amnon Rapoport, Ronald Rivest, Ruth Rosenholtz, Tim Roughgarden, Stuart Russell, Roma Shah, Donald Shoup, Steven Skiena, Dan Smith, Paul Smolensky, Mark Steyvers, Chris Stucchio, Milind Tambe, Robert Tarjan, Geoff Thorpe, Jackson Tolins, Michael Trick, Hal Varian, James Ware, Longhair Warrior, Steve Whittaker, Avi Wigderson, Jacob Wobbrock, Jason Wolfe, and Peter Zijlstra.

Thanks to the King County Public Library, the Seattle Public Library, the Northern Regional Library Facility, and the UC Berkeley libraries for backstage passes into their operations.

Thanks to those with whom we corresponded, who pointed us in the direction of research worth knowing, including Sharon Goetz, Mike Jones, Tevye Krynski, Elif Kuş, Falk Lieder, Steven A. Lippman, Philip Maughan, Sam McKenzie, Harro Ranter, Darryl A. Seale, Stephen Stigler, Kevin Thomson, Peter Todd, Sara M. Watson, and Sheldon Zedeck.

Thanks to many of those with whom conversation led in short order to many of the insights herein, and of whom the following is an incomplete list: Elliot Aguilar, Ben Backus, Liat Berdugo, Dave Blei, Ben Blum, Joe Damato, Eva de Valk, Emily Drury, Peter Eckersley, Jesse Farmer, Alan Fineberg, Chrix Finne, Lucas Foglia, John Gaunt, Lee Gilman, Martin Glazier, Adam Goldstein, Sarah Greenleaf, Graff Haley, Ben Hjertmann, Greg Jensen, Henry Kaplan, Sharmin Karim, Falk Lieder, Paul Linke, Rose Linke, Tania Lombrozo, Brandon Martin-Anderson, Sam McKenzie, Elon Musk, the Neuwrite group at Columbia University, Hannah Newman, Abe Othman, Sue Penney, Dillon Plunkett, Kristin Pollock, Diego Pontoriero, Avi Press, Matt Richards, Annie Roach, Felicity Rose, Anders Sandberg, Claire Schreiber, Gayle and Rick Shanley, Max Shron, Charly Simpson, Najeeb Tarazi, Josh Tenenbaum, Peter Todd, Peter van Wesep, Shawn Wen, Jered Wierzbicki, Maja Wilson, and Kristen Young.

Thank you to some of the fine free and open-source software that made the work possible: Git, LaTeX, TeXShop, and TextMate 2, for starters.

Thanks to those who lent their skills and efforts on various fronts: to Lindsey Baggette, David Bourgin, and Tania Lombrozo for bibliographic and archival sleuthing.

Thanks to the Cambridge University Library for permission to print Darwin's wonderful diary page, and to Michael Langan for a crisp restoration thereof.

Thanks to Henry Young for a sharp portrait.

Thanks to those who read drafts and offered invaluable feedback along the way: to Ben Blum, Vint Cerf, Elizabeth Christian, Randy Christian, Peter Denning, Peter Eckersley, Chrix Finne, Rick Fletcher, Adam Goldstein, Alison Gopnik, Sarah Greenleaf, Graff Haley, Greg Jensen, Charles Kemp, Raphael Lee, Rose Linke, Tania Lombrozo, Rebekah Otto, Diego Pontoriero, Daniel Reichman, Matt Richards, Phil Richerme, Melissa Riess James, Katia Savchuk, Sameer Shariff, Janet Silver, Najeeb Tarazi, and Kevin Thomson. The book is immeasurably better for their saccades and thoughts.

Thanks to our agent, Max Brockman, and the team at Brockman Inc. for being astute and exuberant champions of the work.

Thanks to our editor, Grigory Tovbis, and the team at Henry Holt for their perspicacious, tireless, enthusiastic work at making the book the best it could be and for bugling it forth proudly into the world.

Thank you to Tania Lombrozo, Viviana Lombrozo, Enrique Lombrozo, Judy Griffiths, Rod Griffiths, and Julieth Moreno, who picked up the slack on the childcare front on multiple occasions, and to the Lombrozo Griffiths family, and the members of the UC Berkeley Computational Cognitive Science lab, and to all who exhibited grace and patience with book-induced scheduling constraints.

Thanks to the various institutions who offered direct and indirect support. Thank you first to the University of California, Berkeley: to the Visiting Scholar Program in the Institute of Cognitive and Brain Sciences for a productive two-year stint, and to the Department of Psychology for its ongoing support. Thank you to the Free Library of Philadelphia, the University of California, Berkeley Library, the Mechanics' Institute Library, and the San Francisco Public Library for both space and tomes. Thank you to the University of Pennsylvania Fisher Fine Arts Library for allowing a nonstudent in off the streets day after day. Thank you to the Corporation of Yaddo, the MacDowell Colony, and the Port Townsend Writers' Conference for beautiful, inspirational, and fertile residencies. Thank you to the USPS Media Mail rate for making a peripatetic ink-and-pulp lifestyle possible. Thank you to the Cognitive Science Society and the Association for the Advancement of Artificial Intelligence for invitations to attend their annual conferences, at which many connections were made: interpersonal, interdisciplinary, and interhemispheric. Thank you to Borderlands Cafe for being the one place we know in San Francisco that serves coffee without music. May you always prosper.

Thank you to Rose Linke—

Thank you to Tania Lombrozo—

—as readers, as partners, as supporters, as inspirations, as ever.

Index

Page numbers in *italics* refer to figures and tables.

About the Authors

BRIAN CHRISTIAN is the author of *The Most Human Human*, a *Wall Street Journal* bestseller, *New York Times* editors' choice, and a *New Yorker* favorite book of the year. His writing has appeared in *The New Yorker*, *The Atlantic*, *Wired*, *The Wall Street Journal*, *The Guardian*, and *The Paris Review*, as well as in scientific journals such as *Cognitive Science*, and has been translated into eleven languages. He lives in San Francisco.

TOM GRIFFITHS is a professor of psychology and cognitive science at UC Berkeley, where he directs the Computational Cognitive Science Lab. He has published more than 150 scientific papers on topics ranging from cognitive psychology to cultural evolution, and has received awards from the National Science Foundation, the Sloan Foundation, the American Psychological Association, and the Psychonomic Society, among others. He lives in Berkeley.